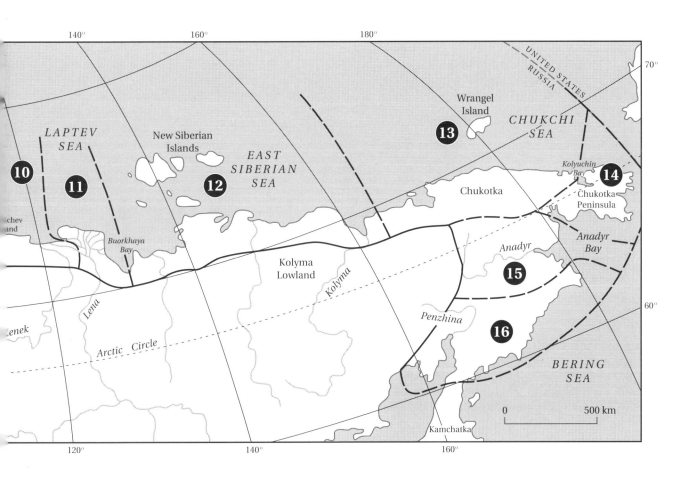

1. Murman District
2. Kanin-Pechora District
3. Polar Ural District
4. Yugorskiy District
5. Novaya Zemlya
6. Franz Josef Land
7. Ob-Tazovskiy District
8. Yenisey District
9. Taymyr District
10. Anabar-Olenek District
11. Lena District
12. Yana-Kolyma District
13. North Chukotka District
14. Beringian Chukotka District
15. Anadyr District
16. Koryak District

FLORA OF THE RUSSIAN ARCTIC
VOLUME II

A.I. TOLMACHEV
Russian Edition Editor

J.G. PACKER
English Edition Editor

G.C.D. GRIFFITHS
Translator

Flora of the Russian Arctic

A Critical Review of the Vascular Plants Occurring in the Arctic Region of the Former Soviet Union

Russian Edition Authors
V.V. Petrovskiy
A.I. Tolmachev
T.V. Yegorova
B.A. Yurtsev

Volume II

Cyperaceae—Orchidaceae

The University of Alberta Press

First English edition published by
The University of Alberta Press
Athabasca Hall
Edmonton, Alberta
Canada T6G 2E8

Copyright © English edition by The University of Alberta Press 1996
Permission to translate and publish this edition of *Arkticheskaya Flora SSSR (Flora Arctica URSS)* was granted by the Severo-Zapadnoye Agentstvo po Avtorskim Pravam (SZAAP) on behalf of the original publishers, USSR Academy of Sciences, V.L. Komarov Botanical Institute.

Originally published as *Arkticheskaya Flora SSSR (Flora Arctica URSS),* Volumes III and IV, by the USSR Academy of Sciences, V.L. Komarov Botanical Institute.

ISBN 0–88864–270–9

Canadian Cataloguing in Publication Data

Main entry under title:

Flora of the Russian Arctic

    Translation of: Arkticheskaya flora SSSR (Flora arctica URSS)
    Contents: Vol. 1. Polypodiaceae—Gramineae — v. 2. Cyperaceae—Orchidaceae.
    Includes bibliographical references and index.
    ISBN 0–88864–269–5 (V.1) — 0–88864–270–9 (V.2)

    1. Botany—Russia, Northern. 2. Botany—Arctic regions.
I. Griffiths, Graham C.D. II. Packer, John G.
QK676.F66 1995   581.9472   C95–910815–7

All rights reserved.

No part of this publication may be produced, stored in a retrieval system, or transmitted in any forms or by any means, electronic, mechanical, photocopying, recording, or otherwise, without the prior permission of the copyright owner.

Printed on acid-free paper. ∞
Printed and bound in Canada by Best Book Manufacturers, Louiseville, Quebec.

COMMITTED TO THE DEVELOPMENT OF CULTURE AND THE ARTS

Contents

- vii   Acknowledgements
- ix   Translator's Preface
- xiii   Preface to Volume III of the Russian edition, Cyperaceae
- xv   Preface to Volume IV of the Russian edition, Lemnaceae—Orchidaceae
- xvii   Abbreviations Used in Citing Floristic and Systematic Literature

1   FAMILY XIV / Cyperaceae—*Sedge Family*

- 3   GENUS 1 / Eriophorum—*Cotton Grass*
- 18   GENUS 2 / Trichophorum—*Deer Grass*
- 20   GENUS 3 / Scirpus—*Bulrush*
- 22   GENUS 4 / Eleocharis—*Spike Rush*
- 25   GENUS 5 / Kobresia—*Kobresia*
- 30   GENUS 6 / Carex—*Sedge*

141   FAMILY XV / Lemnaceae—*Duckweed Family*

- 141   GENUS 1 / Lemna—*Duckweed*

143   FAMILY XVI / Juncaceae—*Rush Family*

- 143   GENUS 1 / Juncus—*Rush*
- 160   GENUS 2 / Luzula—*Woodrush*

181   FAMILY XVII / Liliaceae—*Lily Family*

- 183   GENUS 1 / Tofieldia—*False Asphodel*
- 186   GENUS 2 / Zygadenus—*Camas*
- 187   GENUS 3 / Veratrum—*False Hellebore*
- 192   GENUS 4 / Gagea—*Yellow Star-of-Bethlehem*
- 193   GENUS 5 / Allium—*Onion*
- 197   GENUS 6 / Fritillaria—*Fritillary*
- 198   GENUS 7 / Lloydia—*Lloydia*
- 199   GENUS 8 / Smilacina—*False Solomon's Seal*
- 200   GENUS 9 / Maianthemum—*May Lily*
- 200   GENUS 10 / Paris—*Herb Paris*

201 **FAMILY XVIII** / Iridaceae—*Iris Family*

201 GENUS 1 / Iris—*Iris*

203 **FAMILY XIX** / Orchidaceae—*Orchid Family*

205 GENUS 1 / Cypripedium—*Lady's Slipper*
206 GENUS 2 / Corallorhiza—*Coralroot*
207 GENUS 3 / Listera—*Twayblade*
208 GENUS 3A / Epipactis—*Helleborine*
208 GENUS 4 / Goodyera—*Rattlesnake Plantain*
209 GENUS 5 / Chamaeorchis—*Alpine Orchid*
209 GENUS 6 / Coeloglossum—*Frog Orchid*
212 GENUS 7 / Leucorchis—*Small White Orchid*
212 GENUS 8 / Lysiella—*Blunt-leaved Orchid*
213 GENUS 9 / Gymnadenia—*Rein Orchid*
213 GENUS 10 / Dactylorchis—*Spotted Orchid*

215 **APPENDIX I** / Summary of Data on the Geographical Distribution of Vascular Plants of the Soviet Arctic

217 TABLE 3 / Distribution of Vascular Plants of the Soviet Arctic, *Cyperaceae*
223 TABLE 4 / Distribution of Vascular Plants of the Soviet Arctic, *Lemnaceae—Orchidaceae*

227 Index of Plant Names

❦ Acknowledgements

PERMISSION TO TRANSLATE and publish the 10-volume original *Arkticheskaya Flora SSSR (Flora Arctica URSS)* was granted by the Severo-Zapadnoye Agentstvo po Avtorskim Pravam (SZAAP) on behalf of the original publishers, the USSR Academy of Sciences, V.L. Komarov Botanical Institute. The Press acknowledges the assistance of B.A. Yurtsev of the V.L. Komarov Botanical Institute in obtaining these rights.

In 1994 a successful application was made to the Selection Committee for Scientific Publication Grants, Natural Sciences and Engineering Research Council of Canada (NSERC) for a grant to assist in the publication of the first three Russian volumes of *Flora of the Russian Arctic*. The University of Alberta Press would like to thank the Natural Sciences and Engineering Research Council of Canada for its commitment to this project.

The Press would like to thank Dr. Graham Griffiths, who provided the translation, and Dr. John Packer for his editorial guidance. Dr. Keith Denford's involvement and support of this project since its inception is gratefully acknowledged. The Press also acknowledges the people involved in preparing the manuscript for publication, especially Alan Brownoff, designer, Mary Mahoney-Robson, editor, Karen Chow and Heidi Betke, editorial assistants, and Michael Fisher, cartographer, who prepared the new endpaper maps. We would also like to thank retired Press Director, Norma Gutteridge, for her initial involvement in the project.

Translator's Preface

IN THIS TRANSLATION OF *Arkticheskaya Flora SSSR*, I have tried to preserve the flavour of the original edition. Some of the Russian authors (especially Tolmachev and Tsvelev) tend to write in much longer sentences than approved in current English style manuals. However, I do not think it the task of a translator to rewrite material merely to conform with current stylistic canons. I have followed the original sentence structure wherever possible, only breaking down or drastically reorganizing sentences when this seemed necessary to achieve clarity in English. Since this is a translation, not a revised edition, new nomenclatorial proposals are presented in the same form as in the original Russian edition, indicated by "sp. nova," "comb. nova," etc.

References to the Soviet Union (USSR) in the text naturally refer to the former Soviet Union within its postwar boundaries prior to its dissolution in 1991. With respect to the Arctic Region, the terms "Soviet Arctic" and "Russian Arctic" are equivalent since all arctic territory of the former Soviet Union lies within the present boundaries of the Russian Federation. However, references to distribution within the Soviet Union as a whole may include occurrence outside the present boundaries of the Russian Federation.

The transliteration system used for Russian personal and place names in this translation is the "popular system" that has evolved in English-language newspapers. It is the same as that used on maps produced by the National Geographic Society of the USA except for omission of the apostrophe used to represent the Cyrillic hard and soft signs. Such use of the apostrophe means nothing to English-speaking readers unfamiliar with the subtleties of Slavic pronunciation, and complicates printing for little advantage.

The following points will assist the reader with no knowledge of Russian with the pronunciation of the Russian words transliterated according to this system. The letter **y** does double service both as a vowel and consonant in this transliteration system (hence in words ending "**-yy**" the first **y** is to be read as a vowel). This vowel sound is pronounced more or less as in the French "feuille." While this double use of the letter **y** is perhaps a disadvantage of the system, the same double usage is present in English spelling. Differentiation of the consonantal **y** as **j** (according to German convention) would likely cause it to be mispronounced by unilingual English readers. Note that the vowel **i** is always to be pronounced as in "ravine," not as in "fine" (i.e., it represents the sound also represented in English by **double e**, as in "bee"). The Russian **e** is represented by **e** following a consonant, but by **ye** at the beginning of words and following a vowel. The consonant **ch** is to be pronounced as in "chalk," while the sound represented by the **ch** in "loch" is represented in transliterated Russian words by **kh**. The consonant **zh** represents the voiced equivalent of **sh**, a sound not normally used in English but roughly the same as the French pronunciation of **j**.

In rendering place names I have translated words with obvious English equivalents, such as bay for *guba* and *zaliv*, cape for *mys*, island for *ostrov*, lake for *ozero*, mountain for *gora*, peninsula for *poluostrov* and range for *khrebet*, rather than merely transliterated such words as on National Geographic maps. There is a general problem in rendering Russian place names in English arising from the highly inflected structure of the Russian language; this is whether to retain all the inflectional elements attached as suffices. I have adopted a pragmatic (though inconsistent) approach. In the case of major geographical features, I have used the shortened forms in common use on English maps (for example, "Barents Sea" rather than "Barentsovo Sea"). I have also used shortened forms in cases where the English reader can more readily relate these to a neighbouring town or river (for example, the "Kara Sea" rather than "Karskoye Sea" or the "Verkhoyansk Range" rather than "Verkhoyanskiy Range"). But for many localized features, which the reader may be able to locate only on Cyrillic maps, I have spelled out the fully inflected form in transliteration since that is the form to be found on such maps. In all cases I cite fully inflected place names in the nominative case, and wherever possible have checked the orthography against the 1986 edition of the *Atlas of the USSR*. There remains a possibility that a few errors may have been made in citing the nominative form of local names not included in the Atlas. It should be appreciated that many place names in the Russian North derive from aboriginal languages, and it is not always obvious (even to Russians) how they should be declined or whether they are declinable at all in Russian. The terms *kray* and *oblast* are transliterated when they refer to units of regional administration (such as *Primorskiy Kray* and *Leningrad Oblast*), since these terms are not exactly equivalent to any terms used for regional administrative units in western countries; but they are translated (as "territory" and "region") when not used in an administrative sense.

There are various discrepancies between the English transliterations of Russian personal names here used and the abbreviations of these names used in citing authorship of the Latin names of plants. This arises because the abbreviations for Russian authors given in standard lists generally follow German or Eastern European spelling conventions. I have decided to accept these discrepancies and to copy all abbreviations in the form given in the Russian edition. It would not be realistic to try to revise the abbreviations in this translation, since botanists are unlikely to accept such changes unless made in the context of publishing a new standardized list after widespread consultation. Examples of such discrepancies affecting the abbreviations for the name of authors of this Flora include Czer. for Cherepanov, Egor. for Yegorova, Jurtz. for Yurtsev and Tzvel. for Tsvelev.

A few unusual words for landform or vegetational types may need clarification. The term "placorn tundra" is used by Russian ecologists for level tundra with fine soil. *Baydzharakh* is a Yakutian word for small relict mounds enclosing polygonal ice blocks (normally arranged in groups or rows). These are not the same as pingos (*bulgunnyakhs* in Russian), which are larger isolated mounds resulting from the eruption of underground ice lenses. The term *khasyr* (presumably an aboriginal word, since I can not find it in any Russian dictionary) refers according to the context to small temporary lakes that support terrestrial (but flood-tolerant) vegetation on their beds. This description applies to what the Irish call a "turlough," so I have used this word in translation. I consider "barrens" to be an appropriate translation of the Russian term *goltsy* (the poorly vegetated high alpine zone of mountains), since the root of the Russian word means "bare." The Russian word *stlanik* is retained for vegetation dominated by cedar pine, a small tree or shrub with sprawling growth form. This term is sometimes translated as "krummholz," but not entirely appropriately since this growth form of cedar pine is genetically determined and does not require extreme environmental conditions.

Descriptive botanical terms have been checked in cases of doubt in the *Russian-English Botanical Dictionary* by P. Macura (1982, Slavica Publishers Inc.).

With respect to the summaries of distribution in the North American Arctic, it should be noted that the authors often use the term "Labrador" to include arctic parts of Quebec (the Ungava Pensinsula, etc.), as well as Labrador in the present political sense.

The precise date of final approval for printing (the "imprimatur") is given at the end of each volume of the Russian edition. The date of effective publication in the sense of the International Code of Botanical Nomenclature, the date that the work became available, was presumably very shortly after that date.
A list of the dates can be found in Table P-1 in Volume I.

Maps in this English edition are numbered by the original Russian volume number and the corresponding original number of the map. For example, in this volume, Map III-4 is the fourth map in the Russian volume III on Cyperaceae.

I hope that this translation will facilitate further collaboration between Russian and western botanists in elucidating the fascinating present and past interrelationships within the circumpolar arctic flora.

G.C.D. Griffiths
Translator

Preface to Volume III of the Russian Edition
Arkticheskaya Flora SSSR
Cyperaceae

THE PRESENT VOLUME of the *Arkticheskaya Flora USSR* completes the treatment of material of the monocotyledonous plants of the Soviet Arctic. The role of monocotyledons in the composition of the arctic flora is extremely important. They include two of the families of higher plants most richly represented in the Arctic, the grasses and sedges. A series of representatives of these families, together with the Juncaceae, are very characteristic of the Arctic and play an extremely substantial role in the formation of particular vegetational communities. For this reason and also because there are numerous critical forms among arctic monocotyledons, the study of these plants possesses special importance for understanding the arctic flora as a whole.

The present volume of the *Flora* is entirely devoted to the sedge family, Cyperaceae. Several authors have taken part in reviewing material for this volume. The largest share of the work was assumed by T.V. Yegorova, a specialist on the extensive genus *Carex*, which is the most richly represented genus in the Arctic. She also treated the genera *Eleocharis* and *Trichophorum*, and prepared the key for identification of the genera of the family. Material of the genus *Kobresia* was treated by V.V. Petrovskiy, of *Scirpus* by B.A. Yurtsev, and of *Eriophorum* by A.I. Tolmachev, who also wrote the general characterization of the family.

The ecological characterizations of *Carex* species were prepared in the light of data available in herbaria and in the literature, and above all on the basis of direct observations in the Arctic mainly by B.A. Yurtsev.

Maps of plant distributions were prepared by N.V. Matveyeva and O.V. Rebristaya.[1]

Leningrad, June 1964
A. TOLMACHEV

---

[1] For technical reasons we have been unable to indicate on the maps a series of new localities discovered in 1965 (mainly in the district of the Chaun Plain and the Chukotka Mountains). These localities have been included only in the text.

Preface to Volume IV of the Russian Edition
Arkticheskaya Flora SSSR
Lemnaceae–Orchidaceae

THE PUBLICATION OF the present volume of this work deviates somewhat from the intended sequence of publication. Review of material of two large families of monocotyledons, Gramineae and Cyperaceae, has proved to be too laborious to be completed according to the original schedule. At the present time this task is progressing through the efforts of several of my young colleagues with my direct participation. Meanwhile I have succeeded in completing the review of the remaining families of monocotyledons,[1] and therefore publish data on them earlier.

The central position in the contents of the present volume is occupied by the treatment of the Juncaceae, a family characteristic of the Arctic although quantitatively not very abundantly represented in its flora. Many of the species of Juncaceae are of high interest for investigators of the arctic flora, and we have been able to contribute something new to the treatment of some of them. This has been accomplished through more detailed study of more extensive material in comparison with the work of our predecessors. Certain deviations from the standard treatment of certain well-known forms in the systematic literature are attributable to our concept of the existence of species not only as morphologically and geographically defined assemblages of individuals but also as *distinct in nature*. Accordingly, we consider series of forms joined to one another by gradual transitions to be series of *races of a single polymorphic species*, not as separate species. Therefore, with respect to the sometimes extensive category of plants that are scarcely identifiable accurately (by structure and appearance) under too narrow a species concept and are arbitrarily referable to one or other species mainly on the basis of geographical data, these find their place in our system as transitional forms between incompletely differentiated subspecies within a given species.

As a result of this approach, the total number of species recognized in this work is somewhat reduced in comparison with certain other works, especially the "Flora of the USSR." At the same time, certain forms whose distinctness as species has not been universally recognized or was simply undiscovered are recognized by us as species.

What has just been said about the species of Juncaceae also applies to some degree to the other groups of plants included in the present volume of this Flora. But since these groups are not only more poorly represented in the arctic flora but contain a smaller proportion of representatives truly characteristic of this flora, the investigator of the arctic flora can only contribute something substantially new to the elucidation of these groups in a few special cases.

---

[1] This statement clearly implies that Tolmachev is the sole author of the treatments of families *Lemnaceae* to *Orchidaceae*. — *Translator*

The present volume of the *Arkticheskaya Flora SSSR* contains data on 63 plant species and a whole series of subspecies and varieties. The order of presentation of all data remains the same as that adopted in the first volume. Certain improvements in the format of references to literary sources have been introduced on the advice of colleagues.

As in the preparation of the first volume, I have relied continuously on the help of my very close collaborators, B.A. Yurtsev and O.V. Rebristaya. Both did much in finding and recording material on arctic plants, and in preliminary analysis of their characters. O.V. Rebristaya undertook the major task of inserting localities on the majority of the species distribution maps, including all published up to the present time. Many problems that arose in the course of completing the work were discussed by us and the decisions finally reached are to some degree really the fruits of collective work. To both my assistants I express my heartfelt gratitude.

Leningrad, June 1961
A. Tolmachev

Abbreviations Used in Citing Floristic and Systematic Literature

The list presented in this English edition is a consolidation of the lists and addenda published in volumes II-VIII of the Russian edition. Cyrillic titles are given both in transliteration and in English translation. The classification follows that of the last list on pages 36-40 of volume VIII(2) of the Russian edition.

1. Northern European Part of USSR

Fl. Murm.	Flora Murmanskoy oblasti. [Flora of Murmansk Oblast]. Moscow; Leningrad, 1953, vol. I. 254 pp.; 1954, vol. II. 290 pp; 1956, vol. III. 450 pp.; 1959, vol. IV. 394 pp.; 1960, vol. V. 549 pp.
Fl. sev.-vost yevrop. ch. SSSR	Flora severo-vostoka yevropeyskoy chasti SSSR. [Flora of the North-East European part of the USSR]. Leningrad, 1974, vol. I. 272 pp.; 1976, vol. II. 305 pp.; 1976, vol. III. 295 pp.; 1977, vol. IV. 312 pp.
Fl. yevrop. ch. SSSR	Flora yevropeyskoy chasti SSSR. [Flora of the European part of the USSR]. Leningrad, 1974-1981, vols. 1–5.
Govorukhin, Fl. Urala	Govorukhin V.S. Flora Urala. Oprodelitel rasteniy, obitayushchikh na gorakh i v yevo predgoryakh ot beregov Karskovo morya do yuzhnykh predelov lesnoy zony. [Flora of the Urals. Key to plants living in the mountains and foothills from the shores of the Kara Sea to the southern limits of the forest zone]. Sverdlovsk, 1937.
Hanssen & Lid, Flow. pl. Franz. Josef L.	Hanssen O., Lid J. Flowering plants of Franz Josef Land. Skrifter om Svalbard og Ishavet, 1932, No. 39, pp. 1–42.
Igoshina, Fl. Urala	Igoshina K.N. Flora gornykh i ravninnykh tundr i redkolesiy Urala. [Flora of alpine and lowland tundras and open forests of the Urals]. In book: Rasteniya severa Sibiri i Dalnevo Vostoka. [Plants of Northern Siberia and the Far East]. Moscow; Leningrad, 1966, pp. 135–223.
Leskov, Fl. Malozem. tundry	Leskov A.I. Flora Malozemelskoy tundry. [Flora of the MalozemelskayaTundra]. Moscow; Leningrad, 1937. 106 pp.
Lynge, Vasc. pl. N.Z.	Lynge B. Vascular plants from Novaya Zemlya. Kristiania, 1923. 151 pp.
Perfilev, Fl. Sev.	Perfilev I.A. Flora Severnovo kraya. [Flora of the Northern Territory]. Arkhangelsk, 1934, part I. 160 pp.; 1936, parts II–III. 398 pp.
Rebristaya, Fl. Bolshezem. tundry	Rebristaya O.V. Flora vostoka Bolshezemelskoy tundry. [Flora of the Eastern Bolshezemelskaya Tundra]. Leningrad, 1977. 334 pp.
Ruprecht, Fl. samojed. cisur.	Ruprecht F.J. Flores samojedorum cisuralensium. In: Ruprecht F.J. Symbolae ad historiam et geographiam plantarum rossicarum. St. Petersburg, 1846, pp. 1–67.

Tolmatchev, Contr. fl. Vaig.	Tolmatchev A. Contributions to the flora of Vaigats and of mainland coast of the Yugor Straits. Tr. Botan. muzeya AN SSSR, 1926, vol. 19, pp. 121–154.
Tolmachev, Mat. fl. Mat. Shar	Tolmachev A.I. Materialy dlya flory rayona polyarnoy geofizicheskoy observatorii Matochkin Shar i sopredelnykh chastey Novoy Zemli. [Materials for the flora of the district of the Matochkin Shar Polar Geophysical Observatory and neighbouring parts of Novaya Zemlya]. Tr. Botan. muzeya AN SSSR, 1932, vol. 24, pp. 275–299; 1932, vol. 25, pp. 101–120.
Tolmachev, Obz. fl. N.Z.	Tolmachev A.I. Obzor flory Novoy Zemli. [Review of the flora of Novaya Zemlya]. Arctica, 1936, No. 4, pp. 143–178.

Additional Sources

Aleksandrova, Nov. dan. fl. Yuzh. o. N.Z.	Aleksandrova V.D. Novyye dannyye o flore Yuzhnovo ostrova Novoy Zemli. [New data on the flora of the South Island of Novaya Zemlya]. Biol. MOIP. Otd. biol., 1950, vol. IV, No. 4, pp. 76–85.
Andreyev, Mat. fl. Kanina	Andreyev V.N. Materialy k flore severnovo Kanina. [Materials towards the flora of Northern Kanin]. Tr. Botan. muzeya AN SSSR, 1931, vol. 23, pp. 148–196.
Andreyev, Rast. vost. Bolshezem. tundry	Andreyev V.N. Rastitelnost i prirodnyye rayony vostochnoy chasti Bolshezemelskoy tundry. [Vegetation and natural districts of the eastern part of the Bolshezemelskaya Tundra]. Moscow; Leningrad, 1935.
Dahl & Hadač, Bidr. Spitzb. Fl.	Dahl O.C., Hadač E. Et bidrag til Spitzbergens flora. Norges Svalbard- og Ishavetsundersøkelser. Meddelelser, No. 63, Oslo, 1946.
Feilden, Fl. pl. N.Z.	Feilden H.W. The Flowering Plants of Novaya Zemlya. London, 1898.
Floderus, Nov. Sem. Salic.	Floderus B. Bidrag till kännedomen om Novaja Semljas Salices. Sv. bot. tidskr., VI, 1912.
Gorchakovskiy, Rastit. vysokogor. Urala	Gorchakovskiy P.L. Rastitelnyy mir vysokogoriy Urala. [Vegetational world of the high alpine Urals]. Moscow, 1975, pp. 82–120.
Holm, Nov. Zem. Veg.	Holm T. Novaia-Zemlia's Vegetation. Saerligt dens Phanerogamer. Copenhagen, 1885.
Katenin & al., Fl. Siv. Maski	Katenin A.E., Petrovskiy, V.V., Rebristaya O.V. Sosudistyye rasteniya. [Vascular plants]. In book: Ekologiya i biologiya rasteniy vostochnoyevropeyskoy lesotundry. [Ecology and biology of plants of the Eastern European forest-tundra]. Leningrad, 1970, pp. 37–48.
Kjellman & Lundström, Phanerogam. N.Z. Waig.	Kjellman F.R., Lundström A.N. Phanerogamen von Novaja-Semlja, Waigatsch und Chabarova. In book: Nordenskiöld A.E. Die wissenschaftlichen Ergebnisse der Vega-Expedition. Bd. I. Leipzig, 1883.
Lundström, Weiden Now. Sem.	Lundström, A.N. Kritische Bemerkungen über die Weiden Nowaja Semljas. Acta Reg. Soc. Sci. Upsal., 3, 1877.
Opred. rast. Komi	Opredelitel vysshikh rasteniy Komi ASSR. [Key to the higher plants of the Komi ASSR]. Moscow; Leningrad, 1962.
Perfilev, Mat. fl. N.Z. Kolg.	Perfilev I.A. Materialy k flore ostrovov Novoy Zemli i Kolguyeva. [Materials towards the flora of the islands of Novaya Zemlya and Kolguyev]. Arkhangelsk, 1928. 73 pp.
Sambuk, K fl. sev. yevrop. ch. SSSR	Sambuk, F.V. K flore severa yevropeyskoy chasti SSSR. [Towards the flora of the north of the European part of the USSR]. Zhurn. Russk. botan. obshch., vol XIV, No. 1, 1929.

Schrenk, Enum. pl.	Schrenk, A.G. Enumeratio plantarum in itinere per plages samojedorum cisuralensium per annum 1837 observatarum. In: Schrenk. Reise nach dem Nordosten des europäischen Russlands, Bd. 2, 1854.
Tolmachev, Fl. Kolg.	Tolmachev A.I. Floristicheskiye rezultaty Kolguyevskoy ekspeditsii Instituta po izucheniyu Severa. [Floristic results of the Kolguyev Expedition of the Institute for Study of the North]. Leningrad, 1930. 50 pp. (Tr. Polyarnoy Komissii; vol. 2).
Tolmachev, Fl. kr. sev. N.Z.	Tolmachev A.I. K flore kraynevo severa Novoy Zemli. [Towards the flora of the far north of Novaya Zemlya]. Izv. Glav. bot. sada, 1926, pp. 1–4.
Tolmachev, Fl. pober. Karsk. morya	Tolmachev A.I. K flore yugo-zapadnovo poberezhya Karskovo morya. [Towards the flora of the southwestern coast of the Kara Sea]. Botan. zhurn., 1937, vol. 22, No. 2, pp. 185–196.
Tolmachev, Mat. fl. yevr. arkt. ostr.	Tolmachev A.I. Materialy dlya flory yevropeyskikh arkticheskikh ostrovov. [Materials for the flora of the European Arctic Islands]. Zhurn. Rus. botan. obshch., 1931, vol. XVI, No. 56, pp. 459–472.
Tolmachev, Nov. dan. fl. Vayg.	Tolmachev A.I. Novyye dannyye o flore ostrova Vaygach. [New data on the flora of Vaygach Island]. Botan. zhurn., 1936, vol. 21, No. 1, pp. 88–91.
Tolmachev, Blyumental, Mat. fl. N.Z.	Tolmachev A.I., Blyumental I. Kh. Materialy dlya flory Novoy Zemli. [Materials for the flora of Novaya Zemlya]. Tr. Botan. muzeya AN SSSR, 1931, vol. 23, pp. 197–209.
Tolmachev, Tokarevskikh, Issled. rayona More-Yu	Tolmachev A.I., Tokarevskikh S.A. Issledovaniye rayona «lesnovo ostrova» u reki More-Yu v Bolshezemelskoy tundre. [Investigation of the "forest island" district on the River More-Yu in the Bolshezemelskaya Tundra]. Botan. zhurn., 1968, vol. 53, No. 4, pp. 560–566.
Trautvetter, Consp. fl. Now. Sem.	Trautvetter, E.R. Conspectus florae insularum Nowaja Semlja. Tr. SPb. bot. sada, I, 1871.
Vinogradova, Fl. Pym-va-shor	Vinogradova V.M. Flora rayona teplykh istochnikov Pym-va-shor v Bolshezemelskoy tundre. [Flora of the Pym-va-shor hot springs district in the Bolshezemelskaya Tundra]. Vestn. LGU, 1962, No. 9. Biol., vol. 2, pp. 22-34.

2. Siberian Arctic

Karavayev, Konsp. fl. Yak.	Karavayev, M.N. Konspekt flory Yakutii. [Conspectus of the flora of Yakutia]. Moscow; Leningrad, 1958. 190 pp.
Kjellman, Phanerog. sib. Nordk.	Kjellman F.R. Die Phanerogamenflora der sibirischen Nordküste. In: Nordenskiöld A.E. Die wissenschaftliche Ergebnisse der Vega-Expedition. Leipzig, 1883, Vol. I. pp. 94–139.
Krylov, Fl. Zap. Sib.	Krylov, P.N. Flora Zapadnoy Sibiri. [Flora of West Siberia]. Tomsk, 1927–1964, vols. I–XII.
Opred. rast. Yak.	Opredelitel vysshikh rasteniy Yakutii. [Key to the higher plants of Yakutia]. Novosibirsk, 1974. 533 pp.
Petrov, Fl. Yak.	Petrov V.A. Flora Yakutii [Flora of Yakutia], vol. I. Leningrad, 1930.
Schmidt, Fl. jeniss.	Schmidt F. Florula jenisseensis arctica. In: Wissenschaftliche Resultate der zur Aufsuchung eines angekündigten Mammuth-cadavers Expedition. St. Petersburg, 1872, pp. 73–133.

Tikhomirov, Petrovskly, Yurtsev, Fl. Tiksi	Tikhomirov B.A., Petrovskly V.V., Yurtsev B.A. Flora okrestnostey bukhty Tiksi (arkticheskaya Yakutiya). [Flora of the vicinity of the Bay of Tiksi (Arctic Yakutia)].
Tolmachev, Fl. Taym.	Tolmachev A.I. Flora tsentralnoy chasti vostochnovo Taymyra. [Flora of the central part of East Taymyr]. Tr. Polyarnoy komissii, 1932, vol. 8, pp. 1–126; vol. 13, pp. 5–75; 1935, vol. 25, pp. 5–80.
Trautvetter, Fl. rip. Kolym.	Trautvetter E.R. Flora riparia Kolymensis. Acta Horti Petropol., 1877, vol. 5, pp. 495–574.
Trautvetter, Fl. taim.	Trautvetter E.R. Florula taimyrensis phaenogama. In: Phanerogame Pflanzen aus dem Hochnorden. St. Petersburg, 1847, pp. 17–64.
Trautvetter, Pl. Sib. bor.	Trautvetter E.R. Plantas Sibiriae borealis ab A. Czekanowski et F. Müller annis 1874 et 1875 lectas enumeravit. Acta Horti Petropol., 1877, vol. 5, p. 1-146.
Trautvetter, Syll. pl. Sib. bor.-or.	Trautvetter E.R. Syllabus plantarum Sibiriae boreali-orientalis a Dre Alex. Bunge fil. lectarum. Acta Horti Petropol., 1887, vol. 10, pp. 481–546.

Additional Sources

Aleksandrova, Fl. B. Lyakhovsk.	Aleksandrova V.D. Flora sosudistykh rasteniy ostrova Bolshovo Lyakhovskovo (Novosibirskiye ostrova). [Vascular plant flora of Bolshoy Lyakhovskiy Island (New Siberian Islands)]. Botan. zhurn., 1960, vol. 45, No. 11, pp. 1687–1693.
Drobov, Predst. sekts. *Ovinae* v Yakut.	Drobov, V.P. Predstaviteli sektsii *Ovinae* Fr. roda *Festuca* L. v Yakutskoy oblasti. [Representatives of section *Ovinae* of the genus *Festuca* L. in Yakutsk Oblast]. Petrograd, 1915.
Fl. Putorana	Flora Putorana. [Flora of Putorana]. Novosibirsk, 1976. 246 pp.
Fl. Stanov. nagorya	Vysokogornaya flora Stanovovo nagorya. [High alpine flora of the Stanovoye Highland]. Novosibirsk, 1972. 272 pp.
Hämet-Ahti, Cajand. vasc. pl. Lena R.	Hämet-Ahti L. A.-K. Cajander's vascular plant collection from the Lena River, Siberia, with his ecological and floristic notes. Ann. Bot. Fenn., 1970, vol. 7, pp. 255–324.
Korotkevich, Rastit. Sev. Zemli	Korotkevich E.S. Rastitelnost Severnoy Zemli. [Vegetation of Severnaya Zemlya]. Botan. zhurn., 1958, Vol. 43, No. 5, pp. 644–663.
Malyshev, Fl. Vost. Sayana	Malyshev L.I. Vysokogornaya flora Vostochnovo Sayana. [High alpine flora of the Eastern Sayan]. Moscow; Leningrad, 1965. 368 pp.
Middendorff, Gewächse Sibiriens	Middendorff, A.T. Die Gewächse Sibiriens. In: Middendorff, Sibirische Reise, Bd. IV, Theil 1, 1864.
Polozova, Tikhomirov, Rast. Tarei	Polozova T.G., Tikhomirov B.A. Sosudistyye rasteniya rayona Taymyrskovo statsionara (pravoberezhe Pyasiny bliz ustya Tarei. Zapadnyy Taymyr). [Vascular plants of the district of the Taymyr Station (right bank of the Pyasina near the mouth of the Tareya. West Taymyr)]. In book: Biogeotsenozy Taymyrskoy tundry i ikh produktivnost. [Biogeocenoses of the Taymyr tundra and their productivity]. Leningrad, 1971, pp. 161–184.
Scheutz, Pl. jeniss.	Scheutz N.J. Plantae vasculares jeniseenses. Kongl Svenska vetenskaps Akad. Handling., Bd. 22, No. 10, Stockholm, 1888.
Tikhomirov, Fl. Zap. Taym.	Tikhomirov B.A. K kharakteristike flory zapadnovo poberezhya Taymyra. [Towards the characterization of the flora of the west coast of Taymyr]. Petrozavodsk, 1948. 85 pp. (Tr. Karelofinskovo un-ta; vol. 2).
Tolmachev, Fl. o. Benneta	Tolmachev A.I. K flore ostrova Benneta. [Towards the flora of Bennet Island]. Bot. Zhurn., 44, 4, 1959.

Tolmachev, O fl. nakh. v tsentr. chasti Taym.	Tolmachev A.I. O neskolkikh neozhidannykh floristicheskikh nakhodkakh v tsentralnoy chasti Taymyrskovo polyostrova. [On some unexpected floristic discoveries in the central part of the Taymyr Peninsula]. DAN SSSR, ser. A, 1930, No. 5.
Tolmachev, Raspr. drev. porod	Tolmachev A.I. O rasprostranenii drevesnykh porod i severnoy granitse lesov v oblasti mezhdu Yeniseyem i Khatangoy. [On the distribution of tree species and the northern limit of forest in the region between the Yenisey and the Khatanga]. Tr. polyarnoy komissii, vol. 5, 1931.
Tolmachev, Rast. o. Sibiryakova	Tolmachev A.I. Obzor sosudistykh rasteniy ostrova Sibiryakova v Yeniseyskom zalive. [Review of the vascular plants of Sibiryakov Island in the Bay of Yenisey]. Tr. Botan. muzeya AN SSSR, 1931, vol. 23, pp. 211–218.
Tolmachev, Pyatkov, Obz. rast. Diksona	Tolmachev A.I., Pyatkov P.P. Obzor sosudistykh rasteniy ostrova Diksona. [Review of the vascular plants of Dikson Island]. Tr. Botan. muzeya AN SSSR, 1930, vol. 22, pp. 147–179.
Trautvetter, Fl. boganid.	Trautvetter E.R. Florula boganidensis phaenogama. In: Phanerogame Pflanzen aus dem Hochnorden. St. Petersburg, 1847, pp. 144–167.
Yurtsev, Fl. Suntar-Khayata	Yurtsev B.A. Flora Suntar-Khayata [Flora of Suntar-Khayata]. Leningrad, 1968, 235 pp.

3. Far East, Chukotka

Hultén, Fl. Kamtch.	Hultén E. Flora of Kamtchatka and the adjacent islands. Stockholm, 1927–1930, vols. I–IV.
Kjellman, Phanerog. as. K. Ber.-Str.	Kjellman F.R. Die Phanerogamenflora an der asiatischen Küste der Bering-Strasse. In: Nordenskiöld A.E. Die wissenschaftliche Ergebnisse der Vega-Expedition. Leipzig, 1883, vol. I, pp. 249–379.
Komarov, Fl. Kamch.	Komarov V.L. Flora poluostrova Kamchatki. [Flora of the Kamchatka Peninsula]. Leningrad, 1927-1930, vols. 1–3.
Opred. rast. Kamch. obl.	Opredelitel sosudistykh rasteniy Kamchatskoy oblasti. [Key to the vascular plants of Kamchatka Oblast]. Moscow, 1981, 411 pp.
Petrovskiy, Rast. o. Vrangelya	Petrovskiy V.V. Spisok sosudistykh rasteniy o. Vrangelya. [List of the vascular plants of Wrangel Island]. Botan. zhurn., 1973, vol. 58, No. 1, pp. 113–126.
Tikhomirov, Gavrilyuk, Fl. Bering. Chuk.	Tikhomirov B.A., Gavrilyuk V.A. K flore Beringovskovo poberezhya Chukotskovo poluostrova. [Towards the flora of the Beringian coast of the Chukotka Peninsula]. In book: Rasteniya severa Sibiri i Dalnevo Vostoka. [Plants of Northern Siberia and the Far East]. Moscow; Leningrad, 1966, pp. 58–79.
Trautvetter, Fl. Tschuk.	Trautvetter E.P. Flora terrae Tschuktschorum. St. Petersburg, 1878. 40 pp. (See also: Acta Horti Petropol., 1879, vol. 6, pp. 1–40).
Vasilev, Fl. Komand. ostr.	Vasilev V.N. Flora i paleogeografiya Komandorskykh ostrovov. [Flora and paleogeography of the Commander Islands]. Moscow; Leningrad, 1957.
Voroshilov, Opred. rast. D. Vost.	Voroshilov V.N. Opredelitel rasteniy sovetskovo Dalnevo Vostoka. [Key to plants of the Soviet Far East]. Moscow, 1982. 672 pp.

Additional Sources

Derviz-Sokolova, Fl. Dezhn.	Derviz-Sokolova T.G. Flora kraynevo vostoka Chukotskovo poluostrova (poselok Uelen–mys Dezhneva). [Flora of the far east of the Chukotka Peninsula (Uelen village–Cape Dezhnev)]. In book: Rasteniya severa Siberi i Dalnevo Vostoka. [Plants of Northern Siberia and the Far East]. Moscow; Leningrad, 1966, pp. 80–107.
Filin, Yurtsev, Rast. o. Ayon	Filin V.R., Yurtsev B.A. Sosudistyye rasteniya o. Ayon (Chaunskaya guba). [Vascular plants of Ayon Island (Chaun Bay)]. In book: Rasteniya severa Sibiri i Dalnevo Vostoka. [Plants of Northern Siberia and the Far East]. Moscow; Leningrad, 1966, pp. 44–57.
Floderus, Salic. Anadyr.	Floderus B. Salices peninsulae Anadyrensis. Arkiv f. bot., 25A, 10, 1933.
Floderus, Salix Kamtch.	Floderus B. On the Salix-flora of Kamtchatka. Arkiv f. bot., 20A, 6, 1926.
Fl. Sib. Daln. Vost.	Flora Sibiri i Dalnevo Vostoka. [Flora of Siberia and the Far East]. Moscow; Leningrad, 1966.
Kharkevich, Buch, Rast. Sev. Koryak.	Kharkevich S.S., Buch T.G. Sosudistyye rasteniya Severnoy Koryakii. [Vascular plants of Northern Koryakia]. Botan. zhurn., 1976, vol. 61, No. 8, pp. 1089–1102.
Khokhryakov, Fl. p-ova Taygonos	Khokhryakov A.P. K flore poluostrova Taygonos i severnovo poberezhya Gizhiginskoy guby. [Towards the flora of the Taygonos Peninsula and the north coast of the Bay of Gizhiga]. In book: Biologiya rasteniy i flora severa Dalnevo Vostoka. [Plant biology and flora of the Northern Far East]. Vladivostok, 1981, pp. 8–11.
Khokhryakov, Fl. r. Omolon	Khokhryakov A.P. K flore srednevo techeniya reki Omolon. [Towards the flora of the middle course of the River Omolon]. In book: Flora i rastitelnost Chukotki. [Flora and vegetation of Chukotka]. Vladivostok, 1978, pp. 53–75.
Khokhryakov, Mater. fl. yuzhn. ch. Magad. obl.	Khokhryakov A.P. Materialy k flore yuzhnoy chasti Magadanskoy oblasti. [Materials towards the flora of the southern part of Magadan Oblast]. In book: Flora i rastitelnost Magadanskoy oblasti. [Flora and vegetation of Magadan Oblast]. Vladivostok, 1976, pp. 3–36.
Khokhryakov, Yurtsev, Fl. Olsk. plato	Khokhryakov A.P., Yurtsev B.A. Flora Olskovo bazaltovovo plato (Kolymsko-Okhotskiy vodorazdel). [Flora of the Olskoye basalt plateau (Kolyma-Okhotsk watershed)]. Byul. MOIP. Otd. biol., 1974, vol. 79, No. 2, pp. 59–70.
Kitsing, Koroleva, Petrovskiy, Fl. b. Rodzhers	Kitsing L.I., Koroleva T.M., Petrovskiy V.V. Flora sosudistykh rasteniy okrestnostey bukhty Rodzhers (o. Vrangelya). [Vascular plant flora of the vicinity of Rodgers Bay (Wrangel Island)]. Bot. zhurn., 59, 7, 1974.
Kudo, Fl. Paramushir	Kudo Y. Flora of the island of Paramushir. J. Coll. Agric. Hokkaido Univ., 9, 2, 1922.
Kurtz, Fl. Tschuktsch.	Kurtz F. Die Flora der Tschuktschenhalbinsel. Nach den Sammlungen der Gebrüder Krause. Engler's Bot. Jahrb., 1895, Bd. 19, pp. 432–493.
Polezhayev & al., Fl. Bering. r-na	Polezhayev A.N., Khokhryakov A.P., Berkutenko A.N. K flore Beringovskovo rayona Magadanskoy oblasti. [Towards the flora of the Beringian district of Magadan Oblast]. Botan. zhurn., 1976, vol. 61, no. 8, pp. 1103–1110.
Schmidt, Reisen Amurl.	Schmidt Fr. Reisen im Amurlande und auf der Insel Sachalin. Mém. Acad. Sci. SPb., sér VIII, 12, 1, 1869.
Sugawara, Ill. fl. Saghal.	Sugawara S. Illustrated flora of Saghalien. II. 1939. Tokyo.
Vorobev, Mat. fl. Kuril.	Vorobev D.P. Materialy k flore Kurilskykh ostrovov. [Materials towards the flora of the Kurile Islands]. Tr. Dalnevost. filiala AN SSSR, ser. botan., vol. III(V), 1956.
Voroshilov, Fl. D. Vost.	Voroshilov V.N. Flora sovetskovo Dalnevo Vostoka. [Flora of the Soviet Far East]. Moscow, 1966, 476 pp.

4. American Arctic

Hultén, Comments Fl. Al.	Hultén E. Comments on the Flora of Alaska and Yukon. Stockholm, 1967. 147 pp. (Ark. Bot. Ser. 2, Bd. 7, Heft 1).
Hultén, Fl. Al.	Hultén E. Flora of Alaska and Yukon. Lund, 1941-1950, vols. I–X.
Hultén, Fl. Al. & neighb. terr.	Hultén E. Flora of Alaska and neighbouring territories. Stanford, 1968. 1008 pp.
Porsild, Ill. fl. Arct. Arch.	Porsild A.E. Illustrated flora of the Canadian Arctic Archipelago. Ottawa, 1957. 209 pp. (Nat. Mus. of Canada Bull. No. 146); 2nd edition, revised. Ottawa, 1964.
Porsild, Vasc. pl. W Can. Arch.	Porsild A.E. The vascular plants of the western Canadian Arctic Archipelago. Ottawa, 1955. 226 pp. (Nat. Mus. of Canada Bull. No. 135).
Porsild & Cody, Checklist pl. NW Canada	Porsild A.E., Cody W.J. Checklist of the vascular plants of the Continental Northwest Territories, Canada. Ottawa, 1968. 102 pp.
Porsild & Cody, Pl. continent. NW Canada	Porsild A.E., Cody W.J. Vascular plants of continental Northwest Territories, Canada. Ottawa, 1980. 667 pp.
Simmons, Survey phytogeogr.	Simmons H.G. A survey of the phytogeography of the Arctic American Archipelago with some notes about its exploration. Lund, 1913. 183 pp.

Additional Sources

Anderson, Fl. Al.	Anderson J.P. Flora of Alaska and adjacent parts of Canada. Ames, 1959. 724 pp.
Coville, Willows Alask.	Corille F.V. Willows of Alaska. Proceed. Washingt. Acad. Sci., 3, 1901.
Gjaerevoll, Bot. invest. centr. Alaska	Gjaerevoll O. Botanical investigations in central Alaska, especially in the White Mountains. Kgl. Norske Videnskabers Selskabs Skrifter, 1958, No. 5, pp. 1–74; 1963, No. 4, pp. 1–115; 1967, No. 10, pp. 1–63.
Gröntved, Vasc. pl. Arctic Amer.	Gröntved J. Vascular plants from Arctic North America. Report Fifth Thule exped. II, No. 1. Copenhagen, 1936.
Holm, Contr. morph. syn. geogr. distr. arct. pl.	Holm T. Contributions to the morphology, synonymy and geographical distribution of arctic plants. In: Report of the Canadian Arctic Expedition 1913–1918, vol. 5. Botany, part B. Ottawa, 1922.
Hultén, Fl. Aleut. Isl.	Hultén E. Flora of the Aleutian Islands and Westernmost Alaska Peninsula with notes on the flora of the Commander Islands. Stockholm, 1937. 397 pp.; 2nd edition, revised. Weinheim, 1960. 376 pp.
Hultén, Suppl. Fl. Al.	Hultén E. Supplement to Flora of Alaska and neighbouring territories. A study in the flora of Alaska and the Transberingian connection. Bot. Notis., 1973, vol. 126, No. 4, pp. 459–512.
Macoun & Holm, Vasc. pl.	Macoun J.M., Holm T. The vascular plants of the Arctic coast of America west of the 100th meridian. In: Report of the Canadian Arctic Expedition 1913–1918, vol. 5. Botany, part A. Ottawa, 1921.
Polunin, Bot. Can. E Arctic	Polunin N. Botany of the Canadian Eastern Arctic. National Museum of Canada, Bull. No. 97, Biological ser., No. 26, 1947.
Porsild, Bot. SE Yukon	Porsild A.E. Botany of Southeastern Yukon adjacent to the Canol Road. Bull. Nat. Mus. Canada, 151, 1951.
Porsild, Contrib. fl. Alaska	Porsild A.E. Contributions to the flora of Alaska. Rhodora, 41, 1939.
Porsild, Mat. fl. NW territ.	Porsild A.E. Materials for a flora of the continental Northwest territories of Canada. Sargentia, IV, 1943.

Raup, Bot. SW Mackenz.	Raup H.M. The botany of Southwestern Mackenzie. Sargentia, VI, 1947.
Raup, Willows Huds.	Raup H.M. The willows of the Hudson Bay region and the Labrador Peninsula. Sargentia, IV, 1943.
Raup, Willows W. Amer.	Raup H.M. The willows of boreal Western America. Contrib. Gray Herbar., CLXXXV, 1959.
Rydberg, Cespit. willows	Rydberg P.A. Cespitose willows of Arctic America. Bull. N.Y. Bot. Gard., I. 1899.
Schneider, Amer. willows	Schneider C. Notes on American willows. I, Bot. Gaz., LXVI, 2, 1918; II, ibid., LXVI, 4,1918; III, ibid., LXVII, 1, 1919; VI, J. Arnold Arboret., I, 1919, 67–97; VIII, ibid., 1919, 211–232; X, ibid., II, 1920, 65–90; XI, ibid., 1921, 185–204.
Simmons, Vasc. pl. Ellesm.	Simmons H.G. The vascular plants in the flora of Ellesmereland. Kristiania, 1906.
Welsh, Fl. Al.	Welsh S.L. Anderson's Flora of Alaska and adjacent parts of Canada. Provo, 1974. 724 pp.
Young, Fl. Lawr. Isl.	Young S.B. The vascular flora of St. Lawrence Island with special reference to floristic zonation in the arctic regions. Contrib. Gray Herbar. Harvard Univ., 1971, No. 201, pp. 11–115.

5. Greenland

Böcher & al., Fl. Greenl.	Böcher T.W., Holmen K., Jakobsen K. The flora of Greenland. 2nd edition. Copenhagen, 1968. 312 pp.
Böcher & al., Grønl. Fl.	Böcher T.W., Holmen K., Jakobsen K. Grønlands Flora. Copenhagen, 1957.
Holmen, Vasc. pl. Peary L.	Holmen K. The vascular plants of Peary Land, North Greenland. Meddl. om Grønl., 1957, Bd. 124, No. 9, pp. 1–149.

Additional Sources

Devold & Scholander, Fl. pl. SE Greenl.	Devold J., Scholander P.F. Flowering plants and ferns of Southeast Greenland. Skrifter om Svalbard og Ishavet, 1933, No. 56, pp. 1–209.
Floderus, Grönl. Salic.	Floderus B. Om Grönlands Salices. Meddl. om Grønl., 1923, 63.
Gelting, Vasc. pl. E Greenl.	Gelting P. Studies on the vascular plants of Eastern Greenland. Meddl. om Grønl., 1934, Bd. 101, No. 2, pp. 1–340.
Jorgensen & al., Flow. pl. Greenl.	Jorgensen C.A., Sørensen Th., Westergaard M. The flowering plants of Greenland. A taxonomical and cytological survey. Copenhagen, 1958. 172 pp.
Lagerkranz, Fl. W & E Greenl.	Lagerkranz J. Observations on the flora of West and East Greenland. Nova Acta Reg. Soc. Sci. Upsal., ser. 4, XIV, 6, 1950.
Lange, Consp. fl. groenl.	Lange J. M. C. Conspectus florae groenlandicae. Meddl. om Grønl., Bd. 3, 1880.
Ostenfeld, Fl. Greenl.	Ostenfeld C. H. The flora of Greenland and its origin. Det Kgl. Danske Videnskabernes Selsk. Biologiske Meddelelser, Bd. VI, 3, Copenhagen, 1926.
Polunin, Contrib. fl. SE Greenl.	Polunin N. Contribution to the flora and phytogeography of Southeastern Greenland. J. Linn. Soc., Bot., LII, 1943.
Porsild, Fl. Disko	Porsild M. The flora of Disko Island and the adjacent coast of West Greenland. Meddl. om Grønl., Bd 58, 1920.
Seidenfaden, Vasc. pl. SE Greenl.	Seidenfaden G. The vascular plants of Southeast Greenland from 60° 04' to 64° 30' N lat. Meddl. om Grønl., 1933, Bd. 106, No. 3, pp. 1–129.

Seidenfaden & Sørensen, Summary spec. E Greenl.	Seidenfaden G., Sørensen Th. Summary of all species found in Eastern Greenland. Meddl. om Grønl., Bd. 101, 4, 1937.
Seidenfaden & Sørensen, Vasc. pl. NE Greenl.	Seidenfaden G., Sørensen Th. The vascular plants of North-East Greenland from 74° 30' to 79°N lat. Meddl. om Grønl., Bd. 101, 4, 1937.
Sørensen, Revis. Greenl. sp. Puccinellia	Sørensen Th. A revision of the Greenland species of Puccinellia Parl. Copenhagen, 1953.
Sørensen, Vasc. pl. E Greenl.	Sørensen Th. The vascular plants of East Greenland from 71° 00' to 73° 30'. Meddl. om Grønl., 1933, Bd. 101, No. 3, pp. 1–177.

6. Arctic Western Europe

Fl. europ.	Flora europaea. Cambridge, 1964-1980, vols. 1–5.
Gröntved, Pterid. Spermatoph. Icel.	Gröntved J. The Pteridophyta and Spermatophyta of Iceland. Copenhagen, 1942. 427 pp.
Hultén, Atlas	Hultén E. Atlas of the distribution of vascular plants in NW Europe. Stockholm, 1950. 512 pp.; 2nd edition, revised, 1971. 531 pp.
Löve, Isl. Ferdafl.	Löve A. Islenzk Ferdaflora. Reykjavik, 1970. 428 pp.
Löve & Löve, Consp. Icel. fl.	Löve A., Löve D. Cytotaxonomical conspectus of the Icelandic flora. Acta Horti Gotoburg., 1956, vol. 20, No. 4, pp. 65–290.
Rønning, Svalb. fl.	Rønning O.I. Svalbards flora. Oslo, 1979. 128 pp.
Scholander, Vasc. pl. Svalb.	Scholander P.F. Vascular plants from northern Svalbard. Skrifter om Svalbard og Ishavet, 1934, No. 62, pp. 1–155.

Additional Sources

Andersson, Salic. Lappon.	Andersson N.J. Salices Lapponiae. 1845. Uppsala.
Benum, Fl. Troms	Benum P. Flora of Troms fylke. 1958. Tromsö.
Floderus, Salic. fennoscand.	Floderus B. Salicaceae fennoscandicae. Stockholm, 1931.
Fries, Mantissa	Fries E.M. Novitiarum florae Sueciae mantissa. I. 1832. Lund-Uppsala.
Hadač, Gefässpfl. Sassengeb.	Hadač E. Die Gefässpflanzen des «Sassengebietes» Westspitsbergen. Norges Svalbard- og Ishavetsundersøkelser. Skrifter, 87, Oslo, 1944.
Hadač, Hist. fl. Spitsb.	Hadač E. The history of the flora of Spitsbergen. Preslia, XXXII, 1960.
Hadač, Not. fl. Svalb.	Hadač E. Notulae ad floram Svalbardiae spectantes. Studia Bot. Cechica, vol. 5, 1942.
Hylander, Nord. Kärlväxtfl.	Hylander N. Nordisk Kärlväxtflora, I. Botan. Notiser, 1953, Heft 3.
Lagerberg & al., Pohj. luon. I	Lagerberg T., Kalela A., Väänänen H. Pohjolan luonnon kasvit. I. 1958. Helsinki.
Lid, Fl. Jan Mayen	Lid J. The flora of Jan Mayen. Oslo, 1964. 108 pp. (Norsk Polarinstitutt Skrifter; No. 130).
Lid, Norsk & Svensk fl.	Lid J. Norsk og Svensk flora. Oslo, 1974. 808 pp.
Lindman, Svensk fanerogamfl.	Lindman C.A.M. Svensk fanerogamflora. Utg. 2. 1926. Stockholm.

Rønning, Vasc. Fl. Bear Isl. Rønning O.I. The vascular flora of Bear Island. Tromsø, 1959. 62 pp. (Acta Borealia A. Scientia; No. 15).

Wahlenberg, Fl. lappon. Wahlenberg G. Flora lapponica. 1812. Berlin.

7. General Sources

Cherepanov, Rast. SSSR Cherepanov S.K. Sosudistyye rasteniya SSSR. [Vascular plants of the USSR]. Leningrad, 1981. 509 pp.

Dorogostayskaya, Sorn. rast. Sev. Dorogostayskaya E. Sornyye rasteniya Kraynevo Severa SSSR. [Weeds of the Far North of the USSR].

Endem. vysokogor. rast. Sev. Azii Endemichnyye vysokogornyye rasteniya Severnoy Azii. [Endemic high alpine plants of Northern Asia]. Novosibirsk, 1975. 336 pp.

Fl. SSSR Flora SSSR. [Flora of the USSR]. Moscow; Leningrad, 1934–1960, vols. I–XXX.

Gelert & Ostenfeld, Fl. arct. Gelert O., Ostenfeld C.H. Flora arctica. Part 1. Copenhagen, 1902.

Hultén, Amph-Atl. pl. Hultén E. The Amphi-Atlantic plants and their phytogeographical connections. Stockholm, 1958. 340 pp.

Hultén, Circump. pl. Hultén E. The circumpolar plants. Stockholm, 1962, pt. I; 1971, pt. II.

Löve & Löve, Cytotaxon. atlas arct. fl. Löve A., Löve D. Cytotaxonomical atlas of the arctic flora. Vaduz, 1975. 598 pp.

Polunin, Circump. arct. fl. Polunin N. Circumpolar Arctic Flora. Oxford, 1959. 515 pp.

8. Miscellaneous

Andersson, Monogr. Salic. Andersson N.J. Monographia Salicum. 1867. Stockholm.

Andersson, Nordamer. Salic. Andersson N.J. Bidrag till kännedomen om de i Nordamerika förekommande Salices. Öfversigtat K. Vet. Akad. Förhandl. XV, 1858.

Andersson, Salic. Japon. Andersson N.J. Salices e Japonica. Mem. Amer. Acad., N.S., VI, 2, 1858.

Buchenau, Monogr. Junc. Buchenau F. Monographia Juncacearum. Engler's Botan. Jahrb., Bd. 12. 1890.

Der. i kust. SSSR Derevya i kustarniki SSSR. [Trees and shrubs of the USSR]. Vols. I–VI. Moscow-Leningrad, 1949–1962.

Dylis, Sib. listv. Dylis N.V. Sibirskaya listvennitsa. [Siberian larch]. Mater. k pozn. fauny i flory SSSR, nov. ser., otd. botan., vol. 3. Moscow, 1947.

Fl. Az. Ros. Flora Aziatskoy Rossii. [Flora of Asiatic Russia].

Gilibert, Exerc. phytol. II Gilibert J.E. Exercitia phytologica. II. 1792. Lyons.

Gorodkov, Obz. russk. osok Gorodkov B.N. Obzor russkikh osok. [Review of Russian sedges]. Tr. Botan. muzeya AN SSSR, vol. XX, 1927.

Hagström, Crit. Res. Potamog. Hagström J.O. Critical researches on the Potamogetons. K. Svenska Vetensk. Akad. Handl., Bd. 55, 5, 1916.

Hoffman, Hist. Salic. I Hoffman G.F. Historia Salicum iconibus illustrata. I. 4. 1787. Leipzig.

Holmen & Mathiesen	Holmen K., Mathiesen H. Luzula Wahlenbergii in Greenland. Bot. tidsskr., Bd. 43, Heft 3, 1953.
Host, Salix	Host N.T. Salix. 1828. Vienna.
Keppen, Raspr. khv. der.	Keppen F. Geograficheskoye rasprostraneniye khvoynykh derev v yevropeyskoy Rossii i na Kavkaze. [Geographical distribution of coniferous trees in European Russia and the Caucasus]. St. Petersburg, 1885.
Kimura, Symb. Iteol.	Kimura A. Symbolae Iteologicae. IV, Sci. Repert. Tohoku Univ., Biol., XII, 1937; VI, ibid., XIII, 1938.
Krall & Viljasoo, Eestis pajud.	Krall H., Viljasoo L. Eestis Kasvavad pajud. 1965, Tartu.
Kükenthal, Cyper. Caricoid.	Kükenthal G. Cyperaceae-Caricoideae. In: Der Pflanzenreich, Heft 38, Leipzig, 1909.
Kükenthal, Cyper. Sibir.	Kükenthal G. Cyperaceae Sibiriae. Russk. botan. zhurn., 1911, Nos. 3–6.
Ledebour, Fl. alt.	Ledebour C.F. Flora altaica. IV. 1833. Berlin.
Ledebour, Ic. pl. fl. ross.	Ledebour C.F. Icones plantarum novarum vel imperfecte cognitarum floram rossicam, imprimis altaicam, illustrantes. V. 1834. Riga.
Linné fil., Supplem.	Linné C. fil. Supplementum plantarum systematis vegetabilium...1781. Braunschweig.
Moench, Meth.	Moench C. Methodus plantas horti botanici et agri Marburgensis a staminum situ describendi. 1794. Marburg.
Pallas, Fl. ross.	Pallas P.S. Flora rossica. I, 2. 1788. St. Petersburg.
Polunin, Real Arctic Pterid.	Polunin N. The Real Arctic and its Pteridophyta. American fern journ., vol. 41, No. 2, 1951.
Rasinsh, Ivy Latv.	Rasinsh A.P. Ivy Latviyskoy SSR. [Willows of the Latvian SSR]. Tr. Inst. biolog. AN Latv. SSR, 8, 1959.
Regel, Monogr. Betulac.	Regel E. Monographia Betulacearum hucusque cognitarum. 1861. Moscow.
Ruprecht, Distrib. crypt. vasc.	Ruprecht F.J. Distributio Cryptogamarum vascularium in Imperio Rossico. In: Ruprecht. Symbolae ad historiam et geographiam plantarum rossicarum. St. Petersburg, 1846.
Salisbury, Prodrom. stirp. horto Allerton	Salisbury R.A. Prodromus stirpium in horto ad Chapel Allerton vigentium. 1796. London.
Seemen, Salic. japon.	Seemen O. Salices japonicae. 1903. Berlin.
Skvortsov, Mat. iv.	Skvortsov A.K. Materialy po morfologii i sistematike ivovykh. [Materials for the morphology and systematics of Salicaceae]. I, Byull. MOIP, biol., LX, 3, 1959; II, ibid., LXI, 1, 1956; III, IV, Bot. mat. Gerb. Bot. inst. AN SSSR, XVIII, 1957; V, Sist. zamet. Gerb. Tomsk. univ., 1956, 79–80; VI, Bot. mat. Gerb. Bot. inst. AN SSSR, XIX, 1959; IX, ibid., XXI, 1961; X, Byull. MOIP, biol., XVI, 4, 1961; XI, Tr. MOIP, III, 1960.
Sukachev, Dendrologiya	Sukachev V.N. Dendrologiya s osnovami lesnoy geobotaniki. [Dendrology on the foundations of forest geobotany]. Leningrad, 1938.
Tikhomirov, Kedr. stlanik	Tikhomirov B.A. Kedrovyy stlanik, yevo biologiya i ispolzovaniye. [Cedar pine stlanik, its biology and utilization]. Mater. k pozn. fauny i flory SSSR, nov. ser., otd. botan., vol. 6. Moscow, 1949.
Tikhomirov, Proiskh. ass. kedr. stl.	Tikhomirov B.A. K proiskhozhdeniyu assotsiatsiy kedrovovo stlanika. [On the origin of the cedar pine stlanik association]. Mater. po istorii flory i rastitelnosti SSSR, sb. II. Moscow; Leningrad, 1946.

Tikhomirov, Raspr. papor.	Tikhomirov B.A. Rasprostranenlye paporotnikov v Sovetskoy Arktike. [Distribution of ferns in the Soviet Arctic]. Botan. mater. Gerbariya BIN, vol. 19, 1959.
Tolmachev, Ist. temnokhv. taygi	Tolmachev A.I. K istorii vozniknoveniya i razvitiya temnokhvoynoy taygi. [Towards the history of the origin and development of spruce-fir taiga]. Moscow; Leningrad, 1954.
Trautvetter, Incrementa	Trautvetter E.R. Incrementa florae phaenogamae rossicae. III. Acta Horti Petrop., IX, 1884.
Trautvetter, Salic. frigid.	Trautvetter E.R. De Salicibus frigidis Kochii. Nouv. Mém. Soc. Nat. Mosc., II, 1832.
Trautvetter, Salic. livon.	Trautvetter E.R. De Salicibus livonicis dissertatio. Nouv. Mém. Soc. Nat. Mosc., II, 1832.
Wimmer, Salic. europ.	Wimmer C.F.H. Salices europaeae. 1866. Bratislava.
Wolf, Mat. izuch. iv Yevr. Ross.	Wolf E. Materialy dlya izucheniya iv, rastushchikh diko v Yevropeyskoy Rossii. [Materials for the study of willows growing wild in European Russia]. Izv. SPb. Lesn. inst., 1900, 4–5.

FAMILY XIV

Cyperaceae J. St. Hil.

SEDGE FAMILY

EXTENSIVE FAMILY CONTAINING about 70 genera and more than 3000 (to 4000?) species. Widely distributed from regions at high latitudes in both hemispheres to equatorial countries inclusively. However, the majority of genera are mainly associated either with relatively cool extratropical (partly even alpine) terrain, or with countries of tropical climate. The overwhelming majority of representatives of the family are plants associated with distinctly (often excessively) moist habitats. Rather many species are amphibious plants. The sedge family also includes species adapted to conditions of considerable aridity, but these are in the minority.

The family Cyperaceae is one of the families characteristic of the arctic flora. In numbers of species it usually occupies one of the first places in the floral composition of arctic areas. Moreover, the hygrophilic nature of the majority of its species is reflected in the fact that the role of this family in floral composition is greater in arctic districts with more oceanic climate and greater development of marshes than in districts with more continental climate. Another characteristic of the disposition of species of this family in the Arctic is their diversity in relatively temperate parts of the region and rapid reduction in species numbers with progression to high arctic conditions.

A considerable part of the species of Cyperaceae occurring in the Arctic are characteristic of temperate areas of the Northern Hemisphere, especially the boreal forest zone. Many of them are typically boreal species that do not penetrate the Arctic very deeply. Subarctic species, roughly equally characteristic of more temperate parts of the Arctic and of the north of the forest zone, occupy a prominent place among arctic Cyperaceae. In particular, these include certain species that play a very substantial role in the formation of the vegetational cover of temperate parts of the Arctic. Cyperaceae possessing an arctic-alpine type of distribution are well represented in the arctic flora. Finally, a certain (rather limited) number of species are truly arctic plants whose distribution does not extend outside the Arctic or only to a very limited extent.

The absolute majority of the species of Cyperaceae entering into the composition of the arctic flora are characterized by a wide geographical distribution. Many species possess a circumpolar or almost circumpolar distribution. There are few more or less narrowly localized species in the Arctic. Species localized when only the arctic part of their ranges is considered are sometimes widespread outside our region.

With respect to ecology, the Cyperaceae in the Arctic (as well as outside its limits) are for the most part associated with well moistened habitats. In particular, a series of species of the genera most characteristic of the Arctic, *Carex* and *Eriophorum*, are prominent as the dominant or codominant plants in various types of marshy moss-herb tundras and in tundra marshes.

Species more characteristic of alpine tundra districts often grow abundantly along snowmelt streamlets, while others form beds on the shores of the more significant streams. Certain species of sedges are characteristic of low-herb meadows on sea shores. A limited number of species of Cyperaceae are associated with relatively dry habitats, growing in so-called open communities of the alpine tundra type. Some of these can survive the winter under conditions of extreme lack of snow.

In the Arctic the family Cyperaceae is represented by six genera. Of these the genus *Carex* is especially characteristic of the arctic flora and in many arctic districts occupies first place among the genera of higher plants with respect to numbers of species.

Less numerous but characteristic almost throughout the Arctic are species of the genus *Eriophorum*. Among the other genera, the presence of species of *Kobresia* is characteristic for some parts of the Siberian Arctic. The remaining genera are not at all characteristic of the arctic flora, contributing to its composition a few species that gravitate towards the southern fringes of our region.

1. Florets *unisexual, without perianth*. ....2.
− Florets *bisexual, with perianth* consisting of bristles or hairs. ....3.

2. Pistillate floret and later achene enclosed in modified closed bract (perigynium). Culm more or less leafy above. ....6. **CAREX** L. — **SEDGE**.
− Pistillate floret and achene surrounded by incompletely closed scalelike bract (more or less connate at base, rarely to middle). Culm leafy only at base. ....5. **KOBRESIA** WILLD. — **KOBRESIA**.

3. Perianth consisting of numerous *straight* silky hairs, much elongating after flowering and forming a downy spherical or somewhat oblong ("cottony") head. Inflorescence consisting of one or a few spikelets aggregated in cluster. Spikelets *many-flowered, 10–20 mm long* (excluding the hairs). Scales thinly membranous, grey or dark grey, not keeled, with attenuate soft tip, persistent, about 10 mm long. ....
....1. **ERIOPHORUM** L. — **COTTON GRASS**.
− Perianth *usually* consisting of *4–6 bristles*, not longer than scales. Sometimes perianth consisting of elongate silky hairs ("cottony"), but in that case hairs *flexuous*, spikelets few-flowered (*5–8 mm long* excluding the hairs), and scales not membranous, yellowish brown, the lower keeled, with median nerve prolonged as short blunt awn. ....4.

4. Spikelets several. ....3. **SCIRPUS** L. — **BULRUSH**.
− Spikelets solitary. ....5.

5. All leaf sheaths lacking blades. Perianth bristles serrate. Style usually with much thickened base (stylopodium) retained on fruit. ....
....4. **ELEOCHARIS** R. BR. — **SPIKE RUSH**.
− Upper leaf sheaths with short blades. Bristles or hairs of perianth smooth. Base of style not thickened; style retained on fruit as short mucro. ....
....2. **TRICHOPHORUM** PERS. — **DEER GRASS**.

GENUS 1

Eriophorum L. — COTTON GRASS

SMALL GENUS (15-20 species) distributed mainly in the temperate zone of the Northern Hemisphere. The majority of the species of the genus possess a wide distribution and are normally among the common plants growing in great quantity. Some of them belong to the most characteristic "background" plants of particular vegetational communities. Accordingly, the role of cotton grasses in the formation of the vegetational cover of those lands in which they grow is considerably greater than their role in the composition of the flora as reflected by species numbers. All cotton grasses are moisture-loving plants growing in relatively moist areas and some are typical marsh plants.

Several species of *Eriophorum* belong among the plants characteristic of the Arctic and widespread within its limits. However, the majority of the species characteristic of the Arctic are native to the more southern parts of our region and do not reach high arctic districts. *Eriophorum Scheuchzeri* and next to it *E. angustifolium* extend the furthest north. The species of the genus growing in the Arctic include typical arctic-alpine plants (*E. Scheuchzeri*), plants widespread in the boreal forest zone (*E. angustifolium, E. vaginatum, E. brachyantherum*), and finally mainly (*E. medium, E. russeolum*) or exclusively (*E. callitrix*) arctic species. A single boreal species (*E. gracile*) penetrates the Arctic only on its very edge and is not characteristic of the arctic flora.

In the Arctic species of *Eriophorum* grow mainly in various types of marshy tundra, showing preference either for peaty areas blanketed with dense moss cover or for sites with more open ground lacking a significant development of moss cover (*E. callitrix, E. Scheuchzeri*). The majority of species grow in abundance on suitable sites. Certain species react very positively to the trampling of mosses by reindeer, and certain others to disturbance of the primary vegetational cover associated with human activity. In particular, *E. Scheuchzeri* and *E. angustifolium* become "roadside" plants to an obvious degree along trails constructed across tundra.

1. Spikelets *several*, situated on culms that bear *1–3 leaflike bracts* at the base of the inflorescence. Inflorescence more or less secund; individual spikelets on slender pedicels of varying length, more or less *conspicuously drooping*. . . . . .2.
– Spikelets *solitary*, situated at tip of culm, standing erect or slightly inclined but *not drooping. No leaflike bracts*. . . . . .4.

2. Leaves *narrowly linear, triangular*. Culms slender, smooth or slightly scabrous below inflorescence. Leaflike bract 1 (sometimes 2), short, *not exceeding inflorescence* and scarcely overlapping it. Scales 4–5.5 mm long, 1.5–2.5 mm wide, greenish or brownish with scarious margin, usually with 3–5 longitudinal nerves. Anthers *linear, about 1.5 mm long*. . . . .
. . . . 3. **E. GRACILE** KOCH.
– Leaves *linear, flat or slightly folded lengthwise*, sometimes rather broad. Culms rather thick. Scales 6–9 mm long, 2–5 mm wide, thin, grey, brown

or blackish, the majority with 1 longitudinal nerve (the lower sometimes with 2–5). Anthers *linear, large, 3–4 mm long.* ....3.

3. Leaves always narrow, moderately developed. Leaflike bracts (1 or 2) scarcely rising above inflorescence. Spikelets at flowering time oblong, mostly about 10 mm long. Scales brown, sometimes ferruginous or dark grey, remaining highly conspicuous long after flowering and covering bases of spikelets. ....2. **E. KOMAROVII** V. VASSIL.
– Leaves sometimes rather broad (up to 8 mm), strongly developed. Leaflike bracts 2–3 in number, often conspicuously projecting above the drooping inflorescence. Spikelets at flowering time oval, 10–15 mm long. Scales grey, grey-brown or almost black, scarcely evident during fruiting. ....1. **E. ANGUSTIFOLIUM** HONCK.

4. Plant *with elongate creeping rhizomes, not forming tufts.* ....5.
– Plant *with abbreviated rhizomes, forming* more or less *dense*, often relatively large tussocklike *tufts.* ....7.

5. Culms solitary (although in beds they may be situated very close to one another, usually with roughly uniform spacing), *perfectly erect, smooth, rather thick*, pale green or often yellowish in fruiting plants. Basal leaves usually *reaching one-third of culm height.* Uppermost culm leaf modified into clasping sheath or possessing only narrow vestigial blade. Fruiting inflorescence *spherical*; its base concealed by downwardly directed hairs. Hairs *pure white* without any tint. Anthers *elongate-elliptic, 0.5–1 mm long*. Achene orbicular to tetragonal in cross-section. ....6. **E. SCHEUCHZERI** HOPPE.
– Culms solitary or in groups of 2–3, *relatively thin, erect or slightly arched on upper part*, green or sometimes brownish but not yellowish. Basal leaves usually *reaching half culm height*. Fruiting inflorescence *broadly obovate*, often *slightly declinate*, sometimes almost spherical; *its base not hidden* by hairs. Hairs *white* (*most often sooty white*), *ferruginous* or more or less *pale ferruginous or bright ferruginous orange*. Anthers *linear, 1–3 mm long*. Achene triangular in cross-section. ....6.

6. Spikelets at flowering time *oblong-elliptic*, in fruit broadly obovate. Hairs of fruit *white, ferruginous white or pale ferruginous* (rarely bright ferruginous). Scales *ovoid-lanceolate, acuminate*, dark grey or blackish, on margin with *very narrow scarious border or not bordered*; the lowest of them larger than the rest but not more than 7–8 mm long. Anthers *1–1.7(2) mm long.* ....5. **E. MEDIUM** ANDERSS.
– Spikelets at flowering time *short, ovoid*, in fruit diffuse, laxish, often almost globose but with base unprotected by hairs. Hairs of fruit either *intensely bright ferruginous* or *whitish* (f. *albescens*). Scales *broad, ovoid*, often *obtuse*, brownish or blackish, *with broad conspicuous hyaline border* (due to which the base of the inflorescence appears mottled); lowest scale large, during flowering concealing sides of spikelet, able to reach 1–1.5 cm long. Anthers *linear, 2–3 mm long.* ....4. **E. RUSSEOLUM** FRIES.

7. Tufts *dense*, sometimes *large, forming tussocks*, with bases buried in mossy turf. *Lower parts of culms clothed with dense covering of brownish leaf sheaths.* Scales *membranous, shining grey, often slightly translucent*, almost colourless on margins, *conspicuously reflexed at flowering time* (due to which the inflorescence appears crisped). Anthers linear, 2–3 mm long. . . . . 9. **E. VAGINATUM** L.
– Tufts small, often loose. Covering of brownish leaf sheaths at base of culms moderately developed. Scales *nontranslucent, dingy dark grey* (sometimes almost black), for the most part *closely imbricate*. Anthers narrowly elliptic, 0.5–1.5 mm long. . . . . .8.

8. Tufts *dense*. Culms up to *30–40 cm high, bearing 2* (more rarely 3) *abortive culm leaves*, the upper of which is almost always situated *distinctly above the middle of the culm.* Sheath of lower culm leaf sometimes with constricted narrow blade; *upper sheath broad, with funnel-shaped expansion towards tip*, for the most part *acuminate, but devoid of even vestigial leaf blade*, darkened on margin. Basal leaves narrow, subulate, erect. Lower scales *ovoid-lanceolate*, acuminate, *dark grey*, often *with greenish tinge*. Young inflorescence oblong, distinctly *projecting* above scales. Anthers up to 1.5 mm long. Hairs of fruit sometimes ferruginous. . . . .
. . . . 8. **E. BRACHYANTHERUM** TRAUTV. ET MEY.
– Tufts *loose*. Culms *8–12 (rarely up to 20) cm high*, normally bearing only *one culm leaf* situated *on lower half of culm*, often near its base. Its sheath *somewhat inflated, abruptly constricted apically into long narrow tip*, dark on margin or entirely more or less uniformly pale. Basal leaves not very narrow, often arcuately curved. Lower scales *large, ovoid*, narrowly acuminate at tip, *blackish grey (to almost black), almost concealing* the young inflorescence during its development. Anthers not more than 1 mm long. Hairs of fruit white. .7. **E. CALLITRIX** CHAM. EX C.A. MEY.

1. ***Eriophorum angustifolium*** Honck., Verz. Gew. Deutschl. (1782), 153; Hultén, Atlas, map 281; Chernov in Fl. Murm. II, 13; A.E. Porsild, Ill. Fl. Arct. Arch. 42; Böcher & al., Grønl. Fl. 240.

E. angustifolium Roth — Ledebour, Fl. ross. IV, 254; Schmidt, Fl. jeniss. 123; Scheutz, Pl. jeniss. 173; Krylov, Fl. Zap. Sib. III, 388; Andreyev, Mat. fl. Kanina 162; Yuzepchuk in Fl. SSSR III, 29; Perfilev, Fl. Sev. I, 133; Leskov, Fl. Malozem. tundry 33; Karavayev, Konsp. fl. Yak. 62.

E. angustifolium L. — Hultén, Circump. pl. I, 58.

E. polystachyum L. (pro parte) — Ostenfeld, Fl. arct. 1, 40; Tolmatchev, Contr. Fl. Vaig. 127; Tolmachev, Fl. Kolg. 15; id., Mat. fl. Mat. Shar 284; id., Obz. fl. N.Z. 150.

E. polystachyum f. *triste* Th. Fries — Lynge, Vasc. pl. N.Z. 99.

E. triste (Th. Fries) Löve et Hadač — Böcher & al., Grønl. Fl. 240.

E. subarcticum V. Vassil. in Not. syst. Herb. Inst. Bot. XIII (1950), 58.

E. angustifolium Roth β *triste* Th. Fries, Tillägg Spetsbergens Fanerogam-Flora.

E. angustifolium ssp. *triste* (Th. Fries) Hultén, Circump. pl. I, 58.

E. angustifolium ssp. *subarcticum* (V. Vassil.) Hultén, Circump. pl. I, 58.

Ill.: Fl. Murm. II, pl. I; A.E. Porsild, l. c., fig. 14, a, b.

A common plant of moist moss-herb tundras, often occurring abundantly and able to form beds. It grows on well moistened sites (in many cases intermittently or persistently with surplus moisture), sometimes forming almost pure stands, some-

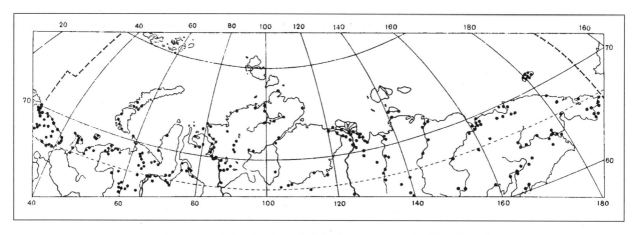

Map III–1 Distribution of *Eriophorum angustifolium* Honck.

times growing mixed with sedges (most frequently *Carex aquatilis*), moisture-loving grasses and other plants. It shows a preference for habitats that are drained to some degree. In northern, truly arctic districts it usually grows on loamy mineral soil at sites with adequate winter snow cover and moistened in summer by melting snowfields or regular atmospheric precipitation. In more southern parts of the tundra zone it sometimes occurs in tundras with flat hummocks, growing in the flarks between peat mounds. In this case the growth of *E. angustifolium* in flarks indicates the presence of some drainage. At sites regularly crossed by reindeer carts in summertime and along tundra trails ("vorg") where the moss cover has to a considerable degree been trampled by reindeer, *E. angustifolium* proliferates and sometimes forms roadside beds in continuous strips. When the cottongrass is in fruit, the trails along which such beds have been formed become especially conspicuous against the brownish green background of the surrounding tundra.

Soviet Arctic. Murman (everywhere); Kanin; Malozemelskaya and Bolshezemelskaya Tundras (common everywhere); Kolguyev Island; Polar Ural; Pay-Khoy and Vaygach Island; Novaya Zemlya (South Island and the south of the North Island); all tundra districts of West Siberia; lower reaches of Yenisey; Taymyr (whole peninsula, common); arctic coast of Yakutia along its whole extent; Bolshoy Lyakhovskiy Island; Wrangel Island; Chukotka; Anadyr Basin; Koryakia (everywhere). (Map III–1).

Foreign Arctic. Arctic Alaska; arctic coast of Canada; Labrador; Canadian Arctic Archipelago; North, West and northern part of East Greenland; Iceland; Spitsbergen; Arctic Scandinavia.

Outside the Arctic. Widespread in Northern and Central Europe, south to Northern Spain, mountainous districts of Italy, and Yugoslavia; found throughout the forest zone of the European part of the USSR and Siberia, in mainland districts of the Far East, on Kamchatka, the Kurile Islands and the Commander Islands; disjunctly in the northern part of the Great Caucasus and in the Tarbagatay; penetrating northern districts of Mongolia, NE China, and the north of the Korean Peninsula; in America throughout Alaska and Canada, and in the NW and NE United States.

Eriophorum angustifolium displays considerable variation within its extensive range, indicative of a tendency towards differentiation of geographical races. However, resolution of this question requires monographic treatment of the total material of *E. angustifolium* and related species, something beyond the task of the present work.

Of greatest interest for us is *E. angustifolium* β *triste* Th. Fries originally described from Spitsbergen, which was considered a high arctic race (subspecies) of this species by Hultén and has sometimes been distinguished from typical (moderately northern) *E. angustifolium* as the separate species *E. triste* (Th. Fries) Löve et Hadač. The basis for such treatment of β *triste* is evidently the considerable constancy of its distinguishing characters in Spitsbergen and again in the northern part of the Canadian Arctic Archipelago. However, it is completely impossible to draw any boundary between *E. angustifolium* s. str. and *E. triste* at all clearly under Soviet conditions. In plants from the Soviet Arctic the characters of β *triste* are rarely found in totality, but more often displayed separately in combination with characters of typical *E. angustifolium*. The most constant character of plants from higher arctic districts is condensation of the inflorescence due to shortening of the spikelet pedicels, a character highly dependent on the direct influence of environmental conditions. Additionally, plants from these districts usually possess very dark, blackish scales. Especially so are almost all plants from Novaya Zemlya, but even in these the intensity of the scale colour is not uniform. If it is considered that the combination of dark scale colour and a compact inflorescence is sufficient to refer plants to β *triste*, then a large portion of the plants from Novaya Zemlya, Central Taymyr and a series of other arctic districts could be considered to belong to that race or variety. But it should be emphasized that the third character of β *triste*, scabrousness of the spikelet pedicels, appears extremely inconstant. In the majority of plants from our European arctic islands this character is expressed weakly or not at all. Plants from the European mainland tundras usually possess the full complex of characters of typical *E. angustifolium* (smooth spikelet pedicels, relatively pale scales, lax inflorescence with many spikelets), but are least constant with respect to the last character. More definite tendencies towards β *triste* are rarely observed here.

The great majority of specimens of *E. angustifolium* from Arctic Siberia possess the complex of characters agreeing with the characterization of the typical species. Plants from the most northern localities (Dikson Island, Yenisey Bay, Central Taymyr, certain sites on the arctic coast of Yakutia) are frequently low-growing and possess a condensed inflorescence with dark blackish scale colour. But, we repeat, they have smooth spikelet pedicels like typical *E. angustifolium*. Only rarely can one find plants with scabrous spikelet pedicels (for example, some plants from the Indigirka Delta and a portion of the plants from Lawrence Bay in Chukotka). There is a complete gamut of transitions between plants with different scale colour and different density of the inflorescence, which suggests that it is inappropriate to separate a high arctic variety (or race) from *E. angustifolium* under our conditions.

Among plants of mainland tundras at some distance from the northern edge of the species' range, there is a rather sharply differentiated form with broader leaves and thickened culm base clothed with the sheaths of old leaves. In some cases a separate spikelet-bearing branch of the inflorescence arises from the axil of the uppermost leaf in such plants. This form agrees with the concept of var. *majus* Schultz (? var. *elatior* Koch), and specimens possessing the lateral branch of the inflorescence also served as the basis for the description of the "species" *E. subarcticum* V. Vassil. *Eriophorum angustifolium* var. *majus* sometimes occurs in temperate arctic districts both in the European North and in extreme NE Asia (more rarely in the greater part of Arctic Siberia). The development of this form (variety?) is associated with habitats where deep mossy turf is converted beneath into a layer of peat (for example, in the flarks of tundra with flat hummocks). This ecological association of var. *majus* is evidently responsible for its development mainly in districts with milder and moister climate, where the growth of mosses and peat development is most intensive.

The relatively broad foliage of *E. angustifolium* var. *majus* has apparently caused plants of this form to be sometimes identified as *Eriophorum latifolium* Hoppe, Bot. Taschenb. (1800) 108. The literature on the arctic flora contains rather numerous records of the growth of the latter species in arctic districts of the USSR. Thus, I.A. Perfilev (Mat. fl. N.Z., Kolg. 52) records it for the vicinity of Bugrino on Kolguyev Island; V.N. Andreyev (l. c. 162) records it for the northern part of Kanin, and this record is repeated in I.A. Perfilev's review of the flora of the Northern Territory (l. c. 133); M.N. Karavayev (l. c. 62) records *E. latifolium* as growing in Arctic Yakutia.

Review of the total material of cottongrasses with many spikelets from the Soviet Arctic confirms the conclusion of Ostenfeld (l. c. 39) that it is inaccurate to identify arctic plants as *E. latifolium*. In particular, the plants from Kanin (Andreyev's collections) really belong to *E. angustifolium* Honck. I.A. Perfilev soon corrected his original identification of the plants from Kolguyev Island. In the tundras near the Pechora, whence abundant material of cottongrasses is available, likewise only one species with many spikelets occurs, *E. angustifolium*.

Records of the possible occurrence of *E. latifolium* in NE Siberia may be ruled out because, as indicated in his time by V.N. Vasilev (Bot. mat. Gerb. Bot. inst. AN SSSR VIII) and more recently emphasized by Hultén (Circump. pl. I, 58), *E. latifolium* does not occur at all either in East Siberia or the Far East. Records for Arctic Yakutia should be referred to *E. angustifolium*.

2. Eriophorum Komarovii V. Vassil. in Not. syst. Herb. Inst. Bot. VIII (1940), 102.
? *E. angustifolium* ssp. *scabriusculum* Hultén, Fl. Al. II (1942), 277; id., Circump. pl. I, 58, map 51.

A critical and still inadequately studied species, undoubtedly very close to *E. angustifolium*. In its typical form occurring on the north coast of the Sea of Okhotsk or in the mountains of the Aldan Basin, it has a somewhat distinctive habit (relatively weak leaf development despite the considerable size of the plant; scales narrow and of a peculiar ferruginous colour; narrow spikelets) which enables it to be distinguished from *E. angustifolium* at first glance. In more northern districts (Koryakia, Anadyr Basin, Chukotka Peninsula), individual characters of *E. Komarovii* are sometimes found in plants in other respects indistinguishable from *E. angustifolium*. On this account the question of the existence of *E. Komarovii* as a species and its relationships with *E. angustifolium* should be considered not yet finally settled. We present data on *E. Komarovii* separately mainly to draw increased attention to this form of cotton grass and thus to encourage the collection of new material needed for critical investigation.

Growing in marshy tundras of various types, apparently not differing from *E. angustifolium* in this respect.

Soviet Arctic. SE Chukotka Peninsula; Anadyr Basin. Doubtful specimens and specimens transitional to *E. angustifolium* exist from the district of Cape Dezhnev and Koryakia (Bay of Korf).

Foreign Arctic. Not reported.

Outside the Arctic. Coast of Sea of Okhotsk and the neighbouring part of the Kolyma Basin; Aldan Mountains. (If the identification of *E. angustifolium* ssp. *scabriusculum* Hult. with *E. Komarovii* V. Vassil. is correct, in the temperate north of North America from Southern Alaska to Newfoundland.)

3. Eriophorum gracile Koch in Roth, Cat. II (1800), 259; Ledebour, Fl. ross. IV, 255; Krylov, Fl. Zap. Sib. III, 390; Perfilev, Fl. Sev. I, 133; Yuzepchuk in Fl. SSSR III, 30; Hultén, Atlas, map 283; Chernov in Fl. Murm. II, 16; Hultén, Circump. pl. I, 96, map 87.

E. coreanum Palla in Oest. Bot. Ztschr. LIX (1905), 190; Yuzepchuk in Fl. SSSR III, 30; Karavayev, Konsp. fl. Yak. 62.

E. gracile ssp. *coreanum* Hultén, Fl. Kamtch. I (1927), 160.

E. asiaticum V. Vassil. in Not. syst. Herb. Inst. Bot. VIII (1940), 104; Karavayev, l. c. 62.

Widespread, almost circumpolar boreal species, weakly differentiated into two (or more?) geographical varieties, the European to West Siberian *E. gracile* s. str. and the Pacific *E. gracile* ssp. *coreanum* (or *E. gracile* ssp. *asiaticum*, if V.N. Vasilev's opinion that more northern plants from East Asia can be separated from *E. coreanum* Palla is confirmed). Just penetrating arctic limits on the northern edge of its range.

Soviet Arctic. East coast of Kola Peninsula north of the mouth of the Ponoy; Nakhodka Bay on the SE coast of Yamal (VII 1912, in flower, Bushevich); basin of River Poluy, margin of marshy pond (23 VII 1914, in fruit, Gorodkov); Anadyr (without precise indication of locality, presented by Bunge). Doubtfully recorded for Arctic Yakutia by M.N. Karavayev; not reaching the forest-tundra on the Yenisey.

Foreign Arctic. Isolated localities in Northern Scandinavia.

Outside the Arctic. Forested region of Europe from the Pyrenees in the southwest and the Balkans in the southeast to Northern Scandinavia; the forest zone of the European part of the USSR, West and East Siberia (in East Siberia and the Far East ssp. *coreanum* or ssp. *asiaticum*); Preamuria; Primorskiy Kray; the north of the Korean Peninsula; Sakhalin; the south of the Kurile Chain; Northern Japan; Kamchatka and the coast of the Sea of Okhotsk; Southern Alaska; forested districts of Canada and the NW and NE United States.

4. *Eriophorum russeolum* Fries in Hartman, Handb. Skand. Fl., ed. 3 (1838), 13; Fries, Novit. Fl. suec. Mant. 3 (1842), 170; Trautvetter, Pl. Sib. bor. 121; Schmidt, Fl. jeniss. 123; Scheutz, Pl. jeniss. 173; Ostenfeld, Fl. arct. 1, 42; Hultén, Fl. Kamtch. I, 162; Tolmatchev, Contr. Fl. Vaig. 127; Tolmachev, Fl. Kolg. 15; Andreyev, Mat. fl. Kanina 163; Yuzepchuk in Fl. SSSR III, 35; Perfilev, Fl. Sev. I, 132; Tolmachev, Obz. fl. N.Z. 150; Leskov, Fl. Malozem. tundry 33; Hultén, Atlas, map 286; Gorodkov in Sp. rast. Gerb. Fl. SSSR 3020a.

E. intermedium Chamisso in litt., pro maxima parte.

E. Chamissonis C.A. Mey. in Ledebour, Fl. Alt. I (1829), 70, pro parte; id. in Mém. Ac. Sci. St.-Pétersb. Sav. Étr. 1 (1831), 204; Lynge, Vasc. pl. N.Z. 98; Lindman, Sv. Fanerogam. fl. 112.

We adopt for the species under consideration the name given to it by Fries, which reflects one of the fundamental characters of the typical form of the species. The earliest name of this species is that given to it by Chamisso (*Eriophorum intermedium*), but this was not published by the author and on the occasion of publication (as Meyer already indicated) could not be retained on account of homonymy with the identical name of another earlier described species (*E. intermedium* Bast.). Meyer included the plants named *E. intermedium* by Chamisso within the species he described as *E. Chamissonis*. But along with them he also referred to this species Altaic plants possessing only limited relationship to the northern. Because these Altaic plants are mentioned by Meyer with precise references to herbarium specimens, it is appropriate to consider these the formal type of *E. Chamissonis* C.A. Mey., not the plants described by Chamisso under the name *E. intermedium*. The uniting of Altaic plants with *E. intermedium* Cham. (with mention only of specimens from Unalaska which in their turn are nonhomogeneous!) forces one to consider the name *E. Chamissonis* as a typical case of a nomen confusum.

The question of the correct name for this species already received lengthy clarification at the time by Hultén (Fl. Kamtch.), but became confused again after publication of the treatment of the genus *Eriophorum* in the Flora of the USSR,

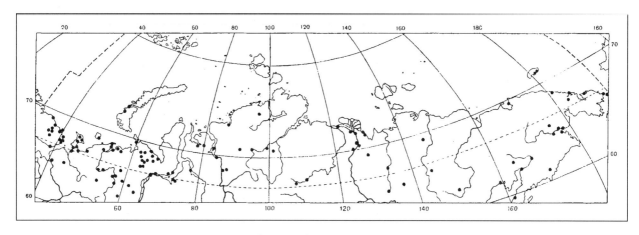

Map III–2 Distribution of *Eriophorum russeolum* Fries.

where S.V. Yuzepchuk attempted to attach the name *E. Chamissonis* to another northern form of cotton grass to which only a minority of the plants from Unalaska seen by Meyer belong. To the contrary, the majority of the northern plants seen by Meyer and referred by him to *E. Chamissonis* (part of the plants from Unalaska directly mentioned in the diagnosis; Mertens' collections from Senyavin Strait; Chamisso's collections from Shishmaref Inlet) consist of confirmed specimens of *E. russeolum* Fries, not of the form intermediate between it and *E. Scheuchzeri* Hoppe to which the name *E. Chamissonis* C.A. Mey. was attached in the Flora of the USSR. Authors who applied the name *E. Chamissonis* to the plants described as *E. russeolum* by Fries were undoubtedly closer to the truth. But the preference accorded by them to Meyer's name as being older was associated with lack of study of the Altaic plants, which cannot be excluded from the ranks of legitimate bearers of this name.

On account of this whole nomenclatural muddle, it is better not to use the name *E. Chamissonis* at all but to recognize *E. russeolum* Fries as the legitimate species name for the bulk of the northern plants identified as *E. Chamissonis* C.A. Mey. The plants treated under the name *E. Chamissonis* in the Flora of the USSR are more correctly (to prevent misunderstanding) called *E. medium* Anderss.

Eriophorum russeolum is a common characteristic plant of moist moss-herb tundras of temperate districts of the Soviet Arctic. Most often it is found in the shrubby (dwarf birch) and mossy tundra subzones. In the more northern subzones it becomes relatively rare (but hardly ever missed in botanical collecting on account of its conspicuous coloration). *Eriophorum russeolum* grows, often forming discontinuous beds, in the flarks between peat mounds and on low shores of tundra lakes, always on sites with continuous moss carpet, often with surplus moisture. It does not shun sites with standing water.

The distribution of *E. russeolum* is of a generally subarctic character. As well as in the Arctic, this species sometimes occurs in the northern forest zone of Eurasia and in Siberia even reaches southern districts where it remains a relatively common plant. However, it is absent from the mountains of the southern fringe of Siberia.

Soviet Arctic. Murman; Kanin; Timanskaya, Malozemelskaya and Bolshezemelskaya Tundras (more or less ubiquitous); Kolguyev Island; Polar Ural; Vaygach Island; South Island of Novaya Zemlya; lower reaches of Ob and district of Ob Sound and Tazovskaya Bay; Gydanskaya Tundra; lower reaches of Yenisey; Taymyr (north to the mouth of the Pyasina and the Mamontovaya in the central part of the Peninsula); Arctic Yakutia (especially the lower reaches of the Lena); Chukotka; Wrangel Island; Anadyr Basin; Koryakia. (Map III–2).

Foreign Arctic. Arctic Alaska; arctic coast of Canada; Labrador; Baffin Island; disjunctly in NW Iceland; Arctic Scandinavia.

Outside the Arctic. Occurring occasionally in Northern Scandinavia, and in the northern part of the forest zone of the European part of the USSR especially the Pechora Basin (south to the Ukhta-Vym watershed and the mouth of the River Shchugor); in Siberia south to Tobolsk, the Vakh and Ket Rivers, the middle course of the Yenisey, the Angara region, southern Transbaikalia, and the Aldan Basin; in the Far East occurring on Kamchatka and on the coasts of the Sea of Okhotsk as far as Ayan and the Shantar Islands. In Preamuria and Sakhalin replaced by the closely related species or race *E. mandshuricum* Meinsh. (= *E. russeolum* var. *majus* Somm.). Found in North America in Alaska and in the north and extreme east of the Canadian forest zone. It is possible that the species is distributed considerably more widely there, but this is difficult to establish because the interpretation of *E. russeolum* and *E. Chamissonis* by American authors differs from ours.

In the lower reaches of the Yenisey, plants combining characters of *E. russeolum* and *E. angustifolium* have been reported. It is possible that we are dealing with hybridization of the two species, although they are very distantly related to one another systematically.

5. ***Eriophorum medium*** Anderss. in Bot. Notis. (1857), 62; Krylov, Fl. Zap. Sib. III, 385; Tolmachev, Obz. fl. N.Z. 150; Hultén, Atlas, map 285.

E. intermedium Cham. in litt., pro minima parte.

E. Chamissonis C.A. Mey. in Ledebour, Fl. Alt. I (1829), 70, pro minima parte; Yuzepchuk in Fl. SSSR III, 36; Raymond in Sv. Bot. Tidskr. 48, 79, pro parte; Leskov, Fl. Malozem. tundry 33.

E. intercedens Lindb. fil., Schedae Pl. Finl. exsicc., fasc. I–VIII (1906), 32-33; Lindman, Sv. Fanerogam. fl. 112; Lynge, Vasc. pl. N.Z. 98; Tolmachev, Mat. fl. Mat. Shar 284.

Our adoption of the name *Eriophorum medium* for the present species is conditioned by the obvious unacceptability of the treatment of this matter in the Flora of the USSR and by authors who relied upon that work. The name *Eriophorum Chamissonis*, proposed by Meyer, refers primarily to Altaic plants that cannot be excluded from the content of the species established by him (as S.V. Yuzepchuk and later Raymond attempted to do). The fact that Meyer himself, two years after publication of the original description of *E. Chamissonis*, focussed his main attention not on Altaic specimens but on Chamisso's material from Kamchatka and Alaska does not save the situation. Furthermore, only part of the series of northern plants referred by Meyer to his species belong to the form here under consideration, while the rest belong to *E. russeolum*. In these circumstances use of the name *E. Chamissonis* would only be a source of confusion. On the other hand, Andersson's name *E. medium* has always been applied only to a single determinate form of cotton grass. This was described by Lindberg under the name *E. intercedens*; but his name is not available for use in view of the existence of the other earlier name.

The question of the nature of *E. medium* and its delimitation from *E. russeolum* is obscure in many respects and complicated by the undoubted existence of plants with intermediate characters. The typical form of *E. russeolum* is well differentiated from *E. medium*. But one of the fundamental characters of *E. russeolum* and that which most strikingly catches the eye, the bright ferruginous colour of the hairs, is actually unreliable; according to the form of the fruiting inflorescence and the broadly obtusish shape of the scales with presence of a broad hyaline border, there are also rather numerous plants with pale ferruginous or even almost white hairs that should be referred to *E. russeolum*. On the whole the following are better distinguishing characters of *E. medium* in comparison with *E. russeolum*: the oblong shape of the spikelet at flowering time, the smaller size of the lowest scale

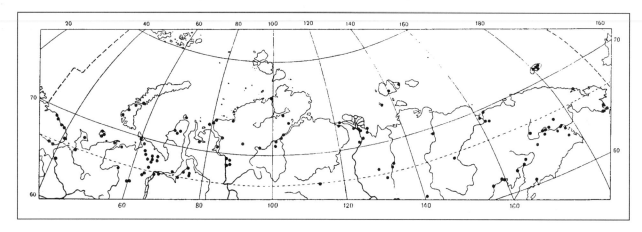

MAP III–3 Distribution of *Eriophorum medium* Anderss.

(only just covering the base of the inflorescence), the narrower acuminate shape of the remaining scales with poor development or lack of a pale hyaline border, the relatively small (not longer than 2 mm) anthers, the more compact (broadly obovoid) shape of the fruiting inflorescence, and the usually pallid hairs (dull ferruginous or whitish, sometimes white).

Additional difficulties in delimiting *E. russeolum* and *E. medium* are caused by the absence of ecological differences between them. Like *E. russeolum*, *E. medium* grows in moist moss-herb tundras, penetrating the moss turf with its rhizomes. It forms beds on level areas of marshy tundra, on shores of tundra lakes, in flarks between peat mounds, and on level mossy shores of tundra streams. It shuns areas of exposed mineral soil. Beds of *E. medium* are usually rather dense (with the species showing clear dominance over other plants), but not entirely closed. When the cotton grass is in fruit, these beds stand out as bright patches against the brownish-greenish background of marshy tundra. The colour of the fruiting inflorescences in each particular bed is more or less uniform, but beds with different colour occasionally occur side by side.

As has been repeatedly noted, *E. medium* in many respects occupies an intermediate position between *E. russeolum* and *E. Scheuchzeri*, in which connection the question of its possible hybrid origin arises. An additional argument in favour of this hypothesis is the fact that *E. medium* occurs almost exclusively where the ranges of *E. russeolum* and *E. Scheuchzeri* overlap. But a few arctic districts (Novaya Zemlya, Northern Taymyr) are exceptions, since here *E. medium* extends further north than *E. russeolum*. It is noteworthy that, for example, in the district of Matochkin Shar on Novaya Zemlya *E. medium* possesses only white hairs and somewhat approaches *E. Scheuchzeri* both morphologically and biologically. (May we not be dealing with populations originating from hybridization between *E. medium* and *E. Scheuchzeri*?) In the south of the range of *E. russeolum* (for example, in the southern half of the Yenisey Basin and in Transbaikalia), where *E. Scheuchzeri* is absent, forms to any degree doubtful with respect to their belonging to *E. russeolum* or *E. medium* have not been observed.

In assessing the likelihood of a hybrid origin of *E. medium*, we must consider the possibility of independent origin of plants of this type in different parts of the expanse over which the ranges of the presumed parent species overlap. Therefore, it is not excluded that more detailed study of local populations of *E. medium* will reveal more or less stable manifestations of the geographical differentiation of this species.

Eriophorum medium is distributed as a more or less common plant in temperate districts of the Arctic. It only reaches high arctic districts at sites uncharacteristic of these districts. It occurs here and there in northern districts of the forest zone.

Soviet Arctic. Murman (rare); Kanin; Malozemelskaya and Bolshezemelskaya Tundras; Kolguyev Island; Polar Ural, Pay-Khoy; Vaygach Island; South Island and the southeast of the North Island of Novaya Zemlya; lower reaches of Ob, Yamal, Obsko-Tazovskiy Peninsula; lower reaches of Yenisey; Taymyr (not ubiquitous!, most northern locality at mouth of Lower Taymyra River); Khatanga River; lower reaches of the Olenek, Lena and Kolyma; Chukotka; Wrangel Island; Anadyr Basin; Koryakia (Penzhina Basin, Bay of Korf). (Map III–3).

Foreign Arctic. Arctic Alaska; arctic coast of Canada; Labrador; the south of the Canadian Arctic Archipelago; Arctic Scandinavia. The data here given are approximate, since the referral of particular data in the literature to *E. medium* is in many cases tentative.

Outside the Arctic. Subarctic Fennoscandia; isolated localities in the lower reaches of the Severnaya Dvina, at Mezen, and in the northern half of the Pechora Basin; very rare in Central Yakutia, more frequent in the Verkhoyansk Range, in northeastern districts of Yakutia and on the northern coasts of the Sea of Okhotsk; isolated localities in Kamchatka. Distribution in North America (except for Alaska where its existence is not in doubt) unclear.

6. ***Eriophorum Scheuchzeri*** Hoppe, Bot. Taschenb. (1800), 104; Ledebour, Fl. ross. IV, 253; Trautvetter, Fl. taim. 22; id., Pl. Sib. bor. 121; Schmidt, Fl. jeniss. 123; Scheutz, Pl. jeniss. 172; Ostenfeld, Fl. arct. 1, 41; Lynge, Vasc. pl. N.Z. 99; Tolmatchev, Contr. Fl. Vaig. 127; Krylov, Fl. Zap. Sib. III, 383; Andreyev, Mat. fl. Kanina 163; Yuzepchuk in Fl. SSSR III, 36; Tolmachev, Fl. Kolg. 15; id., Mat. fl. Mat. Shar 184; id., Obz. fl. N.Z. 150; Perfilev, Fl. Sev. I, 132; Leskov, Fl. Malozem. tundry 33; Hultén, Atlas, map 287; id., Circump. pl. I, 26; Chernov in Fl. Murm. II, 23; A.E. Porsild, Ill. Fl. Arct. Arch. 42; Böcher & al., Grønl. Fl. 240; Karavayev, Konsp. fl. Yak. 63.

E. capitatum Host, Gram. I (1801), 30.

Ill.: Fl. Murm. II, pl. IV.

Circumpolar arctic-alpine species possessing a very wide distribution in the Arctic and usually common starting from the southern fringe of the tundra zone as far as the furthest northern limits of land. Penetrating the northern fringe of the forest zone mainly in river valleys.

Growing in well moistened but not peaty sites, forming dense clusters and beds sometimes spreading over a considerable area. Often growing together with moisture-loving grasses (for example, *Calamagrostis neglecta*). Very characteristic of level mossy sites along streams where the mosses only form a very thin carpet. Also common at sites with silty soil devoid of moss cover. Readily spreading over fresh alluvial areas. Not characteristic of true marshy tundras with hummocks and flarks. In mountainous districts associated with small hollows in flat-topped uplands and with the valleys of rivers and streams. In comparison with other species of cotton grass, it forms the densest beds.

Easily recognized at first glance, especially in the fruiting state, due to its spherical inflorescence with pure white hairs and the relatively short, rather thick, straight culms (leaning but not bending in wind) which often possess a yellowish tint.

Soviet Arctic. Murman; Kanin; Timanskaya and Malozemelskaya Tundras; Kolguyev Island; Bolshezemelskaya Tundra; Polar Ural; Pay-Khoy; Vaygach Island; Novaya Zemlya north to 75°; Yamal with Belyy Island; district of Ob Sound and Tazovskaya Bay; Gydanskaya Tundra; lower reaches of Yenisey and islands of Yenisey Bay; the whole of Taymyr all the way to Cape Chelyuskin; Begichev Island; lower reaches of

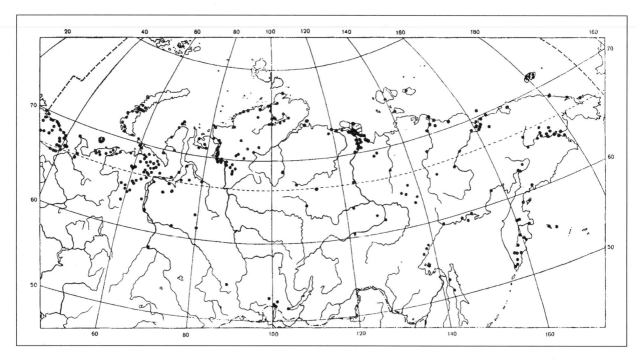

Map III-4 Distribution of *Eriophorum Scheuchzeri* Hoppe.

Olenek and Lena, Lena Delta; New Siberian Islands; lower reaches of Indigirka and Kolyma; Chukotka (not found in the southeast of the Chukotka Peninsula); Wrangel Island; Anadyr Basin; Koryakia (Gizhiga, Penzhina, Bay of Korf). (Map III–4).

Foreign Arctic. Arctic Alaska; arctic coast of Canada; Labrador; Canadian Arctic Archipelago north to the northern part of Ellesmere Island; Greenland from the southern extremity to Peary Land; Iceland; Spitsbergen; Arctic Scandinavia.

Outside the Arctic. Norway, Northern Sweden and Finland; Kola Peninsula, Northern Karelia; northern fringe of the forest zone of the northeast of the European part of the USSR; Urals (south to 59°N); West Siberian Lowland north from the middle course of the Ob; Yenisey Basin north of Turukhansk; considerable part of Yakutia; coasts of Sea of Okhotsk; Northern Sakhalin; Kamchatka; Alaska; northern districts of Canada; mountains of Central Europe. Rarely reported in the mountains of Southern Siberia, where it is mainly replaced by the closely related species *E. altaicum* Meinsh.

7. ***Eriophorum callitrix*** Cham. ex C.A. Mey. in Mém. Ac. Sci. St.-Pétersb. Sav. Étr. I (1831), 203; Ledebour, Fl. ross. IV, 253, pro parte; Fernald in Rhodora 7, 208; Seidenfaden & Sørensen in Meddl. om Grønl. 101, 1, 10; Raymond in Natur. Canad. LXXVIII, 9, 286; Ostenfeld, Fl. arct. 1, 41, pro parte; Macoun & Holm, Vasc. pl. 8a; Krylov, Fl. Zap. Sib. III, 386; Yuzepchuk in Fl. SSSR III, 35; Karavayev, Konsp. fl. Yak. 62; Seidenfaden & Sørensen, Vasc. pl. NE Greenl. 87, 178; Sørensen, Vasc. pl. E Greenl. 126; A.E. Porsild, Ill. Fl. Arct. Arch. 43; Jørgensen & al., Fl. pl. Greenl. 32; Böcher & al., Grønl. Fl. 240; Hultén, Circump. pl. I, 12, map 6.

Ill.: C.A. Meyer, l. c., tab. 2; Seidenfaden & Sørensen in Meddl. om Grønl., fig. 1, pl. 1, 3; Seidenfaden & Sørensen, Vasc. pl. NE Greenl., fig. 31; A.E. Porsild, l. c., fig. 14, g.

Species of Arctic Siberia and North America, just penetrating subarctic districts locally on the southern edge of its range. Forming small, not very dense tufts on moist level or sloping areas of tundra, normally on mineral soil. It can grow both

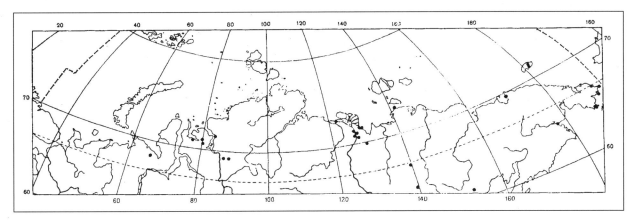

MAP III–5 Distribution of *Eriophorum callitrix* Cham. ex C.A. Mey.

within a continuous carpet of moss and herbs and in loamy areas with semiexposed soil surface. Areas with strong peat development are specifically avoided. Although locally not uncommon, it does not occur in vast numbers. The culms that form the tufts are not closely contiguous to one another at their bases, but often somewhat divergent and arcuate basally.

The distribution of *E. callitrix* is inadequately known. The available data give the impression that the range is somewhat fragmented. But it cannot be excluded that this species may sometimes be overlooked by collectors because it grows in less abundance than other species of cotton grass.

Soviet Arctic. Karskaya Tundra (collected by Sukachev according to Krylov); Gydanskaya Tundra (near Gyda-Yam Bay, on Lake Khasseyn-To, etc., collected by Schmidt, 1866 and Gorodkov, 1927); lower reaches of Yenisey (Golchikha, collected by Kuznetsov and Reverdatto, 1914); forest-tundra beyond the Yenisey (on the River Kargachnaya, collected by Kuznetsov and Reverdatto, 1914); between the Lena and the Olenek; on the southern edge of the Lena Delta from Olenek Bay to Buorkhaya Bay; the lower reaches of the Lena south to the forest-tundra, and mountains of the right bank of the Lena; Cape Svyatoy Nos between the mouths of the Yana and the Indigirka[1]; Wrangel Island; Chukotka (River Chegitun, district of Cape Dezhnev, Lawrence, Olga and Provideniye Bays); mouth of Anadyr. (Map III–5).

Foreign Arctic. St. Lawrence Island; Arctic Alaska; arctic coast of Canada; Labrador; southern part of Canadian Arctic Archipelago (from Banks Island to Baffin Island); East Greenland (between 70° and 78°N).

Outside the Arctic. Barren zone of Verkhoyansk Range; mountains on upper reaches of right-bank tributaries of the Upper Kolyma; the Evota Barren in the Aldan Basin; mountains of Alaska; scattered localities in the Rocky Mountains south to the NW United States; southern part of Hudson Bay; northern tip of Newfoundland.

Certain plants referred by us to *E. callitrix* tend towards *E. brachyantherum* in some characters. This especially applies to certain specimens from the lower reaches of the Lena, the Verkhoyansk Range and the Kolyma Basin, and to a portion of the plants from the Gydanskaya Tundra.

8. *Eriophorum brachyantherum* Trautv. et Mey., Fl. ochot. phaen. (1856), 98; Meinshausen, Cyper. Fl. Russl. 49; Yuzepchuk in Fl. SSSR III, 34; Raymond in Natur. Canad. LXXVIII, 9, 286–287; A.E. Porsild, Ill. Fl. Arct. Arch. 43; Karavayev, Konsp. fl. Yak. 62; Hultén, Atlas, map 282; id., Circump. pl. I, 34, map 28.

E. callitrix auct. non Cham. — Kjellman & Lundström, Phanerogam. N.Z., Waig. 156; Kjellman, Phanerogam. fl. N.Z. 166; Scheutz, Pl. jeniss. 173; Ostenfeld, Fl. arct. 1,

---

[1] Included on the basis of examination of the herbarium of the Botanical Institute of the USSR Academy of Sciences in 1932; the relevant specimen is now missing from the herbarium. Records of the occurrence of *E. callitrix* at Osinin on the Indigirka (Birkenhof) and at Lake Yessey in the Khatanga Basin (I.P. Tolmachev) based on my *(A. T.)* provisional identifications have proved to be erroneous. The relevant specimens must be referred to *E. brachyantherum*.

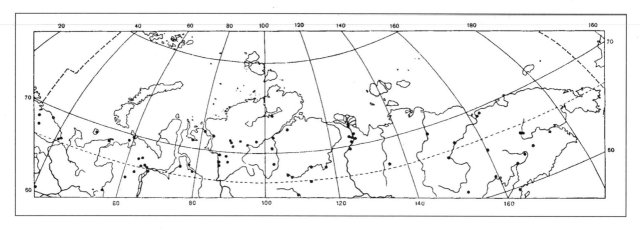

Map III–6 Distribution of *Eriophorum brachyantherum* Trautv. et Mey.

   41, pro parte; Lynge, Vasc. pl. N.Z. 136; Tolmatchev, Contr. Fl. Vaig. 127; Krylov, Fl. Zap. Sib. III, 386, pro maxima parte.
E. vaginatum var. *opacum* Björnstr., Grunddr. Pitea Lapp. växtfys. (1856), 35.
E. opacum Fern. in Rhodora 7, 77 (1905), 85; Seidenfaden & Sørensen in Meddl. om Grønl. 101, 1, 12; Perfilev, Fl. Sev. I, 132; Tolmachev, Obz. fl. N.Z. 150.
E. vaginatum var. *brachyantherum* Kryl., Fl. Alt. (1914), 1437; id., Fl. Zap. Sib. III, 388.
E. vaginatum auct. non L. — Tolmachev, Fl. Taym. I, 102, pro parte.
Ill.: Seidenfaden & Sørensen, l. c., fig. 2, pl. 2, 4.

   Species with distribution of a generally subarctic type. *Eriophorum brachyantherum* penetrates arctic limits on the northern fringe of its range, locally reaching districts rather far from the forest boundary (Southern Novaya Zemlya, the mouth of the Yenisey, the district of Lake Taymyr), but occurs regularly only in the more temperate part of the Arctic. Here *E. brachyantherum* is not uncommon, but does not occur everywhere. The similarity of its appearance to *E. vaginatum* is possibly responsible in part for material of this relatively rarer species (in comparison with *E. vaginatum*) sometimes not being collected where the well-known *E. vaginatum* is common.

   Unlike the latter species, *E. brachyantherum*, even where it is common, does not grow in great abundance and does not become a species that determines the appearance of vegetational communities. It forms small, rather dense tufts but apparently never forms tussocks like those often observed in *E. vaginatum*. It grows both in open clayey areas of tundra and at sites with continuous moss cover, and does not shun areas with surplus moisture. It is absent wherever a rather thick peat layer develops (especially on peat mounds). In general, *E. brachyantherum* apparently has more exacting requirements for mineral nutrition than has *E. vaginatum*.

   Where the ranges of *E. brachyantherum* and *E. vaginatum* overlap, plants with intermediate characters (probably hybrids) are sometimes reported. Particularly characteristic of these is the dingy dark grey colour of the scales and their somewhat reflexed tips, due to which the surface of the inflorescence is rough, almost as in *E. vaginatum*.

Soviet Arctic. Murman (rare); the north of the Malozemelskaya Tundra; Kolguyev Island (apparently rare); eastern fringe of the Bolshezemelskaya Tundra; Polar Ural; Dolgiy and Vaygach Islands; the south of Novaya Zemlya; lower reaches of Ob, district of Ob Sound and Tazovskaya Bay; lower reaches of Yenisey (rather common); the south of the Taymyr Tundra (reaching the SE shore of Lake Taymyr); lower reaches of the Khatanga, Lena (rather common), Indigirka and Kolyma; Anadyr Basin; Koryakia. (Map III–6).

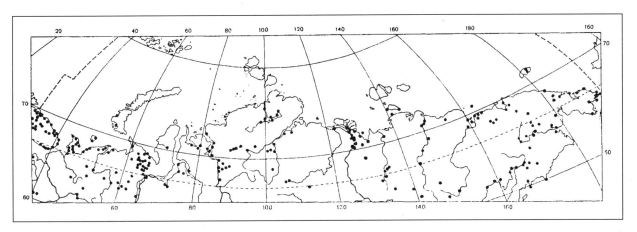

MAP III–7 Distribution of *Eriophorum vaginatum* L.

Foreign Arctic. Arctic Alaska; arctic coast of Canada; Labrador; Baffin Island (single locality on east coast); Arctic Norway.

Outside the Arctic. Mountains of Scotland (rare, reliability of record cast in doubt by Hultén!); Northern Scandinavia (mainly Northern Sweden); remote districts of Northern Finland; here and there in Karelia and on the Kola Peninsula; rarely in the forested part of Arkhangelsk Oblast; Northern Ural; forest zone of East Siberia (distributed more or less throughout and common in many places); Northern Preamuria; NW coast of Sea of Okhotsk; interior districts of Kamchatka; greater part of Alaska and Canada.

9. *Eriophorum vaginatum* L., Sp. pl. (1753), 521; Ledebour, Fl. ross. IV, 252; Trautvetter, Fl. taim. 22; id., Pl. Sib. bor. 121; Schmidt, Fl. jeniss. 123; Scheutz, Pl. jeniss. 172; Ostenfeld, Fl. arct. 1, 40; Lynge, Vasc. pl. N.Z. 100; Tolmatchev, Contr. Fl. Vaig. 127; Krylov, Fl. Zap. Sib. III, 387; Andreyev, Mat. fl. Kanina 163; Tolmachev, Fl. Kolg. 15; id., Obz. fl. N.Z. 150; id., Fl. Taym. I, 102, pro parte; Yuzepchuk in Fl. SSSR III, 33; Perfilev, Fl. Sev. I, 132; Leskov, Fl. Malozem. tundry 34; Hultén, Atlas, map 288; id., Circump. pl. I, 172; A.E. Porsild, Ill. Fl. Arct. Arch. 43; Karavayev, Konsp. fl. Yak. 62.
E. spissum Fern. — Böcher & al., Grønl. Fl. 241.

Species with distribution of a generally subarctic type, widespread in the north of the forest zone of Europe and Asia where it is mainly associated with oligotrophic peat bogs and boggy coniferous forests, growing in abundance at suitable sites. Extremely common in more temperate parts of the tundra zone, where it also often occurs in high abundance. It grows in mossy peaty tundras, forming dense tufts which stand out against a background of more or less smooth moss carpet or peaty bare ground. Usually occurring in great quantity and conspicuous where it grows, often determining the appearance of vegetational communities. Where this cotton grass grows in abundance, the surface of the tundra acquires a tussocky nature.

Distributed in the north all the way to the middle part of Novaya Zemlya, the coasts of Yenisey Bay, the district of Lake Taymyr, and the arctic coast of Yakutia and Chukotka with adjacent islands. Absent from truly High Arctic districts and near the northern limit of its range occupying sites of very restricted area, but remaining a more or less numerous plant at these sites. Much more common in the south of the tundra zone. It grows especially abundantly in the tundras of Northern Yakutia, where it is often one of the dominant plants. When *E. vaginatum* fruits in abundance, wide areas of the surface of the tundra gleam shining whitish from afar due to the mass of cottongrass heads. *Eriophorum vaginatum* is also very characteristic of boggy peaty tundras of Chukotka and Koryakia.

Soviet Arctic. Murman (ubiquitous); Kanin and the Timanskaya and Malozemelskaya Tundras (ubiquitous); Kolguyev Island; Bolshezemelskaya Tundra (frequent almost everywhere); Polar Ural; Pay-Khoy (common); Vaygach Island (rather rare); South Island and southern part of North Island of Novaya Zemlya (relatively rare and not everywhere); Yamal; district of Ob Sound and Tazovskaya Bay; Gydanskaya Tundra; lower reaches of Yenisey (ubiquitous); Southern Taymyr Tundra (frequent); lower reaches of the Upper Taymyra and the south of Lake Taymyr (localized in conjunction with the distribution of suitable peaty areas); lower reaches of the Khatanga, Olenek and Lena and the coast east of the Lena Delta (also growing in great abundance in mainland tundras on the southern edge of the delta, but not reaching the northern edge of the delta); lower reaches of the Yana, Indigirka and Kolyma; Bolshoy Lyakhovskiy Island; Chukotka; Wrangel Island; Anadyr Basin; Koryakia. (Map III–7).

Foreign Arctic. Arctic Alaska; arctic coast of Canada; Labrador; southern islands of Canadian Arctic Archipelago; extreme SW Greenland; Arctic Scandinavia.

Outside the Arctic. Northern and Central Europe; forest zone of West Siberia and the north of the forest zone of East Siberia (absent over a considerable expanse of the middle portion of the forest zone from the Ob to the Vilyuy and the upper course of the Lena); coasts of Sea of Okhotsk; Kamchatka; Kurile Islands; Sakhalin; Amur Basin; Northern Japan; in North America in a large part of Alaska, Canada and the NE United States. Represented in Eastern North America by a distinct race, *E. vaginatum* ssp. *spissum* (Fern.) Hult. (= *E. spissum* Fern.).

A peculiar feature of the geographical distribution of *E. vaginatum* is the absence of this species from a broad zone south of the Arctic Circle to the east of the Ob. In the south of Central Siberia as well as in a considerable part of the Lena Basin, only isolated localities are known for *E. vaginatum*. As we proceed further east, this again becomes a common plant on the coasts of the Sea of Okhotsk and in the Amur Basin. However, in Preamuria *E. vaginatum* is represented by a form differing from the type in having more slender culms, narrow sheaths, maculate scales and sometimes ferruginous hairs. The range of this form partly overlaps the range of *E. humile* Turcz., which grows over the expanse from the Altay Mountains to Ayan on the Sea of Okhotsk and consists of low-growing plants not forming tufts (or at least not large tussocklike tufts of the same type as in *E. vaginatum*) and possessing a rather well developed blade of the culm leaf and a series of other distinguishing characters. Since it is not a form replacing *E. vaginatum*, *E. humile* Turcz. should obviously be considered a separate species, not a race or variety of the former. As for plants from Kamchatka, Sakhalin and northern districts near the Sea of Okhotsk, here *E. vaginatum* is represented by a more or less typical form; but only in these plants is trifurcation of the mouth of the inflated culm sheath sometimes reported.

GENUS 2

Trichophorum Pers. — DEER GRASS

SMALL GENUS CONTAINING about 6 species distributed in the temperate zone of the Northern Hemisphere. In the Arctic represented by two species. One of these, *T. caespitosum*, is rather characteristic for certain arctic districts.

1. Culms sharply triangular, scabrous. No florets in axils of lowest two scales. Scales more or less pale ferruginous brown. Perianth consisting of long hairs that exceed the spikelets by 2–4 times their length and form small

cottony heads. Fruit oblong-ovoid, dark brown. Densely tufted plant with very short creeping rhizomes. .... 2. **T. ALPINUM** (L.)PERS.

— Culms almost cylindrical, smooth. Lowest scales fertile. Scales more or less dark brown. Perianth hairs concealed in axils of scales. Fruit ovoid or obovoid, yellowish. Plant without creeping rhizomes, forming very dense tufts. .... 1. **T. CAESPITOSUM** (L.) HARTM.

1. ***Trichophorum caespitosum*** (L.) Hartm. s. str., Handb. Scand. Fl., 5 ed. (1849), 259; Rozhevits in Fl. SSSR III, 37; Leskov, Fl. Malozem. tundry 34; Karavayev, Konsp. fl. Yak. 63; Hultén, Circump. pl. I, 42, map 35; Clapham, Tutin & Warburg, Fl. Brit. Isl. 1061.

    Scirpus caespitosus L., Sp. pl. (1753), 48; Ledebour, Fl. ross. IV, 246; Ostenfeld, Fl. arct. 1, 43; Krylov, Fl. Zap. Sib. III, 397; Andreyev, Mat. fl. Kanina 163; Perfilev, Fl. Sev. I, 130.

    Trichophorum austriacum Palla in Bericht. Deutsch. Bot. Gesellsch. XV (1897), 468; Leskov, Fl. Malozem. tundry 34.

    Scirpus caespitosus ssp. *austriacus* (Palla) Aschers. et Graebn., Synops. Mitteleurop. Fl. II, 2 (1902), 300; Hultén, Fl. Al. II, 290, map 213; id., Atlas, map 290; Böcher & al., Grønl. Fl. 242; A.E. Porsild, Ill. Fl. Arct. Arch. 44, map 67; Polunin, Circump. arct. fl. 110.

    S. bracteatus Bigel. in New Engl. Med. Journ. V (1816), 355.

    Trichophorum bracteatum (Bigel.) V. Krecz. ex Czernov in Fl. Murm. II (1954), 26, map 6.

    Ill.: Fl. SSSR III, pl. III, fig. 7; Fl. Murm. II, pl. V, fig. 1 (under *T. bracteatum*).

    Arctic-alpine species, rather common in temperate districts of the European and American Arctic. Also occurring in the west (Yamal) and extreme east of Arctic Siberia.

    Growing in dense tufts in the flarks of sedge-cottongrass hummocky tundras, in bogs with mounds and flarks, in marshes with small sedges, in peaty meadows, and on marshy shores of waterbodies.

    Trichophorum caespitosum is divided at the present time into two subspecies. The majority of authors consider plants subsequently described as *T. austriacum* Palla to represent the typical subspecies (ssp. *caespitosum*). The range of this subspecies includes Central Europe, Scandinavia, the northern USSR and North America. The second subspecies, ssp. *germanicum* (Palla) Hegi (= *T. germanicum* Palla) occurs only in the west of Western Europe and in Scandinavia (mainly in its western part). The two subspecies differ in the colour of the margins of the mouths of the sheaths (yellowish in ssp. *caespitosum*, brownish red on account of numerous small tubercles in ssp. *germanicum*), the size of the inflorescence, and the number of florets it contains. Plants usually referred to *T. caespitosum* s. str. were revised (in the herbarium) to *T. bracteatum* by V.I. Krechetovich; for reasons unknown to us he referred ssp. *germanicum* to *T. caespitosum* s. str.

    Soviet Arctic. Murman (everywhere), Kanin, Malozemelskaya Tundra; central part of Yamal (Lake Lymbano-To), Chukotka (marsh on Teakachin Hill in the district of Chaun, and Lorino in the SE part of the Chukotka Peninsula); Anadyr Basin; Koryakia (upper part of Oklan Basin and Bay of Korf).

    Foreign Arctic. Arctic Alaska, arctic coast of Canada (mainly western part), Arctic Labrador, Southern Baffin Island, SW and East Greenland, Iceland, Arctic Scandinavia.

    Outside the Arctic. Great Britain (isolated discoveries); disjunctly in Spain and France; Alps, mountains of Hungary and Bulgaria; Fennoscandia; the Baltic (southeast to the Valday Hills); Prepolar, Northern and Middle Urals; occasionally in West Siberia; Altay; basin of the lower course of the Yenisey; northern fringe of Central

Siberian Plateau (watershed between the rivers Romanikha and Medvezhya), occasionally in Prebaikalia; mountains of Eastern and Southern Yakutia (basins of right-bank tributaries of the Upper Kolyma and of the middle course of the River Maya, Lake Toko); coasts of Sea of Okhotsk; Kamchatka; Kurile Islands; Sakhalin; Japan; Alaska; Canada; Rocky Mountains and NE United States.

2. ***Trichophorum alpinum*** (L.) Pers., Syn. I (1805), 70; Rozhevits in Fl. SSSR III, 38; Chernov in Fl. Murm. II, 27, map 7.

Eriophorum alpinum L., Sp. pl. (1753), 53; Ledebour, Fl. ross. IV, 252; Perfilev, Fl. Sev. I, 132; Hultén, Fl. Al. II, 275, map 201; id., Atlas, map 280; id., Circump. pl. I, 42, map 36.

Scirpus hudsonianus (Michx.) Kryl., Fl. Zap. Sib. III (1929), 399.

Ill.: Fl. SSSR III, pl. III, fig. 9; Fl. Murm. II, pl. V, fig. 2.

Rather rare plant of the taiga zone of Eurasia and North America, also distributed in the mountains of Central and parts of Southern Europe. Absent from the northern half of Siberia east of the Yenisey. Only isolated localities known within the limits of the Soviet Arctic.

Growing in sphagnum and peat bogs.

Soviet Arctic. Murman (lower reaches of Kola River, east coast of Kola Bay, Teriberka); Kanin (Lake Krivoye); lower reaches of Yenisey (between Khantayskoye settlement and the Medvezhiy Kamen Range).

Foreign Arctic. Arctic Scandinavia.

Outside the Arctic. Northern Europe and the mountains of Central and parts of Southern Europe; West Siberia (basins of the Severnaya Sosva and Upper Pur Rivers, Lake Tenis); basin of the upper and middle course of the Yenisey; Central Sayan; Prebaikalia; Southern Yakutia (rare); basins of the Zeya and the lower course of the Amur; coast of Sea of Okhotsk (district of Nagayev Bay); Sakhalin; Kamchatka; Kurile Islands; Northern Japan; in North America in the southern half of Alaska, Canada and the northeastern states of the USA.

GENUS 3 **Scirpus L. — BULRUSH**

THIS GENUS INCLUDES a large number of species, mainly hygrophytes, many of them growing in districts with temperate or cold temperate climate. Only one species occurs in the Arctic and high mountains. This belongs to the monotypic section *Pseudo-Eriophorum* Jurtz. whose systematic position is not entirely clear. The leafiness of the culm, the obliquely divergent lowest involucral bract, the strongly scarious blackish scales, and certain other characters allow this species to be referred to subgenus *Scirpus*, within which it most closely approaches the *S. sylvaticus* L. – *S. radicans* Schkuhr species-group. However, it differs from the species of that subgenus in having considerably larger spikelets, scales and achenes (the latter 2–2.4 mm long). It differs from all other species of the genus *Scirpus* in having shorter leaves (reminiscent of certain broad-leaved species of *Luzula*) and slender flexuous, often drooping outer lateral branchlets of the inflorescence, as well as by the distally serrate perianth bristles scarcely appropriate for a species of *Scirpus*; the blade of the lowest involucral bract is involute, shorter than the sheath (in other species of *Scirpus* much longer than the sheath and usually flat).

The sharp morphological distinction between the present species and other members of the genus *Scirpus* and its resemblance in habitus to cotton grasses with many spikelets caused it for long to be included in the genus *Eriophorum* L.; but it differs from all species of that genus in having only 6 perianth bristles that only slightly lengthen in fruit and are twisted and scabrous with trichomes distally, as well as by the leaves, inflorescences, the median nerve of the scales, the whitish colour of the achenes, and later flowering.

1. **Scirpus Maximowiczii** C.B. Clarke in Kew Bull., Add., ser. 8 (1908), 30; Miyabe & Kudo, Fl. Hokkaido, Saghal. 2, 205; Sugawara, Ill. fl. Saghal. I, 345; Ohwi, Fl. Japan 235; Yurtsev in Byull. MOIP, otd. biol. LXX, 1, 132, map (fig. 4).
   Eriophorum japonicum Maxim. in Bull. Acad. imp. Sci. St.-Pétersb. 31 (1886), 111; Nakai, Fl. Koreana 2, 513; Komarov & Klobukova-Alisova, Opred. rast. DVK I, 257; Sochava in Bot. zhurn. XVII, 2, 118; Yuzepchuk in Fl. SSSR III, 28; Nakai in Bull. Nat. Sci. Mus. 31, 132.
   Scirpus japonicus Fern. in Rhodora 7, 79 (1905), 130, in adnot.; non Franch. et Sav. (1875).
   Eriophorum Maximowiczii Beetle in Amer. Journ. Bot. 33, 8 (1946), 663.
   E. latifolium auct. non Hoppe — Regel & Tiling, Fl. Ajan. 124.
   Ill.: Yurtsev, l. c., fig. 1–3.

   Short plant with extravaginal sprouting. Culms basally with a dense brown clothing of dead fibrous sheaths. Plants in bud are reminiscent of *Eriophorum latifolium* Hoppe in appearance.

   East Siberian species of barrens near the Pacific; in the more southern part of its range it occurs both in oceanic (Japan, the Kuriles, Sakhalin) and suboceanic districts (Korean Peninsula, Primorskiy Kray).

   In the northern part of its range the species is more suboceanic, native to districts influenced in summer by the cool monsoon blowing off the northern part of the Pacific Ocean. In the west it reaches Northern Prebaikalia, the Indigirka slope of the Suntar-Khayata Range and the Penzhina Basin.

   Growing in the high alpine (barren) zone, in the Arctic also spreading to the plains; reaching the upper part of the forest zone in bogs and bog forests, and on moist shady rocks. In the barren zone the species grows in hummocky cottongrass-sedge and wet sedge tundras, in springfed marshes and sphagnum bogs, at sites with subsurface groundwater flow, and among rock debris at the foot of slopes, as if combining the ecological features of *E. vaginatum* and *E. angustifolium*.

   The species penetrates the tundra zone only in the Far East. Here it occurs in hummocky cottongrass-sedge tundras, in shrubby hummocky tundras (with layer of *Alnaster kamtschaticus* and *Betula exilis*), in sphagnum bogs in valleys, as well as in moist patchy moss-sedge tundras and springfed marshes on mountain slopes; everywhere under conditions of good winter snow cover and abundant soil moisture; relatively rare.[2]

   Soviet Arctic. Known from some sites in the Penzhina valley near its mouth, from the Palmatkina Range in the Penzhina Basin, and from the Bay of Krest (Olovyannaya Bay).

   Foreign Arctic. Absent.

   Outside the Arctic. Barrens of the Suntar-Khayata, Dzhugdzhur and Baykal Ranges, Northern Transbaikalia, the Stanovoy Range, the Dusse-Alin and Sikhote-Alin, the Korean Peninsula (Kŭmgang-San), southern and central Sakhalin, the southern Kurile Islands, Hokkaido, and the northern part of Honshu.

---

[2] The paucity of collections of S. *Maximowiczii* is possibly explained by its having been overlooked by collectors due to its resemblance to cottongrasses with many spikelets.

GENUS 4   Eleocharis[3] R. Br. — SPIKE RUSH

GENUS CONTAINING ABOUT 80 species, distributed in diverse climatic regions of the World. Five species widespread in the temperate zone of the Northern Hemisphere just penetrate arctic limits.

1. Base of style (stylopodium) *confluent with tip of achene*; achene obovoid, aculeate-acuminate, without ribs, puncticulate, greyish white. Spikelet ovoid or elongate-ovoid, 3–6 mm long, with 3–7 florets. Scales ovoid, obtuse, brownish; perianth bristles 6, persistent, as long as or somewhat shorter than achene, with reclinate serrations. Plant with short creeping rhizomes. . . . . .2. **E. QUINQUEFLORA** (HARTM.) SCHWARZ.
   - Base of style (stylopodium) *delimited from tip of achene by constriction.* .2.

2. Spikelets *very small, 2–3(5) mm long.* Achene elongate-obovoid, with longitudinal vittae and transverse lines between them. Stigmas 3. Scales ovoid or elongate-ovoid, *greenish.* Culms filiform. Plant 2–10 cm high, with slender creeping rhizomes. . . . . . .1. **E. ACICULARIS** (L.) ROEM. ET SCHULT.
   - Spikelets *5–15 mm long.* Scales *brown.* Stigmas 2. . . . . .3.

3. Only *one sterile scale* at base of spikelet. Stylopodium of achene conical, its length slightly less than, equal to or slightly more than its width. Bristles 4 (rarely 5) or none. Plant (5)10–40 cm high, with creeping rhizomes. . . . . .5. **E. UNIGLUMIS** (LINK) SCHULT.
   - *Two sterile scales* at base of spikelet, each of which envelops about half of it. . . . . .4.

4. Stylopodium *oblong-conical or narrowly conical*, longer than wide (but not more than twice as long), or more rarely of equal length and width. Bristles 4, shorter than, equal to or sometimes slightly longer than achene, with small sparse appressed serrations. Spikelets 6–20 mm long. Scales reddish brown, with whitish hyaline margins. Culms rather thick, sometimes abruptly constricted below spikelet. Plant 10–45 cm high, with creeping rhizomes. . . . . .3. **E. PALUSTRIS** (L.) R. BR.
   - Stylopodium *conical-mammiform (broad at base, suddenly narowing distally) or short-conical*, acute at tip, its length less than, equal to or slightly more than its width. Bristles 4, exceeding achene sometimes by 1½ times or more, very rarely absent, their serrations rather large and reclinate. Spikelets 5–15 mm long. Scales dark brown, with narrow whitish hyaline margins. Plant 10–50 cm high, with creeping rhizomes. . . . . .4. **E. INTERSITA** ZINSERL.

    1. ***Eleocharis acicularis*** (L.) Roem. et Schult., Syst. II (1817), 154; Ostenfeld, Fl. arct. 1, 42 (f. *submersa*); Krylov, Fl. Zap. Sib. III, 394; Perfilev, Fl. Sev. I, 129; Zinserling in Fl. SSSR III, 70; Chernov in Fl. Murm. II, map 9; Böcher & al., Grønl. Fl. 241; A.E. Porsild, Ill. Fl. Arct. Arch. 44, map 66; Polunin, Circump. arct. fl. 105; Hultén, Circump. pl. I, 114, map 105; Karavayev, Konsp. fl. Yak. 63.

---

[3] The correct name of this genus is here adopted instead of the commonly used *Heleocharis*.

Scirpus acicularis L., Sp. pl. (1753), 48; Ledebour, Fl. ross. IV, 243; Hultén, Fl. Al. II, 289, map 211; id., Atlas, map 289.

Ill.: Fl. SSSR III, pl. VI, fig. 4; Fl. Murm. II, pl. IX, fig. 6; Polunin, l. c. 106; A.E. Porsild, l. c., fig. 15, f.

Rather rarely occurring plant distributed in the forest zone of Eurasia and North America. Within the Soviet Arctic only isolated localities are known. It grows on silty or sandy shores of bodies of freshwater, and often on the bottom of shallow waterbodies where it is sometimes almost completely submersed. Forming pure beds.

Soviet Arctic. Murman (Pechenga district according to Hultén, Atlas); shore of the Ob south of Salekhard; mouth of Lena; Anadyr Basin (Lake Krasnoye); Penzhina Basin (one locality on the upper course of the Palmatkina River); Bay of Korf.

Foreign Arctic. Arctic Alaska; lower reaches of Mackenzie River; Labrador; southern Baffin Island; west coast of Greenland south of 72°N; Iceland; Arctic Scandinavia (extremely rare).

Outside the Arctic. Western Europe (almost all); European part of USSR (sporadically in the entire territory except northeastern districts); Caucasus; West Siberia (southern half, in the north in the basin of the Severnaya Sosva); Altay; basin of the upper course of the Yenisey; Prebaikalia; Central Yakutia and mountains of Southern Yakutia (very rare); Amur Basin; Primorskiy Kray; Kamchatka; Northern Mongolia; NE China; Korean Peninsula; Japan; North America (in almost the entire territory north of 30°N).

2. *Eleocharis quinqueflora* (Hartm.) Schwarz in Mitteil. Thüring. Bot. Gesellsch. I, 1 (1949), 89; Polunin, Circump. arct. fl. 106.

Scirpus quinqueflorus Hartm., Prim. lin. Inst. bot. (1767), 85.

S. pauciflorus Lightf., Fl. scot. II (1777), 1078; Ledebour, Fl. ross. IV, 246; Hultén, Atlas, map 299; Böcher & al., Grønl. Fl. 242.

Eleocharis pauciflora (Lightf.) Link, Hort. Berol. I (1827), 284; Krylov, Fl. Zap. Sib. III, 397; Zinserling in Fl. SSSR III, 69; Kuzeneva in Fl. Murm. II, 37, map 9.

Boreal, mainly Eurasian species, penetrating the Arctic only in Northern Europe. Growing on moist and marshy shores of waterbodies, in marshy meadows and lowland marshes.

Soviet Arctic. Murman (on the Rybachiy Peninsula at one locality in the northwest and in the upper reaches of the river flowing into Kachkovskiy Bay).

Foreign Arctic. Arctic Scandinavia (not uncommon).

Outside the Arctic. All Fennoscandia; Northern and Central Europe; European part of the USSR (mainly in the forest zone, absent from the northeast); Urals; Caucasus; southern West Siberia; basin of the upper course of the Yenisey; Prebaikalia; Kamchatka; western districts of North America.

3. *Eleocharis palustris* (L.) R. Br., Prodr. fl. Nov. Holland. I (1810), 80; emend. Roem. et Schult., Syst. II (1817), 151; Ledebour, Fl. ross. IV, 244; Krylov, Fl. Zap. Sib. III, 312 (var. *communis* f. *eupalustris*); Ostenfeld, Fl. arct. 1, 43; Perfilev, Fl. Sev. I, 130; Chernov in Fl. Murm. II, 39, map 9; Hultén, Fl. Al. II, 294, map 218; id., Circump. pl. I, 174, map 165; Böcher & al., Grønl. Fl. 241; Polunin, Circump. arct. fl. 106.

E. eupalustris Lindb. fil. in Acta Soc. Fauna et Flora fennica 23, 7 (1902), 5; Zinserling in Fl. SSSR III, 76.

Scirpus palustris L., Sp. pl. (1753), 47; Hultén, Atlas, map 299.

Ill.: Fl. SSSR III, pl. VI, fig. 12; Fl. Murm. II, pl. IX, fig. 4; Polunin, l. c. 106.

Widespread in the temperate zone of the Northern Hemisphere. Just penetrating arctic limits. It grows in waterbodies and on their very moist shores, sometimes in marshes.

A very variable species, mainly with respect to the form of the stylopodium.

Soviet Arctic. Murman (Pechenga district and the lower reaches of the Kola River); Southern Kanin.

Foreign Arctic. Extreme south of Greenland, Iceland.

Outside the Arctic. Fennoscandia; the whole of Northern and Central and parts of Southern Europe; almost the entire European part of the USSR (except northeastern districts); Urals; Caucasus; West Siberia; East Siberia (very rare, only in the south, east of Irkutsk); Northern Kazakhstan (very rare); Northern Mongolia. Apparently occurring in the eastern part of North America, replaced by closely related species in the west.

4. ***Eleocharis intersita*** Zinserl. in Fl. SSSR III (1935), 76 and 581; Karavayev, Konsp. fl. Yak. 63.

Ill.: Fl. SSSR III, pl. VI, fig. 11, 26.

Plant mainly of the forest zone of Eurasia, penetrating arctic limits in the northeast of its range. Growing in moist or marshy meadows (often on floodplains), in lowland marshes, and on shores of waterbodies.

Soviet Arctic. Lena Delta (Karavayev, l. c.); Anadyr Basin (vicinity of Ust-Belaya, the mouth of the Mayn and south of Snezhnoye settlement); Penzhina Basin (not uncommon); Bay of Korf.

Foreign Arctic. Not occurring.

Outside the Arctic. Central Europe; European part of the USSR (except northern and southern districts); Caucasus; Urals; West Siberia (mainly in the southern half, in the north in the basin of the Severnaya Sosva); Altay; basin of the upper course of the Yenisey; Prebaikalia; basin of the upper course of the Vilyuy; Central Yakutia [reported by M.N. Karavayev (l. c.) for the Kolyma Basin and the Yana-Indigirka district]; Amur Basin; Primorskiy Kray; coasts of Sea of Okhotsk; NE China; North America.

5. ***Eleocharis uniglumis*** (Link) Schult., Mant. II (1824), 88; Ledebour, Fl. ross. IV, 245; Perfilev, Fl. Sev. I, 130; Chernov in Fl. Murm. II, 40, map 10; Hultén, Fl. Al. II, 295, map 219; id., Circump. pl. I, 120, map 111; Böcher & al., Grønl. Fl. 241.

Scirpus uniglumis Link — Hultén, Atlas, map 308.
Eleocharis eu-uniglumis Zinserl. in Fl. Yugo-Vost. III (1929), 278; id. in Fl. SSSR III, 82, 584; Karavayev, Konsp. fl. Yak. 63.

Ill.: Fl. Murm. II, pl. IX, fig. 1.

Plant rather widespread in the temperate zone of the Northern Hemisphere. Just penetrating arctic limits. Growing on the shores of waterbodies and in wet meadows, sometimes in water.

Soviet Arctic. Murman (Pechenga district, lower reaches of the Kola River, Teriberka); Southern Kanin; Anadyr Basin.

Foreign Arctic. Greenland (extreme south); Iceland; Arctic Scandinavia.

Outside the Arctic. Western Europe (but not in Spain or Italy); European part of USSR; Caucasus; West and East Siberia (mainly in the southern half); Kamchatka; Central Asia; Mongolia; the Himalayas; North America.

GENUS 5   **Kobresia** Willd. — KOBRESIA

COMPARATIVELY SMALL GENUS (containing about 40 species), distributed mainly in high alpine regions of the temperate and parts of the tropical zones of the Northern Hemisphere. Fourteen species of *Kobresia* occur within the USSR, mainly in the high mountains of Asian parts of the Union where some of them are the dominants in a series of vegetational associations.

In the Soviet Arctic the genus *Kobresia* is represented by four species, of which one (*K. filifolia*) just penetrates its limits. The Siberian-Mongolian *K. filifolia* is a species with wide ecological amplitude, gravitating towards vegetational communities of the meadow-steppe type in river valleys. The three other species (*K. sibirica*, *K. Bellardii* and *K. simpliciuscula*) are typical arctic-alpine plants. The latter two of these possess almost circumpolar ranges. Their distribution in the Soviet Arctic is principally determined by the availability of suitable habitats.

Kobresia simpliciuscula clearly gravitates towards calcareous substrates, so that the occurrence of this species depends on the presence not only of mountainous relief but also of appropriate minerals. *Kobresia Bellardii* and *K. sibirica* occur mainly on outcrops of soft sandstones and shales and on sandy alluvia. The relatively small number of localities for *Kobresia* species in the Soviet Arctic may be largely explained by inadequate investigation of the territory. This is confirmed by rather careful floristic collections in the Polar Ural, around Lake Taymyr, in the lower reaches of the Lena, and on the east coast of the Chukotka Peninsula. The inconspicuous nature of the plants and their restriction to specific habitats is surely significant, but lack of botanical investigation of Asian districts of the Soviet Arctic can be considered the basic reason for the paucity of localities for *Kobresia*. This applies both to such a widespread species as *K. Bellardii* which plays a prominent role in the vegetational cover of the Verkhoyansk and Cherskiy Ranges, and to the extremely rarely encountered Siberian-American species *K. sibirica*. According to available data, it may be postulated with respect to *K. sibirica* that the arctic part of the range of this species is fundamental.

Inadequate factual data on the distribution of *Kobresia* in Siberia make it impossible to form sufficiently precise concepts of the nature of the distribution of these species. In those districts from which *Kobresia* species are known, they sometimes do not occur everywhere. An example of this is provided by the distribution of *K. Bellardii* in Taymyr. This fact should be a deterrent against excessive extrapolation in composing range maps (cf. Hultén, Circump. pl. I, 47, map 40). Although species like *K. Bellardii* and *K. simpliciuscula* have almost circumpolar distributions, it is nevertheless probable that their ranges are not continuous in a great part of the USSR. With respect to the arctic part of their ranges, the only real fact at the present moment is that both these species are absent from tundras of the European north of the USSR and adjacent islands (Novaya Zemlya, etc.), which have been relatively well investigated.

1. Inflorescence a simple narrow linear spike 10–20 mm long and 1.5–2.5 mm wide. Spikelets with 2 florets (one pistillate, one staminate). Stigmas 3. Culms slender, 0.5–0.7 mm in diameter, densely clustered. Leaves narrow, 0.2–0.5 mm wide, erect, sulcate. Lower sheaths without blades, light brown or brown. Plant forming dense tufts. ....3. **K. BELLARDII** (ALL.) DEGLAND.
– Inflorescence compound or a spike with spikelets consisting of more than 2 florets. ....2.

2. Inflorescence *compact, an oval or ovoid spike 10–20 mm long and 5–10 mm wide*. Spikelets with 3–7 florets (1–2 pistillate, 2–5 staminate). Stigmas 3. Achene distinctly triangular. Culms comparatively thick, 1–2 mm in diameter. Leaves sulcately involute, 1–2 mm wide. Lower sheaths lustrous stramineous brown, 3–6 cm long. Roots *thick*, cordlike, densely covered with *albescent hairs*. ....1. **K. SIBIRICA** TURCZ.
– Inflorescence otherwise; roots *slender, without pubescence*. ....3.

3. Inflorescence a *linearly elongated spike*, sometimes branched near base, *10–20 mm long and 2.5–4 mm wide*. Spikelets with 3–7 florets (one pistillate, the remainder staminate). Stigmas 3 or 2. Culms 7–30 cm long, rounded. Leaves somewhat shorter than culms, erect, sulcately involute, 1–2.5 mm wide. Lower sheaths *without blades*, comparatively tightly appressed to culms. ....2. **K. FILIFOLIA** (TURCZ.) C.B. CLARKE.
– Inflorescence *oblong-ovoid, oblong or rhomboid*, more or less lobed, *10–15 mm long and 4–10 mm wide*. Lateral spikelets usually bisexual with two florets, sometimes with a single female floret; apical florets male. Stamens and stigmas 3. Culms 7–15 cm long, usually triangular. Leaves considerably shorter than culms (about half their height), for the most part curved. All sheaths *with leaf blade*, not tightly appressed to culm. ....4. **K. SIMPLICIUSCULA** (WAHLB.) MACKENZ.

1. ***Kobresia sibirica*** Turcz. ex Bess. in Flora XVII, 1, Beibl. (1834) 26 et in Bull. Soc. Nat. Mosc. XI (1838), 103 (nomen); Boeck in Linnaea XXXIX, 7; Ivanova in Bot. Zhurn. XXIV, 5/6, 480.

   Elyna sibirica Turcz. ex Ledeb., Fl. ross. IV (1852), 262 (descr.).
   E. schoenoides auct. non C.A. Mey. — Ostenfeld, Fl. arct. 1, 44.
   Kobresia schoenoides auct. non Steud. — Krylov, Fl. Zap. Sib. III, 412, pro parte; Sergiyevskaya in Fl. SSSR III, 106, pro parte; Karavayev, Konsp. fl. Yak. 63; Tolmachev & al., Opred. rast. Komi 95.
   K. hyperborea A.E. Porsild in Bull. Nat. Mus. Canada 121 (1951), 103 (nomen).
   K. arctica A.E. Porsild in Sargentia IV (1943), 15 (descr.) non Ivanova — A.E. Porsild, Ill. Fl. Arct. Arch. 45; Polunin, Circump. arct. fl. 110.
   Ill.: A.E. Porsild, Ill. Fl. Arct. Arch., fig. 15, a.

   Occurring in rather moist habitats in moss-lichen, sedge-*Dryas* and *Kobresia* tundras, as well as on riverine alluvia, sandy exposures, and mounds in polygonal bogs. All these habitats are usually continuously snow-covered in wintertime. The depth of snow cover is not less than 3–5 cm.

   Soviet Arctic. Polar Ural (upper reaches of the Sob); middle course of the Pyasina and Anabar; lower reaches of the Olenek and Lena; east coast of Chukotka Peninsula

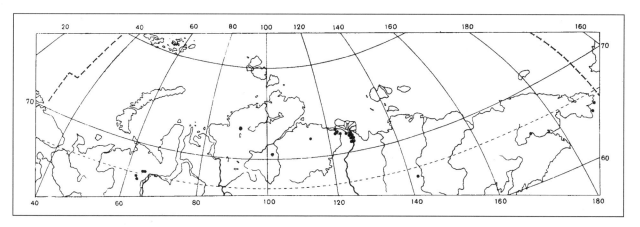

MAP III–8 Distribution of *Kobresia sibirica* Turcz. ex Bess.

(Lawrence Bay); Arakamchechen Island; Anadyr Basin (doubtful specimen from the vicinity of Ust-Belaya settlement). (Map III–8).

Foreign Arctic. Arctic coast of Alaska; lower reaches of Mackenzie River; arctic coast of Canada (Coronation Gulf, Boothia Peninsula); Canadian Arctic Archipelago (Victoria Island).

Outside the Arctic. Northern and Middle Urals; the north of the Central Siberian Plateau (Kotuy Basin); Cherskiy Range; Nukhu-Daban Range and the mountains of Northern Mongolia.

2. *Kobresia filifolia* (Turcz.) C.B. Clarke in Journ. Linn. Soc. XX (1884), 381; Sergiyevskaya in Fl. SSSR III, 109; Ivanova in Bot. zhurn. XXIV, 5/6, 483.

Elyna filicifolia Turcz. in Bull. Soc. Nat. Mosc. XXVIII (1885), 353.

Kobresia capillifolia (Decne.) C.B. Clarke var. *filifolia* (Turcz.) Kük. in Öfvers. Finska Vet.-Soc. Förhandl. XLV, 8 (1902–1903), 1; Krylov, Fl. Zap. Sib. III, 414.

Siberian-Mongolian species penetrating the Arctic in the lower reaches of the Lena, Chukotka and the Anadyr Basin. Occurring in the Arctic exclusively on limestone outcrops (in the Tuora-Sis Range often in communities with *Caragana jubata*). In the rest of its range it occurs in a great diversity of habitat types: on stony slopes, rocks and screes, in *Dryas-Kobresia*-small sedge tundras, on dry steppelike slopes, in moist meadows and willow carr, in dry sunny woods (pine or larch) and in solonetzic meadows, near ice sheets and among floodplain herbage, on riverine alluvia, and on forest edges. Very often in communities of the steppe type. The majority of recorded localities possess persistent snow cover in winter of a thickness usually greater than 5 cm.

Soviet Arctic. Lower reaches of Lena to 72°N, only on the right bank in the system of the Tuora-Sis Range; middle course of the Anadyr; Chukotka (basin of the Kuvet River). (Map III–9).

Foreign Arctic. Not occurring.

Outside the Arctic. Northern part of Central Siberian Plateau (middle course of the River Medvezhya, a tributary of the Kheta); upper reaches of the Olenek; Lena Basin; Cherskiy Range; Altay; the south of Krasnoyarsk Kray (Minusinsk Depression); Transbaikalia; Northern Mongolia.

3. *Kobresia Bellardii* (All.) Degland in Loisel., Fl. Gall. II (1807), 626; Simmons, Survey Phytogeogr. 58; M. Porsild, Fl. Disko 48; Krylov, Fl. Zap. Sib. III, 413; Sergiyevskaya in Fl. SSSR III, 109; Ivanova in Bot. zhurn. XXIV, 5/6, 486; A.E. Porsild, Bot. SE Yukon 102; Karavayev, Konsp. fl. Yak. 64; Tolmachev & al., Opred. rast. Komi 95.

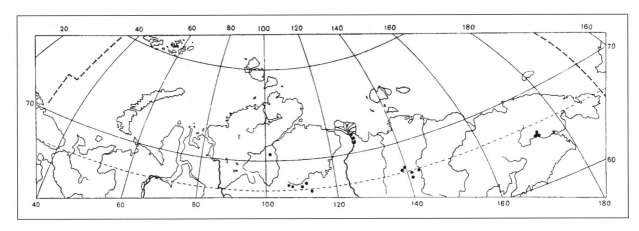

Map III–9 Distribution of *Kobresia filifolia* (Turcz.) C.B. Clarke.

Carex Bellardii All. in Fl. Pedem. II (1785), 264, tab. 92, fig. 2.
Kobresia scirpina Willd. in Sp. pl. IV (1805), 205; Devold & Scholander, Fl. pl. SE Greenl. 129.
Elyna Bellardii (All.) C. Koch in Linnaea XXI (1848), 616; Lange, Consp. fl. groenl. 13; Ostenfeld, Fl. arct. 1, 44; Tolmachev, Fl. Taym. I, 102; Seidenfaden, Vasc. pl. SE Greenl. 86; Sørensen, Vasc. pl. E Greenl. 126; Seidenfaden & Sørensen, Vasc. pl. NE Greenl. 86.
Kobresia myosuroides (Vill.) Fiori et Paol. in Ic. Fl. Ital. (1895), 52; Hultén, Fl. Al. II, 296; id., Atlas, map 314; id., Circump. pl. I, 46; Böcher & al., Grønl. Fl. 242; A.E. Porsild, Ill. Fl. Arct. Arch. 44; Jørgensen & al., Fl. pl. Greenl. 34.
Ill.: A.E. Porsild, Ill. Fl. Arct. Arch., fig. 15, b, c.

In the Arctic and Subarctic growing mainly in communities of the tundra type (*Dryas*- moss, *Dryas*-lichen, dwarf shrub-*Dryas*; locally very abundant in *Dryas-Kobresia*-lichen and *Dryas-Kobresia*-small sedge tundras). Very often found on stony slopes among rocks and rock debris (at sites where fine soil accumulates), on floodplain gravelbars, and on sandy dunes and spits.

In more southern districts usually growing in moss-lichen or stony *Dryas* tundras of the barren zone, and in dwarf shrub or dwarf birch associations of the subalpine zone. In the forest zone the species occurs on stony steppelike slopes, in dry sunny open larch groves, in meadows on the terraces of dry river valleys, as well as in floodplain meadows and on gravelbars along rivers.

All types of habitats of *K. Bellardii* in the Arctic are areas with little snow cover in wintertime (but not without snow!). The depth of snow cover usually does not exceed 4–5 cm.

Soviet Arctic. Rybachiy Peninsula (Hultén, Atlas); Taymyr (the basin of the Lower Taymyra and Lake Taymyr); lower reaches of the Anabar, Olenek and Lena; Tiksi Bay; the north of the Verkhoyansk Arc; Northern Anyuyskiy Range; district of the Bay of Chaun; Wrangel Island; the east of the Chukotka Peninsula (Lorino settlement, Lawrence Bay); Arakamchechen Island; Bay of Krest; lower reaches of the Anadyr. (Map III–10).
Foreign Arctic. West and arctic coasts of Alaska; arctic coast of Canada; Labrador; Canadian Arctic Archipelago (on Ellesmere Island roughly to 81°N); Greenland (north to Peary Land inclusively); Iceland; Arctic Scandinavia.
Outside the Arctic. Mountains of Europe (Pyrenees, Alps, Appenines, mountains of Northern Scandinavia); Urals; Tarbagatay; Dzhungarskiy Alatau; Eastern Tien Shan; the north of the Central Siberian Plateau; Verkhoyansk Range; mountains of South Siberia (Altay, Sayans, Prebaikalia); Yakutia; mountains of Mongolia;

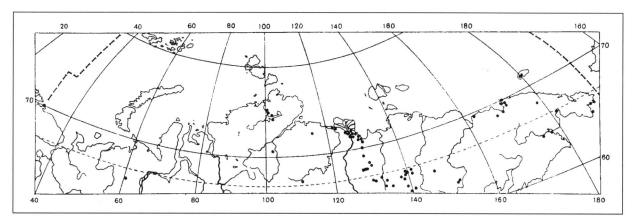

MAP III–10 Distribution of *Kobresia Bellardii* (All.) Degland.

Northern China; Korean Peninsula; Northern Japan; Kurile Islands; North America (Rocky Mountains).

4. ***Kobresia simpliciuscula*** (Wahlb.) Mackenz. in Bull. Torr. Bot. Club L (1923), 349; Sergiyevskaya in Fl. SSSR III, 110; Ivanova in Bot. zhurn. XXIV, 5/6, 495; Hultén, Fl. Al. II, 297; id., Amphi-atl. pl. 232; id., Atlas, map 315; Böcher & al., Grønl. Fl. 243; A.E. Porsild, Ill. Fl. Arct. Arch. 45; Jørgensen & al., Fl. pl. Greenl. 35; Polunin, Circump. arct. fl. 109.
   Carex simpliciuscula Wahlb. in Svensk. Vet. Akad. Nya Handl. XXIV (1803), 141.
   Kobresia caricina Willd. in Sp. pl. IV (1805), 206; Lange, Consp. fl. groenl. 130.
   K. bipartita Dalla Torre in Anleit. wiss. Beob. Alpenreisen (1882) 330; Ostenfeld, Fl. arct. 1, 44; Simmons, Survey Phytogeogr. 59; M. Porsild, Fl. Disko 49; Krylov, Fl. Zap. Sib. III, 415; Sørensen, Vasc. pl. E Greenl. 125; Seidenfaden & Sørensen, Vasc. pl. NE Greenl. 86.

   A species associated in the Arctic basically with outcrops of limestone or other calcareous minerals. Occurring mainly at sites with adequate moisture. Growing in sedge-moss, sedge-*Dryas*, *Dryas*-mixed herb, and *Dryas-Kobresia*-sedge tundras, in openings in larch groves, and in moist places in limestone rock debris. In the Tuora-Sis Range (lower reaches of the Lena) constantly present in communities with *Caragana jubata*. All these habitats possess stable winter snow cover of a depth of 7–10 cm or more.

   Soviet Arctic. Lower reaches of Olenek (Chekanovskiy Range); lower reaches of Lena (Tuora-Sis Range); east and SE coasts of Chukotka Peninsula; basin of the middle course of the Anadyr. (Map III–11).

   Foreign Arctic. West coast of Alaska; arctic coast of Canada; Labrador; Canadian Arctic Archipelago (reaching 80°N on Ellesmere Island); Greenland [on the west coast north to 72°, on the east coast to 78°N(?)]; Spitsbergen; Arctic Scandinavia.

   Outside the Arctic. Mountains of Europe (Scotland and the north of England, Scandinavia, Pyrenees, Alps); Northern and Middle Urals; Altay; the north of the Central Siberian Plateau (middle course of the River Medvezhya, a tributary of the Kheta); Western and Eastern Sayans; Northern Mongolia; mountains of North America; Newfoundland (western part).

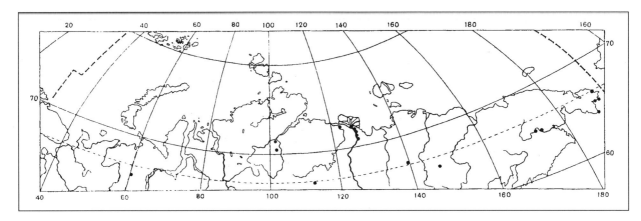

MAP III–11 Distribution of *Kobresia simpliciuscula* (Wahlb.) Mackenz.

GENUS 6   Carex L. — SEDGE

ENORMOUS GENUS CONTAINING about 1000 species according to one source, up to 2000 according to another. Representatives of *Carex* are distributed almost throughout the World from the High Arctic to the southernmost limits of the distribution of angiosperms, occurring in all climatic zones. The majority of species grow in the Northern Hemisphere, mainly in the temperate zone. Within the USSR about 400 species have been found. The genus is represented in the flora of the Soviet Arctic by 103 species, mostly associated with its temperate districts.

About 10% of all the sedges growing in the Arctic are typical arctic species (*C. stans, C. Bigelowii, C. ensifolia* ssp. *arctisibirica, C. lugens, C. rariflora, C. misandra, C. glacialis, C. rotundata, C. membranacea, C. holostoma, C. parallela, C. scirpoidea*). All these listed species range beyond arctic limits, occurring in the Subarctic and sometimes penetrating more southern districts in mountain systems. The most widespread in the Arctic are *C. stans, C. ensifolia* ssp. *arctisibirica, C. rariflora, C. rotundata, C. misandra* and *C. glacialis*. These species usually grow as large numbers of individuals; they play a very substantial role in the formation of vegetational communities and determine their appearance.

Another group of sedges in the arctic flora consists of true arctic-alpine species. Of these the most characteristic of the Arctic are *C. tripartita, C. rupestris* and (in part) *C. atrofusca*.

Nine species growing in the Arctic have distributions associated with northern and polar coasts (*C. subspathacea, C. maritima, C. glareosa, C. Gmelinii, C. recta, C. paleacea, C. salina, C. Mackenziei, C. ursina*). *Carex ursina* is an exclusively arctic species. Among the remainder, *C. maritima* and *C. subspathacea* may be considered predominantly arctic.

The most numerous group of arctic sedges consists of boreal species. A portion of these species are very widespread in the forest zone and only penetrate the Arctic to an insignificant extent near its southern edge (*C. nigra, C.*

acuta, C. lasiocarpa, C. rhynchophysa, C. vesicaria, C. limosa, C. appendiculata, C. diandra, C. loliacea, etc.). Certain boreal species are relatively characteristic of more temperate districts of the Arctic (*C. canescens, C. brunnescens, C. chordorrhiza, C. dioica, C. Redowskiana, C. caespitosa, C. aquatilis, C. wiluica, C. vaginata, C. rostrata, C. angarae*, etc.).

Finally, there are isolated localities situated within the Arctic for steppe species (*C. duriuscula, C. pediformis, C. supina*).

Sedges are especially characteristic of temperate districts of the Soviet Arctic. Only a comparatively small number of species penetrate high arctic latitudes (*C. misandra, C. stans, C. ensifolia* ssp. *arctisibirica, C. saxatilis* ssp. *laxa, C. rupestris, C. rariflora, C. tripartita, C. ursina, C. parallela, C. maritima, C. rotundata*). The islands lying east of Novaya Zemlya are extremely poor in *Carex* species.

The arctic sedges belong to diverse systematic groups. The largest components of the arctic flora with respect to numbers of species are section *Canescentes* of subgenus *Vignea* (14 spp.) and sections *Acutae* (20 spp.), *Atratae* (13 spp.) and *Physocarpae* (8 spp.) of subgenus *Carex*. The section *Dioicae* is represented by almost all its species (except one). The most extensive and abundant distributions in the Arctic are possessed by certain species of section *Acutae*, the most advanced group of subgenus *Carex* in an evolutionary sense.

The majority of arctic sedges are distributed in association with marshy lowland tundras and marshes or with meadow associations. The latter habitats are more characteristic of boreal elements of the flora. A few species grow in maritime meadows or on maritime sands and gravelbars (*C. Gmelinii, C. maritima, C. subspathacea, C. glareosa*, etc.).

Certain xerophilous species occur on dry slopes (rocky, stony or stony with fine soil) and on sandy sites (*C. rupestris, C. glacialis, C. macrogyna, C. duriuscula, C. pediformis, C. sabulosa*, etc.).

1. Inflorescence consisting of only a *single* terminal spikelet. . . . . .2.
   – Inflorescence consisting of *2 or more* spikelets. . . . . .18.

2. Plant *dioecious* (i.e., pistillate and staminate spikelets situated on different individuals). . . . . .3.
   – Plant *monoecious*, usually with androgynous spikelets (i.e., with staminate florets on distal part of spikelet and pistillate florets on basal part), rarely with gynaecandrous spikelets (i.e., with reverse position of florets). . . . . .8.

3. Stigmas *3*; ripe perigynia *appressed to spikelet rachis or slightly divergent*. Culms scabrous; leaves *flat*, 1.5–2 mm wide. . . . . .4.
   – Stigmas *2*; ripe perigynia *horizontal or reclined from spikelet rachis*. Culms more or less smooth; leaves *sulcate-triangular*, no more than 1 mm wide. Plant loosely tufted, with slender creeping rhizomes. . . . . .5.

4. Plant with *more or less long creeping rhizomes covered, like the culm bases, with pale brown scalelike sheaths*. Scales *castaneous, not ciliate*, on lower

part of spikelet often awned and 2–3 times as long as the perigynia. Perigynia *lanceolate*, 3–4 mm long, *glabrous*. . . .7. **C. ANTHOXANTHEA** PRESL.

— Plant densely tufted, with *very short creeping rhizomes*; culm bases entirely surrounded by *dark purple scalelike sheaths*. Scales *blackish purple, usually ciliate on margins*. Perigynia *obovoid*, 2.5–3(3.5) mm long, *very densely covered with yellowish brown hairs*. Pistillate and staminate spikelets 1.5–2.5(3) cm long. .6. **C. SCIRPOIDEA** MICHX.

5. Culms *cylindrical*, very finely and inconspicuously furrowed. Perigynia *ovoid or elliptical*, with 10–13 *thick nerves, thick-walled, with short* more or less scabrous beak. . . . . .6.

— Culms *ridged*, deeply furrowed. Perigynia *ovoid-lanceolate, rarely ovoid, thin-walled*, with 10–12 *thin nerves*, mostly *with elongate* smooth beak. . . . . .7.

6. Pistillate spikelets *dense*, more or less elongate-ovoid, 0.7–1.2(1.5) cm long, *with (10)15–18* florets; their scales usually entirely dark brown, *without whitish hyaline margins or with very narrow* margins, normally broadly rounded or obtuse distally, more rarely acute. Perigynia *ovoid*, (2.5)3–3.5(3.7) mm long, planoconvex or more rarely unequally biconvex, brown when ripe, *gradually* tapering to scabrous beak with whitish hyaline borders. 1–2 pistillate florets occasionally present at base of staminate spikelets, or very rarely spikelets androgynous. . . . . . . .17. **C. DIOICA** L.

— Pistillate spikelets *loose*, elongate, 0.7–1.5 cm long, *with 5–10* florets which are often abortive distally; scales pale ferruginous or more rarely dark ferruginous, with more or less *broad whitish hyaline margins*, obtuse or acute. Perigynia *elliptical* or more rarely elongate-ovoid, (2.5)3–3.7(4) mm long, unequally biconvex (more strongly so behind), *abruptly* tapering to short (0.2–0.3 mm long), slightly scabrous or sometimes smooth beak with whitish hyaline borders. As in the preceding species, plants with androgynous spikelets occur very rarely. . . . . .18. **C. GYNOCRATES** WORMSK.

7. Perigynia *ovoid-lanceolate, 3.5–4.3(4.5) mm long*, planoconvex or unequally biconvex, *slightly bent* when ripe, *gradually tapering without incurvation to elongate beak*. Pistillate spikelets *loose*, with perigynia somewhat spaced; scales ovoid or broadly ovoid, obtuse, brown with *broad whitish hyaline margins, usually half as long as* (more rarely *slightly shorter than*) perigynia. Individuals with androgynous spikelets occur occasionally. . . . . .19. **C. REDOWSKIANA** C.A. MEY.

— Perigynia *ovoid or elongate-ovoid, 3–3.5(4) mm long, straight* when ripe, *abruptly tapering to short beak*; pistillate spikelets *compact*; scales ovoid, *entirely dark brown or with narrow whitish hyaline margins, almost equalling* perigynia. . . . . .20. **C. PARALLELA** (LAEST.) SOMMERF.

8. Stigmas *2*; achenes *biconvex*. . . . . .9.

— Stigmas *3*; achenes *usually triangular*. . . . . .14.

9. Spikelets *gynaecandrous*. . . . . .28. **C. URSINA** DEW. *(see couplet 30).*

– Spikelets *androgynous*. .... 10.

10. Perigynia *with well expressed nerves, leathery*, almost sessile, *without rachilla* at base of achene (plants of *C. dioica* group with androgynous spikelets). .... *(see couplet 5)*.
– Perigynia *without nerves, membranous, with rachilla* at base of achene; if without rachilla, then perigynia on slender elongate pedicels (up to 1 mm long). .... 11.

11. Perigynia *without rachilla*, elongate-ovoid or ovoid, 3–3.5 mm long, *on pedicels 0.7–1 mm long*, when ripe *horizontal or reclined from spikelet rachis*; scales soon falling. Culms smooth; leaves flat or folded lengthwise. .... 3. **C. MICROPODA** C.A. MEY.
– Perigynia *with rachilla* at base of achene; ripe perigynia *imbricate, sessile*. .... 12.

12. Spikelets ovoid or elongate-ovoid, *without bract*. Bases of shoots surrounded by long *pale yellowish brown* sheaths. Culms very slender, *smooth*; leaves setiform-involute, 0.25–0.5 mm wide, stiff, almost equalling culm. Perigynia scarious, elliptical or more rarely obovoid, (3)3.5–5 mm long, rounded distally, abruptly tapering to short beak (0.2–0.3 mm long); margins of perigynia scabrous distally. .... 8. **C. HEPBURNII** BOOTT.
– Base of the almost globose spikelets *with a scalelike bract*, in whose axil (unlike the scales) there is no floret. Bases of shoots surrounded by *purple or reddish brown* sheaths. Culms slender, firm, *scabrous above*; leaves setiformly involute, stiff, 0.5–0.7 mm wide. Perigynia more or less abruptly tapering to beak 0.5–0.8 mm long. .... 13.

13. Spikelets *usually pale*, with brownish or light ferruginous scales that are half as long as and narrower than the perigynia. Perigynia *ovoid*, (3)3.3–3.7(4) *mm long, with smooth* beak 0.6–0.8 mm long. .... 9. **C. CAPITATA** L.
– Spikelets *dark*, with reddish brown scales that are shorter than the perigynia only by the length of their beaks and almost equally wide. Perigynia *orbicular-ovoid or broadly elliptical, 2.5–3 mm long*, with sparsely *scabrous* margins and beak about 0.5 mm long. .... 10. **C. ARCTOGENA** H. SMITH.

14. Perigynia *ovoid-lanceolate* (lanceolate-conical), *gradually* tapering
(8) without incurvation to elongate smooth beak. Rhizomes and culm bases covered with *light brown* scalelike sheaths. .... 15.
– Perigynia *ovoid, broadly ovoid or elongate-obovoid, abruptly* tapering to short smooth beak. Rhizomes and culm bases covered with more or less *castaneous or purple* scalelike sheaths. .... 17.

15. Scales *falling early, mature perigynia always without scales*. Perigynia almost subulate, *deflexed* when ripe. Rhizomes slender, creeping. ..... 16.

– Scales *not falling when the perigynia ripen, the latter scarcely divergent from the rachis of the spikelet.* . . .7. **C. ANTHOXANTHEA** PRESL *(see couplet 4).*

16. Culms *cylindrical,* smooth. Leaves *cylindrical or compressed-cylindrical,* 0.8–1.5 mm wide. Perigynia 5–9, 4–5 mm long, with compressed-cylindrical *rachilla* exserted by 1.5–1.8 mm (perigynia together with rachilla 5.5–7 mm long); stigmas *displaced sideways by rachilla.* . . . .
. . . . .5. **C. MICROGLOCHIN** WAHLB.
– Culms *triangular,* smooth or slightly scabrous. Leaves *flat or folded lengthwise,* 1–1.5 mm wide. Perigynia (2)3–4(5), 6–7 mm long, *without rachilla;* stigmas in normal position *at centre of tip of perigynium.* . . . .
. . . . .4. **C. PAUCIFLORA** LIGHTF.

17. Plant with long creeping rhizomes that are covered, like the culm bases, with *purple scalelike sheaths.* Leaves 1.5-2 mm wide, almost straight. Perigynia *ovoid or broadly ovoid, almost orbicular,* 2.5–3.5 mm long, blackish brown, *shining, rather strongly divergent* from spikelet rachis when ripe. *Achene with rachilla of about equal length situated at its base.* . .
. . . . .1. **C. OBTUSATA** LILJEBL.
– Plant with long or sometimes short creeping rhizomes, which are covered with *light brown or castaneous scalelike sheaths.* Leaves 2–2.5 mm wide, yellowish and *flexuous* (*often recurved*) *distally.* Perigynia *elongate-obovoid, bluntly triangular,* 3–4 mm long, rusty brown, *not shining, appressed* to spikelet rachis when ripe. *Rachilla absent.* . . . .
. . . . .2. **C. RUPESTRIS** BELL. EX ALL.

18. Spikelets in inflorescence *almost uniform* in shape and size, all or at least
(1) one or more of the upper spikelets bisexual (occasionally spikelets in inflorescence heterogeneous: the upper and lower pistillate or bisexual, the middle bisexual or staminate). . . . . .19.
– Spikelets *of two types:* the upper narrower and staminate, the lower thickened and pistillate (rarely pistillate spikelets with a certain number of staminate florets distally). . . . . .53.

19. Stigmas 2; achenes biconvex. . . . . .20.
– Stigmas 3; achenes triangular. . . . . .42.

20. Spikelets usually sessile; no sheathlike bracteole in the axil of the scalelike spikelet bracts. . . . . .21.
– Spikelets on short or more rarely elongate pedicels; sheathlike bracteole surrounding the base of the pedicel present in the axil of the (leaflike or scalelike) bracts. . . . . .41.

21. Plant of sphagnum bogs *with flagelliform prostrate leafy shoots,* from whose nodes arise erect vegetative and generative shoots. Inflorescence capitate, 0.8–1.5 cm long, consisting of 3–5 spikelets so closely aggregated that they appear to form a single spikelet. Perigynia broadly elliptical or

ovoid, brown when ripe, shining, with dark nerves, with short entire beak. Culms smooth; leaves 1.5–2 mm wide. . . . . .13. **C. CHORDORRHIZA** EHRH.
- Plant *not forming flagelliform prostrate leafy shoots.* . . . . .22.

22. Inflorescence either with all spikelets *androgynous* (i.e., with staminate florets on distal part of spikelet and pistillate florets on basal part) *or with admixture of entirely staminate and entirely pistillate spikelets.* Plant loosely tufted, with more or less *long creeping rhizomes.* . . . . .23.
- Inflorescence with all spikelets *gynaecandrous* (i.e., with pistillate florets on distal part of spikelet and staminate florets on basal part). Plant more or less densely tufted, *with abbreviated noncreeping rhizomes;* culms aggregated in clusters. . . . . .27.

23. Spikelets in inflorescence *of two types:* the upper and lower pistillate or androgynous, the middle androgynous or staminate. Perigynia more or less elongate-ovoid, 4.5–5.5 mm long, *with ciliately serrated wing on margins, often tuberculate and with short dense appressed pubescence* on both sides. Culms sharply triangular and scabrous; leaves flat, 3–5 mm wide. . . . . .12. **C. PALLIDA** C.A. MEY.
- All spikelets in inflorescence *androgynous.* Perigynia *glabrous, not winged* on margins. . . . . .24.

24. Inflorescence consisting of *2–4(5) small pale green spikelets separated by 1–2 cm;* spikelets with few florets (1–2 staminate and 2–4 pistillate). Perigynia elliptical, 2.5–3 mm long, thickened, unequally biconvex, leathery, shining, brown when ripe, abruptly tapering to short entire beak, with thin conspicuous nerves and with marginal ribs in the form of two thickened nerves displaced to its posterior side. Culms *filiform, weak, lodging,* sharply scabrous above. Leaves flat, soft, 1–2 mm wide. Rhizome very slender, 0.5–0.8 mm in diameter. . . . . .35. **C. DISPERMA** DEW.
- Inflorescence consisting of *more numerous, closely aggregated brown spikelets.* Culms *strong, erect.* . . . . .25.

25. Inflorescence *spikelike, with short branches below, 2–3.5 cm long,* consisting of *numerous* small spikelets. Perigynia ovoid, 2.5–3 mm long, nonuniformly biconvex, castaneous when ripe, shining, with elongate scabrous beak which is cleft anteriorly. Culms *scabrous,* sharply triangular, on their lower part surrounded by long dark brown sheaths. Leaves sulcate, 1.5–3 mm wide. Plant 30–80(100) cm high. . . . .
. . . . .11. **C. DIANDRA** SCHRANK.
- Inflorescence *capitate, ovoid or almost globose, not branched, 0.7–1.5 cm long,* consisting of *3–10* spikelets. Perigynia with entire or slightly notched beak. Culms *usually smooth,* bluntly triangular or almost cylindrical. Plant 5–15(20) cm high. . . . . .26.

26. Rhizome *slender, 0.8–1(1.5) mm in diameter,* with firm cortex *not peeling during drying.* Culms slender, 0.6–1(1.2) mm in diameter; leaves usually

folded lengthwise, setiform. Perigynia (2)2.5–3 mm long, planoconvex, thick-walled, with *slightly scabrous* short beak, *sessile*. Achene *completely* filling perigynium. Plant of steppelike habitats. ....
....14. **C. DURIUSCULA** C.A. MEY.

– Rhizome *thickened, 1.5–2 mm in diameter*, with cortex *peeling* during drying. Culms 1.5–2 mm in diameter; leaves sulcate or folded lengthwise. Perigynia (3.7)4–5.5 mm long, somewhat turgidly biconvex, usually with *smooth* beak, borne on well differentiated *slender pedicel*. Achene *not completely* filling perigynium. ....15. **C. MARITIMA** GUNN.

27. Perigynia *concave-convex, thickened*, 4–5 mm long, anteriorly with 8–10
(22) conspicuous nerves and deeply cleft beak, from base with more or less *broad marginal wing* (up to 0.3 mm wide) that is finely serrate distally. Achene *half as wide as and shorter* than perigynium. Inflorescence spikelike or capitate, 2–4 cm long, consisting of 4–6 aggregated obovoid spikelets *0.7–1 cm long*. Culms scabrous above, leafy almost to half their height; leaves flat, 2–3(4) mm wide. ....16. **C. LEPORINA** L.

– Perigynia *planoconvex, not thickened, without marginal wing*; achene *completely or almost completely* filling perigynium. Spikelets usually *not more than 0.5 cm long*, rarely up to 1 cm. ....28.

28. Inflorescence *capitate*, 1–2 cm long, consisting of 2–4(5) usually *closely aggregated spikelets* (occasionally the lowest spikelet separated by 5 mm).
....29.

– Inflorescence *not capitate*, mostly *spikelike*, with more or less *separated spikelets*. ....35.

29. Plant *loosely tufted, with slender stolons*, more rarely densely tufted. Leaves flat or semifolded, 1–1.5(2)mm wide. Scales *pale green or rarely pale ferruginous*. Perigynia ovoid or elliptical, (2.5)2.7–3.2(3.5) mm long, thin-walled with whitish tint, acute or sometimes obtuse distally, *almost beakless*, with 4–8 thin nerves which are conspicuous when brunnescent.
....30. **C. TENUIFLORA** WAHLB.

– Plant *rather densely tufted, without stolons*. Scales *dark brown, cinnamon brown or ferruginous*. Perigynia with *clearly expressed beak*. ....30.

30. Plant *3–5(10) cm high; leaves almost equalling culm*. Inflorescence with *no more than 2 spikelets*, of which the terminal one is larger (4–7 mm long) and the second lateral one smaller. Scales broadly ovoid or broadly obovoid, broadly rounded distally, castaneous, slightly shorter than perigynia. Perigynia broadly ovoid or ovoid, *2–2.3 mm long*, greyish green, with 5–6 thin *inconspicuous nerves*, with very short (0.2 mm long) smooth beak.
....28. **C. URSINA** DEW.

– Plant *(10)15–40(50) cm high; leaves usually shorter than culm*. Inflorescence consisting of *(2)3–5 spikelets*. Scales mostly ovoid or elongate-ovoid. Perigynia *2.5–3.5 mm long*, usually with *conspicuous nerves*.
....31.

31. *Upper spikelet in inflorescence gynaecandrous, lateral spikelets entirely pistillate.* Perigynia ovoid or obovoid, more rarely elongate-ovoid or lanceolate. Inflorescence ovoid or elongate-ovoid, 1–1.5(2) cm long, consisting of 2–3(4) aggregated or rather loosely arranged ovoid or oblong spikelets. Culms smooth or only scabrous below inflorescence, bent, sometimes drooping. Leaves greyish green, flat or folded lengthwise, 1–2 mm wide. ....27. **C. GLAREOSA** WAHLB.
- *All spikelets in inflorescence gynaecandrous.* ....32.

32. Culms *scabrous*. Spikelets *ovoid* or more rarely *almost globose, 3–8 mm long*, 4–5 mm wide; scales ovoid or elongate-ovoid, acute, brown or light brown. Perigynia with *opaque, greenish-whitish bloom*, later turning brown distally, *gradually tapering to broad beak*, with thin nerves which are conspicuous when brunnescent. Leaves greyish green, folded lengthwise or flat, 1–1.5(2.5) mm wide, stiff, almost as long as culm or slightly shorter. ....33.
- Culms *smooth or slightly scabrous only below inflorescence*. Spikelets *obovoid or oblong, 7–12 mm long*, 3–5 mm wide; scales broadly ovoid or ovoid, mostly obtuse, reddish brown, castaneous, ferruginous or blackish brown. Perigynia *without opaque bloom*, membranous, light ferruginous, more or less *abruptly tapering to narrow beak*. Leaves flat or folded lengthwise, 1.5–2(2.5) mm wide. ....34.

33. Perigynia *ovoid or ovoid-elliptical, with spongy tissue at base weakly developed, with well expressed (0.5–0.7 mm long), slightly scabrous (or more rarely smooth) beak* that is more or less *deeply cleft anteriorly*. Inflorescence 1–2 cm long, consisting of (2)3–4(5) closely aggregated spikelets or at most the lowest separated by 1 cm; spikelets 5–8 mm long (the lower sometimes entirely pistillate). ....31. **C. HELEONASTES** EHRH.
- Perigynia *elliptical, with spongy tissue at base well developed; their beaks very short (just expressed), smooth or more rarely slightly scabrous, slightly bent, entire or slightly cleft anteriorly.* Inflorescence 1–1.5(1.7) cm long, consisting of 2–3(4) slightly separated spikelets 3–5 mm long. ....
....32. **C. AMBLYORHYNCHA** V. KRECZ.

34. Perigynia *obovoid, ovoid or elliptical*, 2.5–3.5 mm long, with smooth margins, *when ripe appressed or slightly divergent from spikelet rachis and scarcely bent, with elongate smooth beak* 0.5–0.8 mm long. Inflorescence ovoid or elongate-ovoid, somewhat lobed (reminiscent of inflorescence of *C. leporina*), 1–2 cm long, consisting of (2)3–4(5) closely aggregated brown spikelets. ....33. **C. TRIPARTITA** ALL.
- Perigynia *broadly ovoid or broadly elliptical, when ripe distinctly divergent from spikelet rachis and bent, with very short beak and distally scabrous margins*. ....34. **C. PRIBYLOVENSIS** J.M. MACOUN.

35. Perigynia *thick-walled, usually broadly elliptical or broadly ovoid*, more
(28) rarely ovoid, 2.8–3(3.5) mm long, *about 1.8 mm wide*, glaucous green, later brunnescent, with 7–10 indistinct nerves which become more conspic-

uous when brunnescent, situated on a short (0.4–0.6 mm long) but well defined pedicel, abruptly tapering to a short (smooth or slightly scabrous) beak. Scales broadly ovoid, obtuse, light brown to reddish brown or brown, equalling perigynia or slightly longer and broader. Culms mostly *smooth*. Leaves *glaucous or yellowish green, sometimes bluish*, flat, 2–3 mm wide. Inflorescence (1.5)2–4.5 cm long, consisting of (3)4–6 spikelets of which the *uppermost spikelet is clavate, 1–1.5 cm long*, normally with staminate florets to middle (very rarely with only 1–2 staminate florets in var. *isostachya* Norman), and the *lateral spikelets ovoid or elongate-ovoid*, 0.5–1.3 cm long, with very few staminate florets or entirely pistillate. . . . . .23. **C. MACKENZIEI** V. KRECZ.

– Perigynia *membranous or thin-walled, ovoid or elliptical, 1–1.5 mm wide*; *uppermost spikelet shorter and scarcely differing in shape from the lateral spikelets*. Culms *scabrous*. . . . . .36.

36. Inflorescence consisting of 2–4 *small (3–5 mm long), almost globose spikelets with few florets (2–4 pistillate and 1–2 staminate)*. Scales broadly ovoid, pale green, half or less as long as perigynia. Perigynia ovoid or elliptical, (2.2)2.5–3.3 mm long, *with thickened raised nerves, beakless*, acute or more rarely obtuse distally, with entire or slightly notched tip, when ripe strongly divergent from spikelet rachis. Plant *loosely tufted, with slender stolons*. Leaves light green, flat, 1–2 mm wide, soft. . . . . . .29. **C. LOLIACEA** L.

– *Spikelets with many florets*. Perigynia *with thin nonprominent nerves, with evident* (though sometimes very short) *beak*. Plant *rather densely tufted, mostly without stolons*. . . . . .37.

37. Perigynia elliptical or obovoid, *1.5–2 mm long*, nerved, *abruptly tapering to a very short (0.1–0.15 mm long) smooth beak*. Inflorescence (2)3–4 cm long, consisting of (3)4–7(8) ovoid or almost globose (more rarely elongate-ovoid) spikelets (3)4–7(9) mm long, 2.5–4(5) mm wide, with ripe perigynia appressed. Leaves greyish green, flat, (1)1.5–2(3) mm wide. . .38.

– Perigynia *(2)2.3–3.5 mm long, gradually tapering to a more elongate scabrous beak*. . . . . .39.

38. Scales *usually whitish green or pale brownish, acute or more rarely obtusish, usually ¾* (rarely only half) *as long as perigynia*. . . . . .25. **C. LAPPONICA** O.F. LANG.

– Scales *brown, obtuse or more rarely acutish, usually about half as long as perigynia*. . . . . .26. **C. BONANZENSIS** BRITT.

39. Perigynia *2.8–3.3 mm long*, ovoid, *ferruginous yellow*, rounded basally, *with well defined slender pedicel about 0.4 mm long, when ripe* more or less *strongly divergent from the spikelet rachis and slightly bent*. Inflorescence 1.5–2.5 cm long, consisting of 4–5 slightly separated spikelets 5–8 mm long; scales ovate, *ferruginous*, with whitish hyaline margins, about ⅔ as long as perigynia. Leaves greyish green, flat, 1.5–2 mm wide, usually half as long as culm. . . . . .21. **C. KRECZETOVICZII** EGOR.

– Perygynia (*2*) *2.2–2.7 mm long, almost sessile, when ripe appressed and straight*; scales *light brownish ferruginous, pale ferruginous or whitish green.* . . . . .40.

40. Perygynia with *elongate* beak which is *cleft anteriorly to its base* (white stripe along beak on margins of fissure). Inflorescence brownish, *(1.5)2–3.5(4–5) cm long*, consisting of (4)5–8(10) *globose-ovoid or ovoid* (more rarely oblong) *spikelets* 3–8 mm long and 3–5 mm wide; spikelets *with oily sheen*. Scales light brownish ferruginous or pale ferruginous. Leaves green, flat, 1.5–2(3) mm wide. Culms *with straight edges, usually strongly spreading* and sometimes lodging.
. . . . .22. **C. BRUNNESCENS** (PERS.) POIR.
– Perygynia with *short* beak that is *not cleft anteriorly but bears only a weak cuneiform notch*. Inflorescence pale green, *2.5–5(8) cm long*, consisting of (3)4–10(12) *oblong or ovoid spikelets* 4–10 mm long and 3–5 mm wide, *without sheen*. Scales pale green or pale ferruginous. Leaves greyish green or light green, flat, 1.5–3(4) mm wide. Culms *with concave edges* (*appearing winged*), *straight and erect*. . . . . .24. **C. CANESCENS** L.

41. Plant *loosely tufted, with creeping rhizomes*. Roots *bare*. Shoots
(20) surrounded at base by *pale brown* leaf sheaths. Leaves 1.5–2.5 mm wide, flat. Inflorescence *1.5–2.5 cm long*; spikelets *0.5–1.2 cm long*, all aggregated or sometimes the lowest separated by 1–2 cm, sessile or the lowest on pedicel 0.5–1.5 cm long and sometimes drooping. Scales *dark brown*, mostly with broadly rounded tip, *with broad green stripe along midrib*, mostly slightly shorter than or equalling perygynia, more rarely distinctly shorter. Perygynia *obovoid or elliptical, very densely covered with minute papillae*, not turning brown, *with 3–4 inconspicuous nerves*, rounded distally, *without beak*. . . . . .50. **C. BICOLOR** BELL. EX ALL.
– Plant *densely tufted, without creeping rhizomes*. Roots *with conspicuous light brown pubescence with greenish tinge*. Shoots surrounded at base by *blackish purple* leaf sheaths. Leaves 2–2.5(3) mm wide, flat. Inflorescence *(2)3–4(5) cm long*; spikelets *2–3(4) cm long*, sessile or the lowest on pedicel 0.3–0.8(1) cm long. Scales *purplish black*, obtuse or acutish, *with pale midrib but without green stripe along it*, slightly shorter than, equalling or just longer than perygynia. Perygynia *elliptical, broadly elliptical or orbicular, without papillae*, sometimes purplish brown distally, *without nerves, with very short but sharply differentiated beak*. . . . .
. . . . .49. **C. ELEUSINOIDES** TURCZ. EX KUNTH.

42. Uppermost spikelet in inflorescence androgynous, with 1–4 pistillate
(19) florets. Leaves narrowly sulcate, 0.5–1 mm wide. . . . .
. . . . .90. **C. WILLIAMSII** BRITT. (*see couplet 91*).
– Uppermost (or 2 uppermost) spikelet(s) gynaecandrous. . . . . .43.

43. Lowest bract *with sheath 1–3 cm long* and blade. Perygynia *lanceolate or elongate-ovoid*, without nerves. Spikelets drooping, on long slender

pedicels 2–4(5) cm long. Plant densely tufted, without creeping rhizomes. ....44.
- Lowest bract normally *without or with very short sheath (2–3 mm long)*, rarely longer (up to 1 cm). Perigynia *not lanceolate*. ....45.

44. Spikelets *ovoid*, (0.5)1–1.5 cm long, *0.6–0.8 mm wide*, 3–5 in number, *not branching*. Scales *blackish brown or purplish brown*. Perigynia *4.5–5 mm long, dark brown distally*, with elongate scabrous beak. Culms surrounded below by *very pale brownish yellow* leaf sheaths. Leaves 2.5–3.5 mm wide, flat or folded lengthwise, *downcurved; many dried up leaves from previous year at base of shoots.* ....60. **C. MISANDRA** R. BR.
- Spikelets *linear*, 1–1.2 cm long, *0.2–0.3 mm wide*, 4–6 in number, *the lowest sometimes with 1–2 branchlets at base*. Scales *light ferruginous brown*. Perigynia (2.5)2.8–3.3(3.5) *mm long, pale ferruginous*, with elongate smooth or slightly scabrous beak. Culms surrounded by *brownish* sheaths. Leaves *straight*, flat, 2–3 mm wide; *almost no old leaves at base of shoots.* ....93. **C. KRAUSEI** BOECK.

45. Culms surrounded at base by brownish sheaths. Spikelets *on slender pedicels, pendant*. Plant with more or less long stolons; roots with very dense yellowish brown or orange pubescence. ....
................................57. **C. MAGELLANICA** LAM. *(see couplet 116).*
- Culms sourrounded at base by dark purple sheaths. Spikelets mostly *upright*. Roots glabrous. ....46.

46. Plant loosely or densely tufted, *with rather long horizontal or obliquely ascending creeping rhizomes.* ....47.
- Plant densely tufted, *without or with very short creeping rhizomes.* .....49.

47. Rhizomes *obliquely ascending*. Culms *smooth*, loosely surrounded at base by numerous long brown or purplish brown sheaths of dead leaves that *do not disintegrate reticulately*. Inflorescence (2)2.5–3.5(5) cm long, consisting of (3)4–5 closely aggregated spikelets or the lowest sometimes slightly separated (by 1 or more rarely 2 cm). Lowest bract mostly *scalelike with short setiform tip or narrowly leaflike, usually shorter than its spikelet* (*rarely longer*). Scales ovate, *acute*, mostly dark purplish brown, slightly longer than perigynia. Perigynia elliptical or obovoid, 4–5(6) mm long, slightly inflated, *yellowish green*, with thin nerves, abruptly tapering *to relatively long, 0.7–1(1.2) mm*, usually distinctly bidentate *beak*. Leaves 2–3.5(4) mm wide, long-acuminate, stiff, with slightly revolute margins.
.....................69. **C. SABULOSA** TURCZ. EX KUNTH *(see couplet 112).*
- Rhizomes *horizontal*. Culms *usually scabrous*, surrounded at base by reddish brown or purplish brown scalelike sheaths that *disintegrate reticulately*. Lowest bract *leaflike, longer than its spikelet and the whole inflorescence*. Scales *acuminate or awned*, with pale stripe along midrib. Perigynia elliptical, *greenish white or yellowish white*, very densely granular, with 4–5 inconspicuous nerves, with *very short beak* (*no more than 0.3 mm long*) *or beakless*, notched distally. ....48.

48. Inflorescence *5–10 cm long*, consisting of *4(5)* more or less separated spikelets. Lowest bract *as long as or longer than inflorescence*. Scales *ovate-lanceolate, brown or purplish brown, with more or less long awn*; lower scales longer than perigynia, upper scales equalling them. Perigynia (3.5)4–4.4 mm long, *with short beak that is broadly indented*. Leaves 2–3 mm wide. . . . . .65. **C. BUXBAUMII** WAHLB.
– Inflorescence *3.5–5 cm long*, consisting of *3(4)* spikelets. Lowest bract usually *shorter than inflorescence, occasionally equalling it*. Scales *ovate, purplish black, acuminate or with very short awn*, shorter than perigynia or more rarely equalling them. Perigynia 3.5–3.8 mm long, *beakless (but with notch distally) or with slightly expressed beak*. Leaves 2.5–3 mm wide. . . . . .66. **C. ADELOSTOMA** V. KRECZ.

49. Scales dark purplish brown, pale along midrib, more or less *abruptly transformed into elongate scabrous awn*, as long as or slightly longer than perygynia. Perigynia elliptical, 4–5 mm long, compressed-triangular, brownish yellow, not granular, with 5–6 distinct nerves, with very short beak 0.1(0.2) mm long that is entire or slightly notched. Inflorescence erect or slightly declinate, (4)7–9 cm long, consisting of (3)4–5 spikelets of which the upper are closely aggregated but the lower more or less separated and borne on pedicels 1–3 cm long (sometimes drooping). Leaves 3–4(5) mm wide. Plant 25–60 cm high. . . . . .70. **C. GMELINII** HOOK. ET ARN. *(see couplet 111)*.
– Scales *acute to obtusish, without awns*. . . . . .50.

50. Inflorescence *3.5–6(7) cm long, sometimes more or less drooping*, consisting of (3)4–5 spikelets *1–3 cm long*; lower spikelets mostly on pedicels *1–2(3) cm long*. Perigynia elliptical, *(3.5)4–4.5 mm long*, without nerves, *not granular or slightly granular distally*. Leaves (3)4–7 mm wide. . . . . .51.
– Inflorescence *always erect, 1–2 cm long*, consisting of 3 (more rarely 4) spikelets *0.8–1.2 cm long*; pedicel of lowest spikelet usually *not more than 0.5 cm long* (extremely rarely 1 cm). Perigynia *2.2–3.6 mm long*, with or without nerves, usually *densely granular*. . . . . .52.

51. Ripe perigynia (3.5)4–4.5 mm long, *ferruginous yellow with purplish black beak*, very rarely slightly brunnescent; scales *elongate-ovate*, purplish black, usually *longer than perigynia*. Inflorescence (3.5)4–6(7) cm long, mostly consisting of *5 spikelets* 1–2(2.5) cm long. *Spikelet almost always present in axil of lowest bract*. (European plant). . . . . .63. **C. ATRATA** L.
– Ripe perigynia (3)3.5–4.2 mm long, *dark purplish*, scarcely differing from scales in colour; scales *ovate*, purplish black or dark brown, *slightly shorter than perigynia*. Inflorescence (3)3.5–4.5 cm long, mostly consisting of *4 spikelets* (1.5)1.8–3 cm long. *Spikelet usually absent from axil of lowest bract, which is usually separated by 2–5 cm*. (Asian plant). . . . . .64. **C. PERFUSCA** V. KRECZ.

52. Scales *purplish black, ⅓–¾ as long as perigynia and almost equalling them in width*. Perigynia *obovoid or elliptical* (rarely broadly elliptical), *2.2–2.5(2.8) mm long, without nerves, early brunnescent*, smooth or with short acicules distally on margins. Spikelets firm. Leaves 2.5–4(5) mm wide, mostly *not more than half as long as culm*. . . .68. **C. NORVEGICA** RETZ.
- Scales *dark brown or more rarely purplish black*, usually *not more than half as long as and narower than perigynia*. Perigynia *elongate-obovoid or elongate-elliptical, (3)3.3–3.6 mm long*, slightly inflated, *with 4–6 distinct* (or rarely inconspicuous) *nerves, pale green, later yellowish* or brownish green, smooth or with isolated acicules. Spikelets less firm. Leaves (2)3–4(5) mm wide, usually *only slightly shorter than culm*. . . . . .67. **C. ANGARAE** STEUD.

53. Stigmas 2; achenes biconvex, flattish. . . . . .54.
(18)
- Stigmas 3; achenes triangular. . . . . .73.

54. Perigynia *shining*, thinly membranous, slightly vesicularly inflated, *3–4(4.5) mm long*, reddish brown or coal black when ripe, with beak *0.5–0.7 mm long*. Style *bent* at extreme base (bent downwards at first, then upwards). Pistillate spikelets 0.8–2.5(3) cm long, 0.6–0.8 cm wide, 2 (more rarely 3) in number, sessile or on long pedicels. Staminate spikelets 1 (more rarely 2). Plant with very short creeping rhizomes. Shoots surrounded basally by reddish brown sheaths which sometimes disintegrate reticulately. . . . . .103. **C. SAXATILIS** L. S. L.
- Perigynia *dull*, planoconvex or biconvex, rarely inflated (and then not vesicularly), usually *2–2.5 (more rarely 3) mm long*, with beak *0.1–0.3 mm long*. Style *straight*. Pistillate spikelets narrower. . . . . .55.

55. Plant loosely tufted (or more rarely densely tufted), with *short or long creeping rhizomes* (i.e., all or some of the shoots in a tuft grow horizontally or almost horizontally for a certain distance). . . . . .56.
- Plant densely tufted, *without horizontal shoots, usually forming tussocks*. . . . . .68.

56. All pistillate spikelets on long pedicels, usually *drooping* or, if erect, then scales of pistillate spikelets pale brown with long awns (5)8–15 mm long. Culms smooth. *Shoots surrounded at base by reddish brown leaf sheaths that are usually scalelike and disintegrate reticulately.* . . . . .57.
- Pistillate spikelets *erect*, almost sessile or on short or more or less long pedicels. *Leaf sheaths surrounding bases of shoots usually prolonged as leaf blades and not disintegrating reticulately.* . . . . .58.

57. Scales *pale brown, gradually prolonged as awns*; on pistillate spikelets awns (5)8–15 mm long, *2.5–5 times* as long as perigynia; awns of staminate spikelets shorter, up to 3 mm long. Perigynia *membranous* (their walls easily shattered when pierced with a needle), with thickened calluslike beak. Achene *deeply pitted* in middle. Inflorescence 10–14 cm long,

consisting of 2 (rarely 3) staminate and 4 (rarely 3) pistillate spikelets. Staminate spikelets 2.5–3.5 cm long, on pedicels like the pistillate; pistillate spikelets 2–4 cm long. Shoots sparsely leaved. Leaves 4–6(10) mm wide, flat or with slightly revolute margins. . . . . .51. **C. PALEACEA** WAHLB.
- Scales usually *dark purplish brown* (or sometimes light brown on staminate spikelets), lanceolate, acute or obtusish, *always without awns, not more than 2.5 times* as long as perigynia, sometimes only slightly longer or very rarely only of equal length. Perigynia *leathery, with thick walls* (shattered with difficulty). Achene usually *not pitted*. Inflorescence consisting of 2 (more rarely 1 or 3) staminate and 3-4 pistillate spikelets. Staminate spikelets 2–2.5(3) cm long, on pedicels; pistillate spikelets 2–3.5 cm long, the uppermost sometimes up among the staminate spikelets. Leaves 3–7 mm wide, flat. . . . . .52. **C. CRYPTOCARPA** C.A. MEY.

58. Scales of pistillate spikelets *prolonged as awn up to 5 mm long* (very rarely awnless), oblong-lanceolate, mostly dark brown, with broad pale stripe along midrib, usually 1.5–2 times as long as perigynia. Perigynia abruptly tapering to short beak. *All pistillate spikelets on pedicels 0.5–5 cm long. Upper and usually lower (sometimes also middle) staminate spikelet on slender pedicel 0.8–2 cm long*. Achene *deeply pitted on side*. . . . . .59.
- Scales of pistillate spikelets *awnless, sometimes with very short mucro at tip. Lateral staminate and upper pistillate spikelets always sessile*, lower pistillate sometimes on short (rarely long) pedicels. Achene *not pitted*. . . . .
 . . . . .60.

59. Staminate spikelets *2.5–3(3.5) cm long*, 3 (rarely 2 or 1) in number. Pistillate spikelets *(2)4–6(7) cm long*, (3)4 in number, often with staminate florets distally; pedicel of lowest spikelet 1–5 cm or more long. Perigynia *elliptical*, 3–3.5(4) mm long, mostly with 3–4 nerves (sometimes nerveless), *covered with minute villous papillae* and sometimes purple spotted, not brunnescent, *with narrow beak (0.2–0.25 mm wide)*. Plant *30–60 cm high*. . . . . .53. **C. RECTA** BOOTT.
- Staminate spikelets *1–1.5(2) cm long*, 1–2 in number. Pistillate spikelets *1–2 cm long*, on pedicel 0.5–3 cm long. Perigynia mostly *ovoid*, 3–3.5 mm long, with or without nerves, usually *without papillae, with beak about 0.5 mm wide*. Plant *10–30 cm high*. . . . . .54. **C. SALINA** WAHLB.

60. Scales of *staminate* spikelets (usually) *and of pistillate* spikelets (sometimes) *with short mucro distally* (if not present on all scales of staminate spikelets, at least on some). Inflorescence 3–4 cm [rarely up to 5(7) cm] long, consisting of 1 (rarely 2) staminate spikelet(s) 0.8–1.3(2) cm long and 2 (rarely 3) pistillate spikelets 0.7–1.2 cm long. Pistillate spikelets with 5–15(20) florets on pedicels 0.5–1.5 cm long. Bracts *folded lengthwise, sheathing the spikelets like scabbards*. Perigynia planoconvex, ovoid or elongate-ovoid, *(2.7)3–3.5(4) mm long*, thickly membranous, nerveless or very rarely with 2–3 nerves, *gradually tapering without incurvation to an elongate, broadly thickened whitish beak*. Achene truncate distally. Leaves 1–2.5 mm wide, folded lengthwise or more rarely flat. Plant 3–15(20) cm

high. Shoots surrounded at base by light brown or more rarely reddish brown leaf sheaths. . . . . .55. **C. SUBSPATHACEA** WORMSK. EX HORNEM.

– Scales of *pistillate and staminate* spikelets *never prolonged as awns or mucros*. Bracts *usually flat*. Perigynia *smaller* or, if large, then beakless or abruptly tapering to very short beak. Plant usually taller. . . . . .61.

61. Perigynia *with distinct nerves*. Culms scabrous. Lowest bract equalling or exceeding inflorescence. . . . . .62.
– Perigynia *without nerves*. . . . . .63.

62. *Large plant, 50–120 cm high; leaves (3)4–8 mm wide*. Inflorescence 10–25(30) cm long, consisting of 2–3 staminate and 3–4(5) pistillate spikelets. Staminate spikelets *2.5–4 cm long*. Pistillate spikelets *(2.5)3–7(9) cm long*, sessile or the lowermost sometimes on a pedicel 1–2 cm long (very rarely pedicel longer and spikelet drooping). Scales of pistillate spikelets *lanceolate* or rarely ovate, *acute* or more rarely obtusish, *purplish black* with pale midrib, narrower than and mostly *1.5 times as long as perigynia* (more rarely slightly longer than or only equalling them, very rarely shorter). Perigynia *biconvex, slightly inflated*, elliptical or obovoid, (2.5)3–3.5 mm long, *without minute papillae*, with sharply differentiated short beak. . . . . .36. **C. ACUTA** L.
– *Smaller plant, 10–40 cm high; leaves (1.5)2–3 mm wide*. Inflorescence (2)3–8(10) cm long, consisting of 1 (rarely 2) staminate spikelet(s) *1–2.5(3) cm long* and 3–4 (rarely 2) pistillate spikelets *1.5–2.5(3) cm long*. Pistillate spikelets usually sessile; their scales *elongate-ovate or ovate, obtuse* or more rarely acutish, *dark brown* with pale midrib, usually *distinctly shorter than perigynia*, more rarely almost as long. Perigynia *planoconvex, not inflated*, ovoid, 2.5–3 mm long, *covered with minute papillae*, with more or less differentiated short beak. . . . . .37. **C. NIGRA** (L.) REICHARD.

63. Culms *smooth*. Bracts of lowest and next lowest (lowest but one) spikelets *exceeding or equalling* inflorescence; bases of bracts *without black scarious auricles or at most with these weakly developed and light brown*. Leaf sheaths *sometimes septate-reticulate due to raised transverse nerves*. Perigynia with short *thickened calluslike beak* (which cannot be split lengthwise with a needle without destroying the whole beak), *sometimes purple spotted over their entire surface, without papillae*. Leaves gradually acuminate. Roots *with very short brown pubescence* just visible beneath a lens, appearing almost bare without a lens. Staminate spikelets *2–4* (rarely 1). . . . . .64.
– Culms usually *slightly scabrous*, sometimes smooth. Bract of lowest spikelet *usually slightly exceeding* its spikelet, *but sometimes equalling or almost equalling* the inflorescence; bract of next lowest spikelet *just equalling or shorter than* its spikelet and always *considerably shorter* than the whole inflorescence; bases of bracts *with black scarious auricles*. Leaf sheaths usually *not septate-reticulate*. Perigynia *not purple spotted, with short beak not thickened and calluslike* (easily split lengthwise with a needle). Leaves mostly abruptly acuminate. Roots usually *with conspic-*

uous yellowish white or greyish white pubescence. Always 1 staminate spikelet. . . . . .65.

64. Plant *large, (40)50–100 cm high*. Leaves *(3)4–6(8) mm wide*, flat or rarely folded lengthwise, *green or greyish green, equalling or slightly longer than culm*. Inflorescence usually *large, (16)18–30(35–40) cm long* (*usually 22–27 cm*), consisting of (1)2–3(4) staminate and (2)3–5(6) pistillate spikelets. Staminate spikelets *(1)2–4(5) cm long* (*most often 3–3.5 cm*), aggregated, mostly *light brown*, with scales broadly rounded distally. Pistillate spikelets *(2)3–7(9) cm long* (*most often 4–6 cm*), separated, firm or lax, sessile or sometimes lowest spikelet on pedicel up to 3 cm long; scales usually obtuse or broadly rounded at tip, rarely acuminate, sometimes with whitish hyaline apex, *brownish, usually with broad green stripe along midrib*, shorter than, equalling or slightly longer than the perigynia and ⅓-⅔ as wide as them. Perigynia obovoid or elliptical, (2)2.5–3(3.5) mm long, *pale, not brunnescent*. . . . . .42. **C. AQUATILIS** WAHLB.

– Plant *not so large, 15–45(60) cm high*. Leaves *(2.5)3–5 mm wide*, flat or folded lengthwise, *yellowish, usually shorter than culm* (*sometimes half as long*). Shoots with (10)15 or more leaves (many from previous year). Inflorescence *(3–5)6–14(15–18) cm long* (*most often 7–12 cm*), consisting of 1–2 (more rarely 3) staminate and (2)3–4 pistillate spikelets. Staminate spikelets *(1)1.5–2.5(3.5) cm long*, mostly *purplish black*, with scales broadly rounded distally. Pistillate spikelets *(1–1.5)2–4(5) cm long* (*most often 3–3.5 cm*), aggregated, the lowermost sometimes on a pedicel up to 2 cm long; scales obtuse or obtusish, *dark brown or purplish black, with pale midrib or entirely dark*, mostly equalling the perigynia in width and more or less equalling them (slightly shorter to slightly longer) in length. Perigynia elliptical or obovoid, (2)2.5–3(3.5) mm long, *brunnescent distally*. . . . . .43. **C. STANS** DREJ.

65. Staminate spikelet *rather narrow (2.5–3 mm wide), usually on pedicel (0.3)0.5–2(2.5) cm long, rarely sessile*. Pistillate spikelets (1)1.5–2.5(3) cm long, 2–4(5) in number, more or less *separated*, lax, the lowermost usually on pedicel 1–3 cm long or longer (sometimes pedicel arising from very base of culm). Scales of pistillate spikelets with *pale* (*green or yellowish*) *midrib, very rarely completely dark*. Leaves (2)3–5 mm wide, mostly flat, sometimes slightly revolute. . . . . .66.

– Staminate spikelet *thickish (3–5 mm wide), usually sessile* (i.e., the next pistillate spikelet situated at its very base), *rarely on pedicel 0.5 cm long* (*very rarely up to 1 cm*). Pistillate spikelets 0.5–2 cm long, 3–4 in number (more rarely 2), lax, *the upper aggregated, the lowermost* more or less *separated* and sometimes borne on a pedicel 0.5–1.5(2) cm long. Scales of pistillate spikelets *entirely black or sometimes with narrow paler or whitish hyaline margins*. Perigynia mostly elliptical, more rarely broadly ovoid or orbicular-ovoid, (2)2.2–2.8 mm long, blackish brown distally, abruptly transforming into short beak. . . . . .67.

66. Scales of pistillate spikelets *purplish black, broadly rounded distally or more rarely acutish, mostly almost equalling the perygynia in length and width*. Perigynia *elliptical, (2.5)3–3.5 mm long, obtusish distally, without beak* or rarely gradually tapering to very short beak; ripe perigynia sometimes purplish brown distally. Staminate spikelet usually *purplish black, 1–1.5(2) cm long*. Shoots surrounded at base by glossy reddish brown or purplish brown leaf sheaths. . . . . .44. **C. BIGELOWII** TORR. EX SCHWEIN.

– Scales of pistillate spikelets mostly *dark brown (castaneous), acutish distally or more rarely obtuse and broadly rounded, mostly longer and narrower than the perigynia*. Perigynia *orbicular-ovoid or rarely elliptical, 2.2–2.7(3) mm long, abruptly tapering to short beak*, usually not brunnescent distally. Staminate spikelet usually *brown or light brown, 1.5–2 cm long*. Shoots surrounded at base by glossy reddish brown or purplish brown leaf sheaths. . . . . .46. **C. RIGIDIOIDES** (GORODK.) V. KRECZ.

67. Plant *loosely tufted; every shoot in the tuft with a rather elongate horizontal portion (rhizome)*. Leaves *(1.5–2)3–4(5) mm wide, abruptly acuminate*, flat, sometimes with slightly revolute margins. Staminate spikelet 0.8–1 cm (more rarely 1.5 cm) long, usually *sessile*. Pistillate spikelets 0.5–2 cm long. . . .45. **C. ENSIFOLIA** (TURCZ. EX GORODK.) V. KRECZ. SSP. **ARCTISIBIRICA** JURTZ.

– Plant *densely tufted*; usually *only certain shoots in the tuft with a short (1–2 cm, more rarely to 4 cm) horizontal portion (rhizome)*, the remaining shoots with such a portion scarcely expressed. Leaves *1.5–2(3) mm wide, gradually acuminate*, flat with slightly revolute margins. Staminate spikelet usually 1–2 cm long, *sessile or on pedicel up to 5 mm long* (rarely longer). Pistillate spikelets (0.5)1–2.5 cm long. . . . . .47. **C. LUGENS** H.T. HOLM.

68. Perigynia *with nerves*. Roots *with dense brownish yellow pubescence that*
(55) *is always conspicuous*. Culms scabrous. . . . . .69.

– Perigynia *without nerves*. Roots *with dense greyish white pubescence or almost glabrous*. . . . . .70.

69. Bract of lowermost spikelet *broad (2–3 mm wide), equalling or slightly exceeding inflorescence*. Inflorescence 6–17(21) cm long (usually *10–12 cm long*), consisting of 1–2 (more rarely 3) staminate and *3 (more rarely 2) pistillate* spikelets. Staminate spikelets (1)2–3(5) cm long, mostly *light brown*. Pistillate spikelets 1–5 cm long (usually *3–4 cm long*); scales elongate-ovate or ovate, mostly *brown (more rarely black), with pale stripe along midrib or more rarely entirely dark*, acutish or obtusish, sometimes whitish hyaline distally, shorter than perigynia (sometimes less than half as long) or more rarely equalling them. Perigynia *(2)2.7–3.5(4) mm long, mostly elliptical (more rarely ovoid), yellowish green when ripe (not blackish brown on distal half)*, usually *abruptly tapering to short beak*. Achene *filling less than two-thirds* of perigynium (so that this seems empty distally). Leaves (2)2.7–3.5(4) mm wide. Shoots surrounded at base by *brown (more rarely reddish brown)* scalelike leaf sheaths. . . . . .39. **C. APPENDICULATA** (TRAUTV. ET C.A. MEY.) KÜK.

– Bract of lowermost spikelet usually *narrow, setiform, 1–1.5 mm wide, slightly exceeding its spikelet but shorter than the inflorescence.* Inflorescence 3.5–6(11) cm long (usually *4–5 cm long*), consisting of 1 (rarely 2) staminate and *2* (*more rarely 3*) *pistillate* spikelets. Staminate spikelets 1.5–2.5(3) cm long, *blackish brown or brown.* Pistillate spikelets usually *1.5–2 cm long* (more rarely 2.5–3 cm long); scales elongate-ovoid or ovoid, *normally entirely blackish brown* (*more rarely with pale midrib*), acutish or with broadly rounded tip, shorter than or more rarely almost equalling perigynia. Perigynia (*2*)*2.5–3 mm long, mostly ovoid* (*more rarely elongate-ovoid or elliptical*), usually with 4–5 distinct nerves (but sometimes these are almost imperceptible or completely absent), *blackish brown on distal half when ripe,* usually *gradually tapering to beak.* Achene *almost completely filling* perigynium. Leaves mostly 1–2 mm (more rarely 3 mm) wide. Scalelike sheaths surrounding bases of shoots *purple, blackish purple or reddish brown* (*more rarely brown*). . . . .
. . . . .38. **C. WILUICA** MEINSH.

70. Shoots surrounded at base by *long* (*up to 10 cm long*), *brownish yellow* (*more rarely brown*), *usually scalelike leaf sheaths that do not disintegrate reticulately.* Roots *thick, up to 2–3 mm in diameter, with long dense greyish white pubescence.* Culms *smooth or scabrous.* Inflorescence 1.5–2(3) cm long, consisting of 3 (more rarely 4) usually densely aggregated spikelets, sometimes with the lowermost spikelet somewhat separated and borne on a pedicel up to 1.5 cm long. Staminate spikelet *sessile*, 1–1.5(2) cm long, thickish (3–5 mm wide), brown or black. Pistillate spikelets 0.5–1.5(2) cm long, 2–3 in number. Lowest bract normally setiform, *with black auricles at base,* just equalling its spikelet or very rarely almost equalling the inflorescence. Scales of pistillate spikelets entirely black, without pale midrib. Perigynia *planoconvex, broadly ovoid* (or more rarely ovoid or elliptical), (2)2.5–3 mm long, purplish black distally, apruptly tapering to short beak.
. . . . .48. **C. SOCZAVAEANA** GORODK.
– Shoots surrounded at base by *short, 1.5–3*(*4*) *cm long, dark purple or brown scalelike leaf sheaths that mostly disintegrate reticulately.* Roots *thin, with short greyish white pubescence or almost glabrous;* or more rarely *thickish with yellowish brown pubescence* (in the latter case the scalelike leaf sheaths rather long). Culms *always scabrous.* Staminate spikelet *on pedicel 1–3 cm long.* Perigynia *biconvex or more rarely planoconvex.* . . . . .71.

71. Perigynia *planoconvex.* Leaf sheaths rather long, *not disintegrating reticulately.* Roots with *dense yellowish brown pubescence.* . . . .
. . . . .38. **C. WILUICA** MEINSH. *(see couplet 69).*
– Perigynia *biconvex, slightly inflated.* Scalelike leaf sheaths short, usually *disintegrating reticulately.* Roots *thin, with greyish white pubescence* or almost glabrous. . . . . .72.

72. Scalelike sheaths surrounding bases of shoots *dark purple, 2–3*(*4*) *cm long.* Leaves (1.5)2–3.5(4) mm wide, flat, sometimes with slightly revolute

margins. Inflorescence normally *3 cm long, consisting of 1 staminate* and 2 (more rarely 3) pistillate spikelets. Staminate spikelet 1–2(3) cm long, light or dark brown, with scales broadly rounded at tip. Pistillate spikelets 1–2(2.5–3) cm long, sessile or more rarely the lowermost on a pedicel up to 1.5 cm long. Lowest bract *setiform, normally shorter or slightly longer than its spikelet* (*very rarely almost equalling the inflorescence*). Scales of pistillate spikelets *ovate or elongate-ovate, obtusish or acutish, entirely blackish brown or more rarely with pale midrib,* narrower and normally shorter than the perigynia (more rarely longer than them). Perigynia sometimes somewhat inflated, ovoid, broadly ovoid, elliptical or more rarely elongate-ovoid, 2–2.5(3) mm long, sometimes blackish brown distally, abruptly or gradually tapering to short *entire pale* beak. Roots *with greyish white pubescence.* . . . . .40. **C. CAESPITOSA** L.

– Scalelike sheaths surrounding bases of shoots *brown, very short, 1.5(-3) cm long*. Leaves 2–3(4) mm wide, flat, shorter than culm. Inflorescence 3–6(*10*) *cm long, consisting of 1–2 staminate* and 1–2(3) pistillate spikelets. Staminate spikelets 1.5–2 cm long, pistillate 1.5–2.5(3.5) cm long, the lowermost sometimes on a pedicel 0.5(-1) cm long. Lowest bract *leaflike, equalling or almost equalling the inflorescence.* Scales of pistillate spikelets *lanceolate, acute or more rarely obtuse, dark brown,* normally longer than or equalling the perigynia (more rarely shorter than them). Perigynia inflated, normally globose or elliptical (more rarely ovoid), (2)2.3–2.5(2.7) mm long, brunnescent, *sometimes brown spotted, mostly with isolated acicules on distal margin,* abruptly tapering to short *emarginate-bidentate* (*or more rarely entire*) *purplish brown* beak. Roots *almost bare, with very short brown pubescence visible only under a microscope.* . . . .
. . . . .41. **C. SCHMIDTII** MEINSH.

73. Lowest bract with tubular sheath 0.5–5 cm long and more or less developed blade. . . . . .74.
(53)
– Lowest bract without or with scarcely expressed sheath (less than 0.5 cm long); or, if sheath longer, then pistillate spikelets very thick (about 1 cm in diameter) and perigynia vesicularly inflated. . . . . .93.

74. Culm leaves and bracts with sheaths and bases of blades more or less *densely pubescent over their entire surface or with pubescence only near the mouths of the sheaths.* Sheaths of bracts 1.5–7 cm long. Pistillate spikelets (*2*)*3–6 cm long, on pedicels 4–10 cm long;* scales usually *with scabrous awns*. Perigynia *6–8 mm long*, with short sparse pubescence on distal half (pubescence better expressed on distal part of spikelet) or sometimes glabrous, with aciculate-ciliate *deeply bidentate beak* (*teeth 1.6–2.2 mm long*) which is reddish brown in its mouth and on the teeth. Culms smooth; leaves 5–7 mm wide. . . . . . .102. **C. SORDIDA** HEURCK ET MUELL. ARG.

– Leaves and bracts with *glabrous* blades and sheaths. Perigynia *smaller, with entire, slightly notched or shallowly bidentate beak.* Scales *awnless or, if with short awns, then pistillate spikelets sessile.* Pistillate spikelets *not more than 3 cm long.* . . . . .75.

75. Perygynia *pubescent* or, if glabrous, then sheath of lowest bract *without blade* and perygynia *2.5–2.7 mm long*. ....76.
– Perygynia *bare, 3–6 mm long*; sheath of lowest bract with more or less developed blade. ....80.

76. Scales *awned-acuminate*; perygynia 4–5 mm long, *very densely pubescent* with greyish brown hairs (walls of perygynium not visible due to the hairs). ....95. **C. LASIOCARPA** EHRH. *(see couplet 94).*
– Scales *awnless*; perygynia *not densely pubescent* (*hairs not entirely concealing surface of perygynium*). ....77.

77. Achene *with disc* distally. Pistillate spikelets on pedicels, that of lowest spikelet (0.5)1–3 cm long; bract of second (lower) pistillate spikelet with short sheath 0.5–1(1.5) cm long. Plant densely tufted, without stolons. ....78.
– Achene *without disc* distally. ....79.

78. Leaf sheaths surrounding bases of shoots *light greyish brown, readily splitting into fibres*. Pedicel of staminate spikelet *0.5* (*more rarely 1*) *cm long* (a pistillate spikelet situated almost at the very base of the staminate spikelet). Sheath of bract of lower (second) pistillate spikelet *with blade 0.5–1.5 cm long*; bract of upper pistillate spikelet scalelike. Scales of pistillate spikelets *ovate or elongate-ovate, scabrous on keel, sometimes with short scabrous mucro*. Perygynia *3.2–3.5(4) mm long, elongate-obovoid, pubescent, with nerves, with elongate beak* (0.6–1 mm long) that is *shallowly bidentate or entire*. Leaves 1.5–2.5 mm wide, standing straight upright, shorter than culm. ....80. **C. SABYNENSIS** LESS. EX KUNTH.
– Leaf sheaths surrounding bases of shoots *brown, entire, not fibrous*. Pedicel of staminate spikelet *1–2.5 cm long*. Sheaths of bracts usually *without or with scarcely expressed blade*. Scales of pistillate spikelets *broadly ovate, broadly rounded distally, often whitish hyaline, smooth on keel*. Perygynia *2.5–2.7 mm long, obovoid, bare, without nerves, with very short entire beak*. Leaves 1.5–2.5 mm wide, stiff or sometimes curved. ......
....81. **C. TRAUTVETTERIANA** KOM.

79. Plant densely tufted, *without stolons*. Perygynia obovoid-conical, *3–3.5 mm long, convex-triangular*, with elongate hairs, *not scabrous on margins*. Staminate spikelet *single, usually not surpassing the pistillate spikelet* next to it, 0.5–1 cm long. Pistillate spikelets 2–3, oblong, 1–2.5 cm long, *sparse-flowered*, on pedicels, *erect*. Sheaths of bracts *without blade*, hyaline and awned-acuminate distally. Culms smooth or slightly scabrous only below inflorescence. Leaves flat or folded lengthwise, 1.5–2.5 mm wide. ....
....85. **C. PEDIFORMIS** C.A. MEY.
– Plant more or less loosely tufted, *with long and rather slender stolons*. Perygynia elongate-ovoid or ovoid, *5–5.5 mm long, planotriangular*, with very short bristly hairs distally, *scabrous on margins*. Staminate spikelet(s) *1–3, surpassing the pistillate spikelets*. Pistillate spikelets (1)2–3, *dense-flowered*, 1.5–2.5 cm long, mostly ovoid or elongate-ovoid, on pedicels

(*sometimes drooping*). *Sheaths of bracts with blades.* Culms smooth. Leaves flat or folded lengthwise, 2–3 mm wide, with long fine tips, flexuous distally. . . . . .61. **C. MACROGYNA** TURCZ. EX STEUD.

80. (75) Plant more or less loosely tufted, with long (more rarely short) creeping rhizomes. . . . . .81.
– Plant densely tufted, without creeping rhizomes. . . . . .86.

81. Pistillate spikelets mostly *sparse-flowered, consisting of 5–10(14) florets, erect*. Perigynia *obtusely triangular or orbicular-triangular in section, greenish yellow* or sometimes brownish, with usually reddish *smooth beak and smooth margins*. Achenes *sessile*. Leaves abruptly acuminate. . . . . .82.
– Pistillate spikelets *densely many-flowered, often drooping*, sometimes erect. Perigynia *planotriangular or almost flat*, entirely or only on distal half *blackish brown or reddish brown, with distal margins and beak more or less scabrous*. Achenes *on slender pedicels*. Leaves gradually long-acuminate. . . . . .84.

82. Inflorescence *2–5(6) cm long*, consisting of *aggregated* spikelets. Perigynia *glaucous green, obtusely triangular, densely covered with granular papillae, almost beakless, with notch distally*. Leaves *almost equalling culm or longer, glaucous, erect, 2–3 mm wide*. . . . . .82. **C. LIVIDA** (WAHLB.) WILLD.
– Inflorescence *(5)8–16 cm long*, consisting of *separated* spikelets. Perigynia *greenish yellow, orbicular-triangular, without papillae*, with inconspicuous nerves, abruptly *tapering to* reddish or brownish *well expressed beak*. Leaves *considerably shorter than culm, bright green, often downcurved, 3–5(7) mm wide*. Staminate spikelet single, 1–1.6 cm long. Pistillate spikelets 2(3), 1–2 cm long. Pedicel of lowest spikelet 3–4 cm long. Bract shorter than inflorescence. Scales entirely brown, dark purplish brown or reddish brown, or with pale stripe along midrib. .83.

83. Perigynia (3.5)4–4.5(5) mm long; beak *0.6–1.2 mm long*, notched or shallowly bidentate anteriorly; depth of notch on beak *0.2–0.3(0.5) mm*. . . . . .83. **C. VAGINATA** TAUSCH.
– Perigynia 4.5–5.5(6) mm long; beak *1–2 mm long*, notched or shallowly bidentate anteriorly; depth of notch on beak *(0.5)0.7–1 mm*. . . . . .84. **C. FALCATA** TURCZ.

84. Plant *with very short stolons* (or sometimes without stolons, see couplet 87). Leaves 3–4 mm wide, half as long as culm or shorter. Perigynia *without nerves*, almost flat, entirely purplish black or purplish brown except for extreme base and margins, their beaks *whitish hyaline distally*. Staminate spikelet single, 0.5–1 cm long; normally 3 pistillate spikelets 1–2 cm long, all on slender pedicels, drooping. Membranous side of bract sheath often *dark coloured*. . . . . .59. **C. ATROFUSCA** SCHKUHR.
– Plant *with long slender stolons*. Perigynia *with nerves*, their beaks *without whitish hyaline border*. Membranous side of bract sheath *pale green*. . . .85.

85. Staminate spikelet(s) *1–3(5)*, *sessile*, 1.5–2(3) cm long; pistillate spikelets 1.2–1.5 cm (more rarely up to 2.5 cm) long, *5–8 mm wide*. Perigynia *5–5.5 mm long, gradually tapering (without incurvation)* to beak. Culms surrounded at base by *light brown* fibrous sheaths. ....
.... .61. **C. MACROGYNA** TURCZ. EX STEUD. *(see couplet 79)*.
– Staminate spikelet *single*, 2–2.5 cm long, *on pedicel 3–4 cm long*. Pistillate spikelets *2*, 2.5 cm long, *3 mm wide*. Perigynia *about 6 mm long, abruptly tapering* to short beak. Culms surrounded at base by *purple* sheaths. .....
....62. **C. KTAUSIPALII** MEINSH.

86. Lowest bract *shorter* than inflorescence. Pistillate spikelets *all on slender*
(80) *pedicels, often drooping*. Perigynia *without nerves, brown, brownish green or purplish black*. ....87.
– Lowest bract *3–4 times as long as* inflorescence. *Upper pistillate spikelets aggregated and sessile, the lowermost* more or less *separated and on a pedicel* sometimes *up to 2 cm long*. Perigynia *with distinct nerves, greenish yellow*. Culms smooth. ....92.

87. Perigynia *almost flat, about 5 mm long, purplish black* (except for extreme base and margins) *like the scales*. Pistillate spikelets *densely many-flowered, ovoid*. Membranous side of bract sheath *often dark coloured*. ....
.... .59. **C. ATROFUSCA** SCHKUHR. *(see also couplet 84)*.
– Perigynia *triangular or orbicular-triangular* in section, *less than 5 mm long, ferruginous or brownish green*. Scales *brown or pale brown*. Pistillate spikelets *few-flowered, lax, oblong*. Membranous side of bract sheath *colourless*. ....88.

88. Uppermost spikelet in inflorescence *staminate*, or very rarely with single pistillate floret at base. Perigynia *without nerves*. ....89.
– Uppermost spikelet in inflorescence *gynaecandrous or androgynous* (with 1–3 pistillate florets), *if androgynous then perigynia with nerves*. ....91.

89. Staminate spikelet *large, (0.8)1–1.5 cm long and 3–5 mm wide, clavate*, brownish orange or brown, usually *rising above the adjacent pistillate spikelet* (*rarely at the same level as it or even slightly below it*). Pistillate spikelets (1)2 (very rarely 3) in number, 1–1.5 cm long; their scales *oblong*, usually tapering towards tip, completely dark brown or with narrow whitish hyaline margins, *slightly shorter than the perigynia*. Perigynia *distinctly triangular, very tightly embracing the achene*, usually strongly strigose-scabrous on distal margins (including beak), rather abruptly tapering to elongate beak, sometimes pubescent with very short bristly hairs. Plant 6–30 cm high; leaves (1.5)2–3 mm wide, flat. ....
.... .94. **C. LEDEBOURIANA** C.A. MEY. EX TREV.
– Staminate spikelet *small, usually 5–7(10) mm long and no more than 2 mm wide, linear*. Pistillate spikelets (1 or 2) more or less *considerably surpassing the staminate spikelet or at the same level as it*; scales of pistillate spikelets mostly *broadly ovate* (more rarely ovate), with broadly rounded tips, ½ - ⅔ *as long as the perigynia*. Perigynia *indistinctly trian-*

gular, almost orbicular in section, not tightly embracing the achene, gradually tapering to slightly scabrous or smooth beak. .90.

90. Staminate spikelet *pale*, with weak brownish or greenish tint, situated *slightly below the adjacent pistillate spikelet* whose base is usually below the tip of the staminate spikelet (or occasionally both spikelets at same level); pedicel of staminate spikelet *exserted from sheath of bract of pistillate spikelet by 0.7–2(2.5) cm*. Scales of pistillate spikelets *pale brown or pale green, with broad whitish hyaline margins*, sometimes entirely whitish. Perigynia *greenish brown*, (3.2)3.5–4 mm long, 1.3–1.5 mm wide, with normally scabrous (more rarely smooth) beak. Plant (5)10–30 cm high; leaves 1.5–2.5(3) mm wide, flat or more rarely folded. . . . . .91. **C. CAPILLARIS** L.

– Staminate spikelet *ferruginous brown*, situated *considerably below the two upper pistillate spikelets* whose bases are usually above the tip of the staminate spikelet; pedicel of staminate spikelet usually *entirely concealed in sheath of bract of pistillate spikelet*, rarely exserted from it by 3–5 mm. Scales of pistillate spikelets *castaneous, usually with narrow whitish hyaline margins*. Perigynia *light ferruginous brown*, (3)3.2–3.5 mm long, 1–1.2 mm wide, usually with smooth beak. Plant 10–25 cm high; leaves 1.5(–2) mm wide, usually flat. . . . . .92. **C. FUSCIDULA** V. KRECZ. EX EGOR.

91. Uppermost spikelet *gynaecandrous*. Pistillate spikelets *many-flowered, 4–5(6) in number*, the lowermost often branched at base (with 1–2 branchlets). Perigynia lanceolate, (2.5)2.8–3.3(3.5) mm long, *without nerves*. Leaves *flat, 2–3 mm wide*. . . . . .93. **C. KRAUSEI** BOECK. *(see couplet 44)*.

– Uppermost spikelet *androgynous*, with (1)2–3(4) pistillate florets. Pistillate spikelets *2–3*, not branched, *few-flowered (with 5–7 florets)*. Perigynia lanceolate, 3–3.5 mm long, *with 3–4(5) nerves*. Leaves *narrowly sulcate, 0.5–1 mm wide*. . . . . .90. **C. WILLIAMSII** BRITT.

92. Perigynia *obovoid, 3–3.5 mm long, with short straight beak that is*
(86) *shallowly notched distally*. Pistillate spikelets globose or ovoid, *0.5–1 cm long*, mostly sessile. Bract *directed upwards*. Culms smooth; leaves *sulcate, 2–3 mm wide*. Plant 5–20 cm high. . . . . .89. **C. OEDERI** RETZ.

– Perigynia *elongate-ovoid, 5–6.5 mm long, with shallowly bidentate elongate downcurved beak*. Pistillate spikelets ovoid, *0.8–2 cm long*, almost sessile or the lowermost sometimes on a pedicel up to 2 cm long. Bract *obliquely downcurved*. Culms smooth; leaves *flat, 3–5 mm wide*. Plant 20–60 cm high. . . . . .88. **C. FLAVA** L.

93. Perigynia pubescent; pistillate spikelets sessile. . . . . .94.
(73)

– Perigynia glabrous; pistillate spikelets on pedicels or sessile. . . . . .100.

94. Perigynia *4–5 mm long*, very densely pubescent with brownish grey hairs, with *acutely bidentate short beak*. Scales elongate-ovate or lanceolate, *awned-acuminate*, brown or reddish brown. Inflorescence *(7)10–22 cm*

long, consisting of *1–3* linear staminate spikelets 2–4(6) cm long and 1–2 ovoid-cylindrical pistillate spikelets 0.8–2.5(3) cm long. Lowest bract surpassing inflorescence. Culms obtusely triangular, smooth or only scabrous below the inflorescence, surrounded at base by reddish brown scalelike sheaths that disintegrate reticulately. Leaves greyish green, *sulcately folded, almost setiform*, 1–2 mm wide. Plant 50–100 cm high, with thick creeping rhizomes. . . . . .95. **C. LASIOCARPA** EHRH.

– Perigynia *1.5–3.5 mm long, with entire beak*. Scales *awnless*. Inflorescence *not more than 5 cm long*; staminate spikelet *always single*. Leaves *flat*. Plant of smaller size, (5)10–30(40–50) cm high. . . . . .95.

95. Pistillate spikelets *sessile*. Sheaths of upper culm leaves and the scalelike sheaths (often split into fibres) on the rhizome and the lower part of the culm *purple*. Achene *without disc or vallate swelling* distally. . . . . .96.

– Pistillate spikelets *on pedicels, that of the lowest spikelet (0.5) 1–3 cm long*. Sheaths of upper (inside) culm leaves *light green*, of lower (outside) leaves surrounding the bases of the shoots *pale brown or brown*. Achene *with disc* distally. Plant densely tufted. . . . . .99.

96. Plant *loosely tufted, with long slender rhizomes* that are covered with scale-like purple sheaths; such scalelike sheaths also present on lower part of culms. Roots *with brown pubescence*. Lowest bract *leaflike*, slightly shorter than inflorescence. Pistillate spikelets *2–3(4)*. Perigynia brownish green, *with nerves*, densely covered with shining stiffish hairs. . . . .
. . . . .79. **C. GLOBULARIS** L.

– Plant *densely tufted, without creeping rhizomes*; if with creeping shoots, these epigeal (stolons) in dense clusters. Roots *glabrous*. Lowest bract *scalelike*. Usually *1–2* pistillate spikelets, very rarely 3. Perigynia *without nerves*. . . . . .97.

97. Scales of pistillate spikelets *acute or with short scabrous mucro*. Clusters of shoots surrounded at base by *purple* leaf sheaths that are *sometimes shredded into fibres*. Culms *scabrous*. Leaves elongate, *straight, rather soft*, 1–2 mm wide. Perigynia orbicular-obovoid, almost orbicular in section, *light green*, sometimes with reddish beak. Inflorescence consisting of (2–)3 aggregated spikelets. .78. **C. VANHEURCKII** MUELL. ARG.

– Scales of pistillate spikelets *broadly rounded distally*. Sheaths surrounding bases of shoots *brown, not fibrous*. Culms *smooth*. Leaves *very stiff, leathery, often curved* and mostly considerably shorter than culm. Perigynia *purplish black distally*. . . . . .98.

98. Scales of pistillate and staminate spikelets *entirely purplish black or with very narrow whitish hyaline margins, with whitish or yellowish pubescence on basal half*. Inflorescence *1.5–2(2.5) cm long*; pistillate spikelet usually single (rarely 2). Staminate spikelet 0.6–1(1.3) cm long, 1.5–2 mm wide; pistillate spikelet(s) *0.5–0.7 cm long, 2–2.5 mm wide*. Perigynia *2 mm long, with short scattered hairs on distal half*, with short (0.2 mm long) differen-

tiated beak that is *whitish hyaline at its tip.* Leaves 1.5–2.5 mm wide, ¼–⅓ as long as culm. Plant without stolons. ....76. **C. MELANOCARPA** CHAM.
- Scales *ferruginous brown, glabrous, with broad whitish hyaline margins.* Inflorescence *2–4 cm long, usually 2 (more rarely 3) pistillate spikelets.* Staminate spikelet 1–2 cm long, 3–3.5 mm wide; pistillate spikelets *(0.5)1–1.5 cm long, 4 mm wide.* Perigynia *(2)2.5–3 mm long, densely pubescent over their entire surface,* gradually tapering to short beak, *without whitish hyaline border.* Leaves 3–4 mm wide, slightly shorter than culm. Plant with short stolons. ....77. **C. ERICETORUM** PALL.

99. Leaf sheaths surrounding bases of shoots *light greyish brown, readily shredding into fibres.* Pedicel of staminate spikelet *0.5 cm (more rarely 1 cm) long* (pistillate spikelet situated almost at very base of staminate spikelet). Scales of pistillate spikelets *ovate or elongate-ovate, scabrous on keel, sometimes with scabrous mucro.* Perigynia *3.2–3.5(4) mm long,* elongate-obovoid, *pubescent, with nerves, with elongate beak* 0.6–1 mm long (shallowly bidentate or entire). Leaves 1.5–2.5 mm wide, standing straight upright, shorter than culm. ....
.....................80. **C. SABYNENSIS** LESS. EX KUNTH *(see couplet 78).*
- Leaf sheaths surrounding bases of shoots *entire, not fibrous, brown.* Pedicel of staminate spikelet *1–2.5 cm long.* Scales of pistillate spikelets *broadly ovate, broadly rounded distally, often whitish hyaline, smooth on keel.* Perigynia *2.5–2.7 mm long,* obovoid, *bare, without nerves, with very short entire beak.* Leaves 1.5–2.5 mm wide, stiff, sometimes curved. ....
..........................81. **C. TRAUTVETTERIANA** KOM. *(see couplet 78).*

100. Perigynia *vesicularly inflated,* more or less abruptly tapering *to well*
(93) *expressed bidentate or bidentate-emarginate (more rarely entire) smooth* beak. Style normally *curved* at extreme base or higher. Plant often large, with creeping rhizomes. ....101.
- Perigynia *not vesicularly inflated,* usually *with very short entire* (smooth or scabrous) beak. Style *straight.* ....106.

101. Culms *sharply triangular and strongly scabrous* above. Leaves green, 3–5 mm wide, flat. Sheaths surrounding bases of shoots long, scalelike, keeled, *purple, disintegrating reticulately.* Plant with short stolons. Perigynia divergent from spikelet rachis *at an acute angle,* ovoid or elongate-ovoid, *thin-walled, gradually tapering* to more or less deeply bidentate beak. Roots with yellowish pubescence. ....102.
- Culms *obtusely triangular, smooth* or rarely scabrous but in that case perigynia *spreading horizontally or somewhat reclinately* from the spikelet rachis. Perigynia *membranous, rather abruptly tapering* to beak. Plant forming long stolons. Sheaths surrounding bases of shoots mostly *light brown and usually not disintegrating reticulately.* Roots glabrous. ....103.

102. Pistillate spikelets *3.5–7 cm long;* perigynia *(6)7–8(9) mm long, stramineous when ripe.* ....100. **C. VESICARIA** L.

- Pistillate spikelets *2.5–3(3.5–4) cm long*; perigynia *4.5–6(7) mm long, usually reddish brown when ripe.* .... 101. **C. VESICATA** MEINSH.

103. Leaves *strongly nodose-reticulate*. Inflorescence *(13) 17–30 cm long*; pistillate spikelets *(3) 3.5–9 cm long, the lowermost often on pedicels up to 5 cm long*. Lowest bract always *considerably longer than* inflorescence. Perigynia *4–6 mm long, with subulately bidentate* beak. Large plant, 40–100(150) cm high. .... 104.
- Leaves *not nodose-reticulate, at most very weakly nodose*. Inflorescence *4–9(12) cm long*; pistillate spikelets *1–3 cm long, sessile or on pedicels less than 5 mm long*. Lowest bract *shorter than or equalling* inflorescence. Perigynia *2.5–4 mm long, with shallowly bidentate or almost entire* beak. Plant smaller, 15–40 cm high. .... 105.

104. Leaves *(6) 8–15 mm wide, green*, rather thin, flat. Perigynia *5–6.5 mm long, early brunnescent (reddish brown), always spreading horizontally from spikelet rachis when ripe, glossy*; pistillate spikelets usually very dense. Culms smooth (more rarely scabrous). ..... 96. **C. RHYNCHOPHYSA** C.A. MEY.
- Leaves *(2) 3–5(8) mm wide, glaucous green or bluish green, stiff, thickened*, mostly flat, more rarely sulcately folded lengthwise. Perigynia *4–5(6) mm long, usually yellowish green, very rarely brownish distally, usually directed obliquely upwards when ripe (rarely spreading horizontally)*, somewhat loose, *scarcely glossy*; pistillate spikelets more or less dense. Culms always smooth. .... 97. **C. ROSTRATA** STOKES.

105. Leaves *sulcately folded lengthwise, 1–2 mm wide, greyish green*. Perigynia 2.5–3.5(4) mm long, *spreading horizontally* from the spikelet rachis, *greenish yellow or light reddish brown distally, sessile*. Culms always smooth; rhizome with *pale brown* scales. .... 98. **C. ROTUNDATA** WAHLB.
- Leaves *flat, 3–5 mm wide, bright green*. Perigynia 3.5–4.5 mm long, *spreading horizontally or more or less distinctly reclinately* from the spikelet rachis, *reddish brown or purplish black, on pedicels 0.5–0.8 mm long*. Culms smooth or scabrous; rhizome with *purple* scales. .... 99. **C. MEMBRANACEA** HOOK.

106. Pistillate spikelets 5–8 mm long, *with 2–5 florets*; rachis of spikelets *often (100) geniculate*; all spikelets *closely aggregated, sessile or on very short pedicels up to 5 mm long, erect*. Staminate spikelet 0.4–1 cm long, surpassing the pistillate spikelets. Leaves 1–1.5 mm wide. Shoots surrounded at base by purple sheaths. .... 107.
- Pistillate spikelets *many-flowered, with straight rachis, on rather long pedicels, often drooping* (or, if sessile, then staminate spikelet scarcely evident and surpassed by the uppermost pistillate spikelet). .... 108.

107. Plant with *long creeping rhizomes*, from which aerial shoots arise in dense clusters. Culms *scabrous*. Pistillate spikelets 1–2; rachis of spikelet *almost*

        straight. Perigynia *ovoid, 3–3.5(4) mm long, 1.5 mm wide.* Leaves *straight.*
        .... 86. **C. SUPINA** WAHLB. SSP. **SPANIOCARPA** (STEUD.) HULT.
- Plant *very densely tufted, without creeping rhizomes.* Culms *smooth.* Pistillate spikelets 2–3; rachis of spikelet *geniculate.* Perigynia *orbicular-obovoid, 2–2.5 mm long, 1.2–1.3 mm wide.* Leaves *flexuous distally.* ....
       .... 87. **C. GLACIALIS** MACKENZ.

108. Roots *glabrous.* Beak of perigynia usually *dark distally, without calluslike thickening.* .... 109.
- Roots *with dense yellowish orange or yellowish brown pubescence.* Beak of perigynia *yellowish distally, with calluslike thickening.* .... 115.

109. Staminate spikelet *scarcely evident, 3–5 mm long, normally surpassed by the upper pistillate spikelet* (rarely at the same level). Pistillate spikelets narrowly cylindrical, 2(–3) in number, 0.6–1.2 cm long, 2–3 mm wide; scales of pistillate spikelets broadly ovate, entirely purplish brown or with pale midrib, shorter than perigynia. Perigynia usually obovoid (more rarely elliptical), *2–2.3 mm long,* inflatedly orbicular-triangular in section, without nerves, *with very short beak 0.15 mm long or beakless.* Plant with more or less long creeping rhizomes; leaves (1.5)2–2.5 mm wide. ....
       .... 71. **C. HOLOSTOMA** DREJ.
- Staminate spikelet *1–2.5 cm long, always rising above the pistillate spikelets.* Perigynia *3–6 mm long,* with more or less *expressed beak.* ...110.

110. Scales of pistillate spikelets *with awns up to 1 cm long.* Perigynia *without acicules* on margins. .... 111.
- Scales of pistillate spikelets *without awns or rarely with very short scabrous mucro* (in that case perigynia more or less *scabrous-aciculate* on margins).
       .... 112.

111. Plant *densely tufted, without creeping rhizomes.* Scales (excluding awns) *ovate or elliptical, shorter than perigynia.* Perigynia *elliptical, (3.5)4–5 mm long,* brownish yellow, without purple spots, thin-walled. ....
       .... 70. **C. GMELINII** HOOK. ET ARN. *(see couplet 49).*
- Plant *with short creeping rhizomes.* Scales (excluding awns) *lanceolate or elongate-lanceolate, considerably longer than perigynia.* Perigynia *elongate-elliptical, 5–6(7) mm long,* with thin nerves, often with purple spots, almost entirely purplish brown when ripe, membranous. Spikelets 3–4 in number, somewhat separated, on slender pedicels up to 4–5 cm long, the lower drooping. Leaves (3)4–6 mm long, abruptly acuminate. ....
       .... 74. **C. MACROCHAETA** C.A. MEY.

112. Sand plant *with obliquely ascending long creeping rhizomes.* Inflorescence consisting of (2)3–5 closely aggregated spikelets, the lowermost sometimes separated and borne on a pedicel *up to 5 mm long.* Perigynia slightly inflated, *with bidentate (or more rarely notched anteriorly only) beak 0.7–1.2 mm long.* .... 69. **C. SABULOSA** TURCZ. EX KUNTH *(see couplet 47).*

– Plant *with short or abbreviated creeping rhizomes.* Lower spikelets on pedicels *up to (3–)5 cm long.* Perigynia *with entire or slightly notched beak 0.1–0.4 mm long.* ....113.

113. Plant densely tufted, *without creeping rhizomes.* Leaves *2–3 mm wide, gradually long-acuminate.* Inflorescence *3–5 cm long,* consisting of 3(–4) spikelets 0.8–1.5 cm long. Scales ovate, obtusish, *dark brown,* sometimes with whitish hyaline margins, slightly shorter than perigynia. Perigynia 3–3.5 mm long, elliptical or ovoid, orbicular-triangular in section, *leathery, without nerves,* ferruginous brown, *without purple spots,* with short beak from which the style often protrudes. ....75. **C. STYLOSA** C.A. MEY.
– Rhizomes more or less *creeping.* Leaves *(3)4–5(6) mm wide, very abruptly acuminate.* Inflorescence *3.5–10 cm long,* consisting of 1(–2) staminate spikelet(s) 1–2 cm long and 2–3 pistillate spikelets 1.5–2.5(3) cm long. Scales mostly *purplish black,* without whitish hyaline margins. Perigynia *membranous, with 5–6* (not always conspicuous) *nerves, purple spotted distally,* later completely purplish brown. ....114.

114. *Upper* 2–3 spikelets (staminate and 1–2 pistillate) *aggregated;* pedicel of staminate spikelet *0.3–1(1.5) cm long;* uppermost pistillate spikelet sessile or more rarely on a pedicel *3(–5) mm long;* lower pistillate spikelets on *thickened pedicels 0.5–1.5(3) cm long, normally erect* (rarely drooping). Pistillate spikelets *oblong, rather narrow* (*mostly 4 mm wide*); scales *ovate,* mostly equalling or slightly shorter than perigynia. Perigynia *2.5–3.5(4) mm long, without acicules on margins, with beak 0.3–0.4(0.5) mm long.* Rhizome *distinctly creeping;* shoots surrounded at base by mostly *light brown* sheaths. ....73. **C. TOLMIEI** BOOTT.
– *All* spikelets *separated;* pedicel of staminate spikelet *(1)1.5–2.5(3) cm long; pistillate spikelets on slender pedicels, often drooping; pedicel of uppermost pistillate spikelet 0.7–1 cm long, that of lower (1)2–4 cm long.* Pistillate spikelets mostly *ovoid, 5–6(7) mm wide;* scales *lanceolate or elongate-lanceolate,* acute or slightly acuminate (more rarely obtusish), mostly longer than perigynia. Perigynia *(3.5)4–5 mm long, smooth or more or less scabrous-aciculate on margins, with very short beak 0.1–0.15(0.2) mm long.* Rhizome *very short creeping;* shoots surrounded at base by mostly *dark purple or dark brown* sheaths. ....72. **C. PODOCARPA** R. BR.

115. Culms *smooth;* leaves greyish green, 1.5–2(3) mm wide. Pistillate spikelets (108) *oblong, rather narrow,* 0.8–1.5 cm long, *3–4(6) mm wide, with 5–10 florets;* scales usually elliptical or ovate (very rarely lanceolate), mostly entirely dark purplish brown, usually slightly longer and broader than the perigynia. Perigynia *orbicular-triangular in section, elliptical-fusiform or elongate-lanceolate* (more rarely ovoid), 3.5–4.2 mm long, glaucous green, later slightly brownish, *gradually tapering to short beak with brown border.* ....58. **C. RARIFLORA** (WAHLB.) SMITH.

– Culms *usually scabrous*. Pistillate spikelets *ovoid or elliptical, 6–8(10) mm wide, with large numbers of florets*. Perigynia *planotriangular, elliptical or broadly ovoid, abruptly tapering to very short light yellow beak*. . . . . . . .116.

116. Leaves *1–1.5(2) mm wide, greyish green, shorter than culm*. Lowest bract usually *shorter than the inflorescence, more rarely equalling it*. Staminate spikelet *(0.8)1–3 cm long*. Pistillate spikelets 1–2(3), *1–2.5 cm long*; scales broad, *elliptical or ovate*, light brown to dark purplish brown, usually *slightly longer than the perigynia and of the same width, not deciduous*. Perigynia *elliptical, (4)4.2–4.5(5) mm long*, bluish green, later slightly brunnescent, *with 6–7 distinct nerves*. Plant *loosely tufted, with more or less long creeping rhizomes*; culms *separated*. . . . . .56. **C. LIMOSA** L.
– Leaves *(2)2.5–3(4) mm wide, bright green, almost equalling the culm*. Lowest bract usually *equalling or longer than the inflorescence*. Staminate spikelet *1–1.5 cm long*. Pistillate spikelets 2–3, *0.8–1.5(2) cm long*, sometimes with staminate florets basally; scales *lanceolate*, blackish purple, often *with involute margins, curved like a claw, 1⅓–2 times as long as and half as wide as the perigynia, deciduous*. Perigynia *broadly ovoid*, almost orbicular, *3–3.5 mm long*, bluish, later brunnescent, *with 3–5 nerves that are distinct only basally*. Plant *rather densely tufted, with short creeping rhizomes*; culms *congested*. . . . . . . .57. **C. MAGELLANICA** LAM. *(see couplet 45)*.

1. ***Carex obtusata*** Liljebl. in Kongl. Vet. Akad. Nya Handl. XIV (1793), 69, tab. 4; Treviranus in Ledebour, Fl. ross. IV, 267; Kükenthal, Cyper. Caricoid. 87; id., Cyper. Sibir. 36; Krylov, Fl. Zap. Sib. III, 434; Perfilev, Fl. Sev. I, 108; Krechetovich in Fl. SSSR III, 381; id. in Mat. ist. fl. rastit. I, 177, map 28; Hultén, Fl. Al. II, 305, map 231; id., Atlas, map 327; Karavayev, Konsp. fl. Yak. 70; Hultén, Circump. pl. I, 146, map 137; Polunin, Circump. arct. fl. 79, 104.
    Ill.: Kükenthal, Cyper. Sibir., fig. 14; Polunin, l. c. 78.

    Characteristic plant of continental taiga and forest-steppe districts of Siberia, just penetrating the Arctic. Here it occurs in the zone of open larch forests, growing on dry grassy slopes composed of fine soil or stones and fine soil, most frequently with southern exposure, in association with steppe (*Festuca kolymensis*, etc.) or cryophilic steppe plants.

    Soviet Arctic. Eastern Bolshezemelskaya Tundra (Vorkuta); lower reaches of Yenisey (Khantayka); lower reaches of Lena (Chekurovka); district of Bay of Chaun; Wrangel Island.

    Foreign Arctic. Arctic part of Alaska; extreme western part of arctic coast of Canada.

    Outside the Arctic. Sweden (extreme south); eastern part of Central Europe; here and there in the European part of the USSR (Luzhki village in Moscow Oblast, Plyushchan on the Don, and the Pinega Basin); Middle and Southern Urals; Caucasus; south half of West Siberia; Altay; Kazakhstan (Kentau); Saur Range; the basin of the upper course of the Yenisey, the vicinity of Turukhansk and the Balyuna Mountains; Prebaikalia; Central Yakutia; Verkhoyansk-Kolymsk mountain country; Amursko-Zeyskoye Plateau; Primorskiy Kray; Northern Mongolia; Western North America from Alaska to the southern extremity of the Rocky Mountains.

2. ***Carex rupestris*** Bell. ex All., Fl. Pedem. II (1785), 264, tab. 92, fig. 1; Treviranus in Ledebour, Fl. ross. IV, 267; Scheutz, Pl. jeniss. 174; Ostenfeld, Fl. arct. 1, 86; Kükenthal, Cyper. Caricoid. 86; id., Cyper. Sibir. 35; Krylov, Fl. Zap. Sib. III, 433;

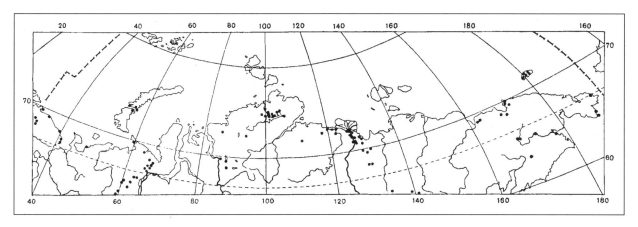

MAP III–12 Distribution of *Carex rupestris* Bell. ex All.

Perfilev, Fl. Sev. I, 108; Krechetovich in Fl. SSSR III, 381; Tolmachev, Fl. Taym. I, 104; id., Obz. fl. N.Z. 150; Leskov, Fl. Malozem. tundry 40; Hultén, Fl. Al. II, 304; id., Atlas, map 326; Kuzeneva in Fl. Murm. II, 127, map 42; A.E. Porsild, Ill. Fl. Arct. Arch. 49, map 75; Böcher & al., Grønl. Fl. 255; Karavayev, Konsp. fl. Yak. 70; Polunin, Circump. arct. fl. 79, 104; Hultén, Circump. arct. fl. 79, 104; Hultén, Circump. pl. I, 26, map 20.

C. petraea Wahlb. in Kongl. Vet. Akad. Nya Handl. XXIV (1803), 139.

Ill.: Ostenfeld, l. c., fig. 63; Polunin, l. c. 78; Fl. Murm. II, pl. XLI, fig. 1.

Arctic-alpine circumpolar species with fragmented distribution. In the Arctic most common in continental mountainous districts; in districts with milder oceanic climate it prefers soils rich in calcium. One of the most xerophilic plants in the Arctic, most frequently colonizing steep (stable or slowly creeping) south-facing slopes composed of stones or stones and fine soil (more rarely rocky), also on dry cliffs, on stony summits and hillocks, and on sandy slopes; found at sites exposed to the winds in winter and therefore retaining little snow; preferring dry, light, well aerated substrates; often growing in turfy tundra soils of its own creation.

Carex rupestris is a turf-former through its formation of short branching rhizomes with vigorous tillering where they break through to the surface; it participates in vegetating steep stony slopes and in some cases dominates in the herb cover of dry stony tundras. Still more frequently it forms a more or less significant admixture in continuous or patchy dry *Dryas* tundras and in so-called turfy tundras (with complete or interrupted network of frost cracks in the tundra turf), including *Dryas*, *Dryas*-sedge, *Dryas-Kobresia*-sedge and even *Kobresia* tundras (in the lower reaches of the Lena and Olenek with *Dryas punctata*, *Carex macrogyna*, *C. misandra*, *C. Trautvetteriana* and other *Carex* spp., *Kobresia Bellardii*, *K. filifolia* and *K. simpliciuscula*). In the more southern subzones of the tundra zone and in high subarctic mountains, this species also occurs not uncommonly in stony *Alectoria* and *Dryas-Alectoria* tundras, and in some variants of *Cetraria* and *Dryas-Cetraria* tundras (with *Cetraria nivalis*, *C. Tilesii*, etc.), often with a significant admixture of dwarf shrubs (*Cassiope tetragona*, *Diapensia* spp., etc. in addition to *Dryas* spp.). In mountainous districts it also enters the subalpine zone of open forests, growing here in exposed portions of dry glades with little snow cover, where the lower vegetational layers have similar composition to the tundras described above.

Cryophilic steppe plants occur frequently in tundras containing *C. rupestris*, and meadow-steppe and rocky steppe plants here penetrate the Arctic.

Soviet Arctic. Murman; Malozemelskaya Tundra (Cape Chaytsyn); Polar Ural; Vaygach Island (Perfilev, l. c.); Novaya Zemlya (northern and southern coasts of

Matochkin Shar); lower reaches of Yenisey (Norilsk Mountains and the upper reaches of the Dudinka); Taymyr; right bank of the Popigay; lower reaches of the Olenek and the district of Olenek Bay; lower reaches and delta of the Lena (frequent); Buorkhaya Bay; middle course of the Kharaulakh; Northern Anyuyskiy Range; district of Chaun Bay; Wrangel Island; Chukotka Peninsula (River Chegitun and vicinity of Chaplino Hot Springs); Anadyr Basin. (Map III–12).

Foreign Arctic. Arctic part of Alaska; north coast of Canada; Labrador; Canadian Arctic Archipelago; Greenland (between 65° and 80°N); Iceland; Spitsbergen; Arctic Scandinavia.

Outside the Arctic. Northern Great Britain; Fennoscandia; mountains of Western Europe; Urals; Caucasus; Altay; Tannu-Ola Range; Sayans; mountains of Transbaikalia; Amursko-Zeyskoye Plateau; Verkhoyansk Range; Southern Primorskiy Kray; coast of Sea of Okhotsk (Ayan district); Sakhalin (mountains along east coast); mountains of Mongolia; Korean Peninsula; Alaska; northern part of Canada; Rocky Mountains.

3. *Carex micropoda* C.A. Mey. in Mém. Ac. Sci. St.-Pétersb. Sav. Étr. I (1831), 210; Treviranus in Ledebour, Fl. ross. IV, 267; Derviz-Sokolova in Bot. mat. Gerb. Bot. inst. AN SSSR XXI, 69.

C. pyrenaica ssp. *micropoda* (C.A. Mey.) Hult., Fl. Al. II (1942), 308, map 235; id., Circump. pl. I, 154, map 146 (*C. pyrenaica* s. l.).

C. pyrenaica auct. non Wahlb. — Kükenthal, Cyper. Caricoid. 104, pro parte; Komarov, Fl. Kamch. I, 224; Hultén, Fl. Kamtch. I, 174.

Ill.: C.A. Meyer, l. c., tab. VI.

Well distinguished from the Western European species *C. pyrenaica* Wahlb. by the presence of 2 (not 3) stigmas, as well as by the smaller ovoid or elongate-ovoid (not ovoid-lanceolate) perigynia.

Growing in moist meadows on the shores of rivers and streams.

Soviet Arctic. Chukotkan coast of Bering Strait (Uelen and Lorino settlements, Chaplino and Senyavin Hot Springs); reported by V.I. Krechetovich (Fl. SSSR III) for the "Anadyr district" in the sense of that Flora, but this record has not so far been confirmed by herbarium material.

Foreign Arctic. Not occurring.

Outside the Arctic. Coast of Sea of Okhotsk (Ola district); Kamchatka; Kurile Islands; Sakhalin; islands of Hokkaido and Honshu; Alaska (south of the Arctic Circle) and adjacent islands; Rocky Mountains.

4. *Carex pauciflora* Lightf., Fl. scot. II (1777), 543, tab. 6, fig. 2; Treviranus in Ledebour, Fl. ross. IV, 268; Kükenthal, Cyper. Caricoid. 110; id., Cyper. Sibir. 40; Komarov, Fl. Kamch. I, 224; Hultén, Fl. Kamtch. I, 175, map 176; Krylov, Fl. Zap. Sib. III, 436; Perfilev, Fl. Sev. I, 107; Krechetovich in Fl. SSSR III, 302; Hultén, Fl. Al. II, 310, map 238; id., Atlas, map 321; id., Circump. pl. I, 85, map 77; Kuzeneva in Fl. Murm. II, 114, map 37; Karavayev, Konsp. fl. Yak. 68.

Ill.: Fl. Murm. II, pl. XXXV, fig. 1.

Circumpolar boreal species just penetrating arctic limits. Growing in sedge-sphagnum bogs.

Soviet Arctic. Murman (Pechenga district, lower reaches of the Kola, southern part of Kola Bay, lower reaches of the rivers Voroney and Iokanga).

Foreign Arctic. Arctic Scandinavia.

Outside the Arctic. Northern and Central Europe; northern half of the European part of the USSR; Northern Ural; southern part of West Siberia; SE Kazakhstan; basin of the upper course of the Yenisey; Prebaikalia; Aldan Basin (Karavayev, l. c.); Amur Basin (River Kur) and the northern Sikhote-Alin; Sakhalin; Kamchatka; Kurile Islands; the north of the Korean Peninsula; Japan; Alaska; Canada; Washington State and northeastern states of the USA.

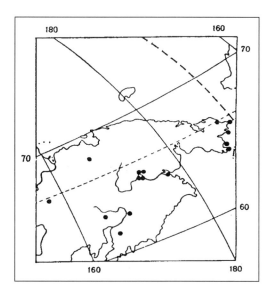

MAP III–13. Distribution of *Carex scirpoidea* Michx.

5. **Carex microglochin** Wahlb. in Kongl. Vet. Akad. Nya Handl. 24 (1803), 140; Treviranus in Ledebour, Fl. ross. IV, 269; Ostenfeld, Fl. arct. 1, 92; Kükenthal, Cyper. Caricoid. 108, pro parte; id., Cyper. Sibir. 39; Krylov, Fl. Zap. Sib. III, 435; Perfilev, Fl. Sev. I, 107; Krechetovich in Fl. SSSR III, 302; Hultén, Fl. Al. II, 310, map 237; id., Atlas, map 322; id., Amphi-atl. pl. 232, map 214; Kuzeneva in Fl. Murm. II, 114, map 37; Roivainen in Ann. Bot. Soc. Vanamo 28, 2, 198; A.E. Porsild, Ill. fl. Arct. Arch. 49, map 76; Böcher & al., Grønl. Fl. 265; Karavayev, Konsp. fl. Yak. 68; Polunin, Circump. arct. fl. 76, 103.
Uncinia microglochin Spreng., Syst. veg. III (1826), 830.
Ill.: Fl. Murm. II, pl. XXXV, fig. 2.

Growing in mossy bogs, in marshy meadows and on river shores, often on carbonate soil.

Soviet Arctic. Murman (Rybachiy Peninsula, between Karlovka and Varzino, Cape Orlov, and the lower reaches of the Ponoy).

Foreign Arctic. Arctic Alaska (one locality); arctic coast of Canada (district of Bathurst Inlet); SE Baffin Island; Labrador; SW and East Greenland; Iceland; Arctic Scandinavia.

Outside the Arctic. Scottish Highlands, Central Europe, Norway and Northern Sweden; the Baltic (near Vilnius); Caucasus (Precaucasus and Dagestan); West Siberia (Tara district); Altay; Central Asia (Pamiroalay, Dzhungarskiy Alatau, Tien Shan); northern edge of Central Siberian Plateau (River Medvezhya); basin of the upper course of the Yenisey (Minusinsk district); Prebaikalia; Vitimskoye Plateau; Cherskiy Mountains (River Boldymba); Northern Mongolia; NW and SW China, Tibet; Himalayas; Alaska; Rocky Mountains; Hudson Bay; Labrador; Newfoundland; Tierra del Fuego.

Until recently it was thought that only *C. oligantha* Boott, a species closely related to *C. microglochin*, grows in South America. According to Roivainen's (l. c.) data, plants indistinguishable from *C. microglochin* also occur in Tierra del Fuego along with the former species.

6. **Carex scirpoidea** Michx., Fl. bor.-amer. II (1803), 171; Ostenfeld, Fl. arct. 1, 82; Kükenthal, Cyper. Caricoid. 81 (incl. var.); id., Cyper. Sibir. 34; Hultén, Fl. Kamtch. I, 173; Krechetovich in Fl. SSSR III, 308; Hultén, Fl. Al. II, 303, map 228; id., Atlas,

map 319; A.E. Porsild, Ill. Fl. Arct. Arch. 49, map 74; Böcher & al., Grønl. Fl. 255; Hultén, Amphi-atl. pl. 188, map 170; Polunin, Circump. arct. fl. 79, 104; Derviz-Sokolova in Bot. mat. Gerb. Bot. inst. AN SSSR XXI, 68.

Ill.: Kükenthal, Cyper. Sibir., fig. 11; A.E. Porsild, l. c., fig. 17; Polunin, l. c. 78.

Growing on rocky and stony slopes in tundra, also on well drained sandy or clayey river shores. Avoiding marshy soils.

Soviet Arctic. Chukotka Peninsula (from the Chegitun River to the SE extremity, Arakamchechen Island); Anadyr and Penzhina Basins. (Map III–13).

Foreign Arctic. Arctic part of Alaska; arctic coast of Canada; arctic part of Labrador; Canadian Arctic Archipelago; NW Greenland (single locality north of Melville Bay); SW and East Greenland.

Outside the Arctic. Yakutia (River Berezovka, a right-bank tributary of the Kolyma), ? Southern Anyuyskiy Range; Magadan Oblast; Kamchatka (River Anauna according to Hultén, Fl. Kamtch. I); northern half of North America; Northern Norway (single locality below 66°N).

7. ***Carex anthoxanthea*** Presl, Reliq. Haenk. I (1830), 203; Kükenthal, Cyper. Caricoid. 97; Mackenzie in N Amer. Fl. XVIII, 1, 25; Krechetovich in Fl. SSSR III, 295; Hultén, Fl. Al. II, 306, map 233.

C. leiocarpa C.A. Mey. in Mém. Acad. Sci. St.-Pétersb. Sav. Étr. I (1831), 208; Treviranus in Ledebour, Fl. ross. IV, 265.

C. anthoxanthea var. *leiocarpa* (C.A. Mey.) Kük., Cyper. Caricoid. (1909), 97.

Ill.: C.A. Meyer, l. c., tab. V.

Soviet Arctic. Chukotka Peninsula (Senyavin Strait).

Foreign Arctic. Absent.

Outside the Arctic. Occurring on the Pacific coast of North America from Alaska and adjacent islands to Vancouver Island.

8. ***Carex Hepburnii*** Boott in Hooker, Fl. bor.-amer. (1840), 209, tab. 207; Gorodkov in Tr. Bot. muz. XX (1927), 201; Krechetovich in Fl. SSSR III, 187.

C. nardina var. *Hepburnii* (Boott) Kük., Cyper. Caricoid. (1909), 70.

C. nardina auct. non Fries — Treviranus in Ledebour, Fl. ross. IV (1853), 265; Ostenfeld, Fl. arct. 1, 48, pro parte; Kükenthal, l. c. (respecting plants from Spitsbergen, Greenland and America); id., Cyper. Sibir. 30; Hultén, Fl. Al. II, 299, map 224; id., Circump. pl. I, 186, map 168, pro parte; Böcher & al., Grønl. Fl. 250; Polunin, Circump. arct. fl. 79, 103, pro parte.

C. nardina var. *atriceps* Kük. in Feddes Repert. sp. nov. VIII (1910), 7; A.E. Porsild, Ill. Fl. Arct. Arch. 47, map 71.

Kobresia Hepburnii (Boott) Ivanova in Bot. zhurn. XXIV, 5/6 (1939), 490.

Ill.: A.E. Porsild, l. c., fig. 16, 1 and 17, a.

Species rather characteristic of the Arctic, distributed mainly within North America where it reaches districts at high latitude; just penetrating the East Asian sector of the Arctic.

Growing in tundra on dry rocky calcareous slopes.

Some authors confuse this species with *C. nardina* Fries, but it is well distinguished from the latter by the elliptical or obovoid perigynia, 1.5–2 mm wide, which are rounded distally (abruptly tapering to a short beak) and have a scarcely differentiated short pedicel; in *C. nardina* the perigynia are mostly ovoid, 1.4–1.6 mm wide, gradually tapering to a more elongate beak, with a sharply defined elongate pedicel. *Carex nardina* was described from Arctic Norway and is restricted in its distribution to Scandinavia and probably also Iceland.

Soviet Arctic. Wrangel Island; Chukotka (River Kuvet, River Chegitun and Arakamchechen Island).

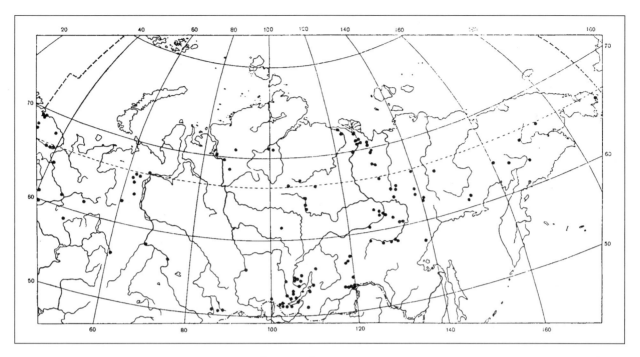

MAP III–14 Distribution of *Carex capitata* L.

Foreign Arctic. Arctic part of Alaska; extreme east of arctic coast of Canada; Labrador; Canadian Arctic Archipelago; all Greenland (to northern extremity); Spitsbergen.
Outside the Arctic. Alaska; western mountains and subarctic districts of North America; Hudson Bay; Labrador.

9. ***Carex capitata*** L., Syst. nat., ed. 10 (1759), 1261; Treviranus in Ledebour, Fl. ross. IV, 266; Ostenfeld, Fl. arct. 1, 49; Kükenthal, Cyper. Caricoid. 70; id., Cyper. Sibir. 30; Gorodkov in Tr. Bot. muz. XXX, 200; Krylov, Fl. Zap. Sib. III, 430; Perfilev, Fl. Sev. I, 108; Krechetovich in Fl. SSSR III, 186; Hultén, Fl. Al. II, 300, map 226; id., Atlas, map 324; id., Amphi-atl. pl. 38, map 20; Kuzeneva in Fl. Murm. II, 74, map 20; Karavayev, Konsp. fl. Yak. 65; Polunin, Circump. arct. fl. 79, 102 (pro parte); Böcher & al., Grønl. Fl. 250; Derviz-Sokolova in Bot. mat. Gerb. Bot. inst. AN SSSR XXI, 68.
Ill.: Fl. SSSR III, pl. XIV, fig. 2; Fl. Murm. II, pl. XIX, fig. 1.

Circumpolar boreal species occurring with large gaps in temperate parts of the Arctic.

Growing in moist, usually sandy sites in moss-small sedge-dwarf shrub tundra, in open larch forest with mosses and small sedges, on polygons, on shores of rivers and streams, and on vegetated sands.

Soviet Arctic. Murman (Pechenga, Rybachiy Peninsula, the lower reaches of the Kola, and the district of Kola Bay); Polar Ural (River Sob); lower reaches of Yenisey; Pyasina Basin (River Dudypta near its mouth); lower reaches of Olenek; lower reaches of Lena (common); middle course of the Kharaulakh River; Chaun Plain (River Milguveyem); Chukotka Peninsula (River Pouten); Anadyr Basin (rare; Nelvti Range, mouth of River Yablonovaya, River Osinovaya) and Penzhina Basin (Oklan and Slovutnaya Rivers); Bay of Korf. (Map III–14).
Foreign Arctic. Arctic part of Alaska; SW Greenland (single locality north of the Arctic Circle); Iceland; Arctic Scandinavia.
Outside the Arctic. Extreme north of Scotland; mountains of Central Europe; Fennoscandia (in the south at Lake Pielisjarvi); basins of rivers flowing into the

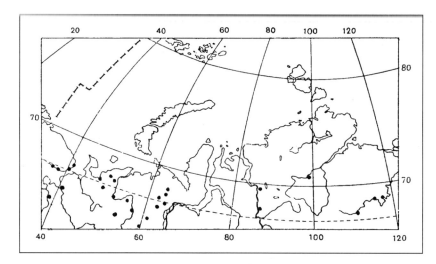

Map III–15 Distribution of *Carex diandra* Schrank.

White Sea; Prepolar and Middle Urals; Precaucasus; West Siberia (the basin of the Severnaya Sosva, the vicinity of Tobolsk, and the basin of the River Tara); Altay (Lake Cheybek-Kol); basin of the upper course of the Yenisey; Central Siberian Plateau north of 62°N; Prebaikalia; Transbaikalia; Aldan Basin; southern part of Verkhoyansk-Kolymsk mountain country; Amur Basin (watersheds of the rivers Nora and Mamyn); Northern Mongolia; Alaska; Canada (mainly subarctic districts); mountains of Western North America and northern Atlantic states of the USA. The North American distributions of *C. capitata* and the closely related *C. arctogena* are inadequately clarified, because these species have usually not been distinguished by investigators of the American flora.

In arctic specimens of *C. capitata* the perigynia are often reddish brown on their distal half.

10. **Carex arctogena** H. Smith in Acta Phytogeogr. Suec. XIII (1940), 193; Hultén, Atlas, map 325; Kuzeneva in Fl. Murm. II, 76, map 20; A.E. Porsild, Ill. Fl. Arct. Arch. 48, map 72; Böcher & al., Grønl. Fl. 250.
    C. capitata f. *arctogena* (H. Smith) Raymond in Contrib. Inst. bot. Univ. Monréal 64 (1949), 38.
    C. capitata var. *arctogena* (H. Smith) Hult., Amphi-atl. pl. (1958) 38, map 19.
    C. capitata auct. non L. — Krechetovich in Fl. SSSR III, 186, pro parte; Polunin, Circump. arct. fl. 79, 102 (pro parte).
    Ill.: Fl. Murm. II, pl. XIX, fig. 2.

    Carex arctogena is well distinguished from *C. capitata* by a complex of characters: smaller perigynia that are sparsely scabrous on their distal margins and taper abruptly to a beak, darker scales, smaller spikelets, and broader leaf sheaths at the bases of the culms. The achenes in *C. arctogena*, unlike in *C. capitata*, almost completely fill the perigynia. Plants possessing the characters of *C. arctogena* fully deserve to be treated as a separate species, not as a form or variety of *C. capitata* according to the opinion of certain authors (see the synonymy). *Carex antarctogena* Roivainen growing on Tierra del Fuego is very closely related to *C. arctogena*.
    Growing on rocky slopes and cliffs.

Soviet Arctic. Murman (Pechenga, Rybachiy Peninsula, Ara-Guba, coast of Kola Bay, Dalniye Zelentsy settlement, and between the rivers Voroney and Kharlovka).

Foreign Arctic. Baffin Island (single locality in extreme southeast); west and east coasts of Greenland south of 72°N (numerous localities); Arctic Scandinavia.

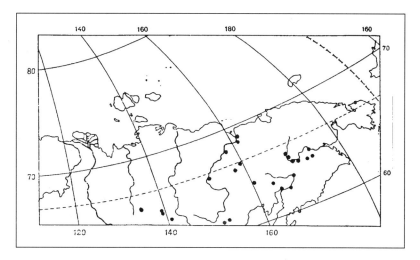

MAP III–16 Distribution of *Carex pallida* C.A. Mey.

Outside the Arctic. Mountains of Sweden and Norway north of 65°N and a few localities in Southern Norway; south coast of Kola Peninsula (Kuzomen). In North America occurring in Northern Manitoba, the south and east of Hudson Bay, and Labrador.

11. ***Carex diandra*** Schrank in Cent. Bot. Anmerk. (1781) 57; id. in Acta Mogunt. 49; Kükenthal, Cyper. Caricoid. 175, pro parte; id., Cyper. Sibir. 5, 57; Komarov, Fl. Kamch. I, 226; Hultén, Fl. Kamtch. I, 177; id., Fl. Al. II, 316, map 244; id., Atlas, map 329; id., Circump. pl. I, 128, map 120; Krylov, Fl. Zap. Sib. III, 446; Krechetovich in Fl. SSSR III, 157; Kuzeneva in Fl. Murm. II, 58, map 11; Karavayev, Konsp. fl. Yak. 65.
    C. teretiuscula Good. in Trans. Linn. Soc. II (1794), 163; Treviranus in Ledebour, Fl. ross. IV, 276; Perfilev, Fl. Sev. I, 110; Leskov, Fl. Malozem. tundry 41.
    Ill.: Fl. SSSR III, pl. X, fig. 14; Fl. Murm. II, pl. XII, fig. 1.

    Characteristic plant of lowland marshes and marshy shores of rivers and lakes, rather widespread mainly in forested regions of Eurasia and North America. Only isolated localities known within the Arctic.
    Soviet Arctic. Murman (Ponoy); Timanskaya Tundra (Pesha village); Malozemelskaya Tundra (River Soyma and between the rivers Indiga and Velt); SE Bolshezemelskaya Tundra (Vorkuta); lower reaches of Yenisey (Talnakh). (Map III–15).
    Foreign Arctic. Alaska (one locality); lower reaches of Mackenzie River; Iceland; Arctic Norway.
    Outside the Arctic. Fennoscandia; almost all Northern and Central Europe; European part of USSR (except southern districts); Caucasus; Urals (including the Prepolar Ural in the basin of the River Voykar); West Siberia (south of 64°N); Altay; East Siberia south of the Arctic Circle (north of it only in the Kheta Basin at the mouth of the River Medvezhya and on the Olenek), east to the Aldan; Amuro-Zeyskoye Plateau; lower reaches of Amur (Lake Orel); Kamchatka; Sakhalin; islands of Hokkaido and Honshu; NE China; Northern Mongolia; India (Himalayas); Turkey; Alaska; Canada; Northern United States (mainly in the east with isolated localities in southern mountains); Canary Islands and New Zealand.

12. ***Carex pallida*** C.A. Mey. in Mém. Ac. Sci. St.-Pétersb. Sav. Étr. I (1831), 215; Treviranus in Ledebour, Fl. ross. IV, 272; Kükenthal, Cyper. Caricoid. 134; id., Cyper. Sibir. 50, pro parte; Krechetovich in Bot. zhurn. XXII, 1, 109.

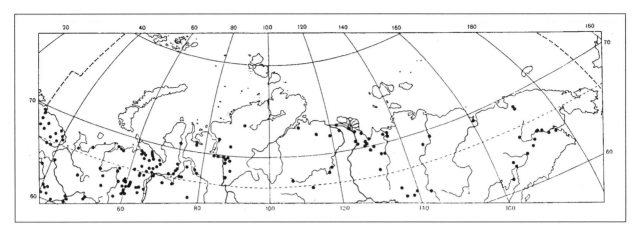

MAP III-17 Distribution of *Carex chordorrhiza* Ehrh.

C. accrescens Ohwi in Mem. Coll. Sci. Kyoto Imp. Univ., ser. B, 6 (1931), 255; Krechetovich in Fl. SSSR III, 136; Karavayev, Konsp. fl. Yak. 64.

Ill.: Kükenthal, Cyper. Sibir., fig. 30; Fl. SSSR III, pl. X, fig. 11.

Meadow-forest species, rather widespread in East Siberia and the Far East. Within the Arctic occurring mainly in river valleys. Growing in shrubbery, open larch forests and mixed herb meadows.

Soviet Arctic. Lower course of the Kolyma (Panteleikha, Nizhniye Kresty); Anadyr Basin (not uncommon); Penzhina Basin; Bay of Korf (Goven Peninsula). (Map III–16).

Foreign Arctic. Not occurring.

Outside the Arctic. Prebaikalia; Transbaikalia; Central and Southern Yakutia; southern part of Verkhoyansk Range; upper course of the Indigirka and the Kolyma; Amur Basin; Primorskiy Kray; coasts of Sea of Okhotsk; Kamchatka; Kurile Islands; Sakhalin; the north of the Korean Peninsula; NE China; Northern Mongolia.

13. *Carex chordorrhiza* Ehrh. in L. fil., Suppl. (1781) 414; id., Beitr. I, 186; Treviranus in Ledebour, Fl. ross. IV, 271; Ostenfeld, Fl. arct. 1, 51; Scheutz, Pl. jeniss. 174; Kükenthal, Cyper. Caricoid. 127; id., Cyper. Sibir. 48; Krylov, Fl. Zap. Sib. III, 440; Perfilev, Fl. Sev. I, 110; Krechetovich in Fl. SSSR III, 194; id. in Mat. ist. fl. rastit. I, 155, map 5; Leskov, Fl. Malozem. tundry 36; Hultén, Fl. Al. II, 315, map 241; id., Atlas, map 336; id., Circump. pl. I, 94, map 85; Kuzeneva in Fl. Murm. II, 78, map 21; A.E. Porsild, Ill. Fl. Arct. Arch. 50, map 79; Karavayev, Konsp. fl. Yak. 66; Polunin, Circump. arct. fl. 80, 102.

Ill.: Fl. SSSR III, pl. XIII, fig. 9; Fl. Murm. II, pl. XXI; A.E. Porsild, l. c., fig. 17, b, c.

Circumpolar species characteristic of sphagnum bogs in the taiga zone. More or less common in temperate parts of the Arctic. Growing in lowland sphagnum bogs or *Hypnum* fens, in flarks and poor peatlands of the forest-tundra and southern tundras, and on vegetated lake shores, at the most marshy sites which thaw considerably in summer. One of the basic plants of wet sinkholes in polygonal bogs.

Soviet Arctic. Murman (almost everywhere but infrequent); southern part of Kanin (Lake Yazhemskoye); Malozemelskaya Tundra (River Sula); Bolshezemelskaya Tundra (not uncommon); basin of the River Kara; Polar Ural (common); lower reaches of the Ob; southern part of Yamal; Obsko-Tazovskiy Peninsula; lower reaches of the River Taz, Gydanskaya Tundra; lower reaches of the Yenisey and islands of Yenisey Bay; Southern Taymyr (River Dudypta); Popigay Basin; lower reaches of the Anabar and Olenek; lower reaches and delta of the Lena; Buorkhaya Bay; lower reaches of the Yana and Kolyma; district of Chaun Bay; Anadyr and Penzhina Basins; west coast of Penzhina Bay. (Map III–17).

Foreign Arctic. Arctic part of Alaska; arctic coast of Canada; the south of Victoria Island and SE Baffin Island; Labrador; Iceland; Arctic Scandinavia.

Outside the Arctic. Scotland; all Fennoscandia; Northern and Central Europe; forest zone of the European part of the USSR (south to the latitudes of Voronezh and Poltava); isolated localities in the Caucasus; all the Urals; the Mugodzhary Mountains; all West Siberia; Altay; basin of the upper course of the Yenisey; isolated localities in Prebaikalia; subarctic districts of the Central Siberian Plateau; the south of the Verkhoyansk Range; Aldan and Indigirka Basins; Northern Preamuria (isolated discoveries in the basins of the rivers Zeya, Tyrma and Amgun); coast of Sea of Okhotsk (valleys of the rivers Yama and Katarba); Southern Kamchatka; the north of the Korean Peninsula; Alaska; Northern and Eastern Canada and the NE United States.

14. ***Carex duriuscula*** C.A. Mey. in Mém. Ac. Sci. St.-Pétersb. Sav. Étr. I (1831), 214, tab. VIII; Krechetovich in Fl. SSSR III, 140; Karavayev, Konsp. fl. Yak. 64; Yegorova in Bot. mat. Gerb. Bot. inst. AN SSSR XIX, 56, map 1; Hultén, Circump. pl. I, 146.

C. stenophylla γ Trev. in Ledebour, Fl. ross. IV (1852), 270.

C. eleocharis L.H. Bailey in Mem. Torr. Bot. Club I (1889), 6.

Ill.: Fl. SSSR III, pl. X, fig. 3.

Siberian and North American steppe species, for which isolated localities are known within the Asian Arctic.

Occurring at steppelike sandy sites.

Soviet Arctic. Chukotka (district of Ust-Chaun, northern foothills and slopes of the Northern Anyuyskiy Range).

Foreign Arctic. Not occurring.

Outside the Arctic. Northern and Eastern Kazakhstan; the south of West Siberia; Altay; the south of East Siberia; Central Yakutia (occurring in the north at Verkhoyansk and at the settlements of Kustur and Tomtor); Amur Basin; Primorskiy Kray; the north of the Korean Peninsula; Northern and NE China; Mongolia; Alaska (one locality); Canada (Yukon and southern districts); the prairies of the USA.

Records of *C. duriuscula* growing in the lower reaches of the Yenisey (Tolstyy Nos) (Scheutz, Pl. jeniss.) in all probability refer to *C. maritima* Gunn.

15. ***Carex maritima*** Gunn., Fl. Norv. II (1772), 131; Krechetovich in Fl. SSSR II, 189; id. in Sov. bot. 4 (1935), 129; Tolmachev, Obz. fl. N.Z. 150; Hultén, Fl. Al. II, 313, map 239; id., Circump. pl. I, 48, map 41; Kuzeneva in Fl. Murm. II, 76, map 20; A.E. Porsild, Ill. Fl. Arct. Arch. 50, map 78; Böcher & al., Grønl. Fl. 252; Polunin, Circump. arct. fl. 80, 103; Karavayev, Konsp. fl. Yak. 65; Derviz-Sokolova in Bot. mat. Gerb. Bot. inst. AN SSSR XXI, 69.

C. incurva Lightf., Fl. Scot. II (1777), 544; Treviranus in Ledebour, Fl. ross. IV, 269 (excl. var. β); Ostenfeld, Fl. arct. 1, 49; Kükenthal, Cyper. Caricoid. 113, pro parte; id., Cyper. Sibir. 42, pro parte; Lynge, Vasc. pl. N.Z. 93; Tolmatchev, Contr. Fl. Vaig. 144; Perfilev, Fl. Sev. I, 13; Leskov, Fl. Malozem. tundry 38; Hultén, Atlas, map 337.

C. incurva var. β *setina* Christ in Scheutz, Pl. jeniss. (1888), 174; Kükenthal, Cyper. Caricoid. 114; id., Cyper. Sibir. 43.

C. bucculenta V. Krecz. in Izv. Bot. sada XXX (1932), 130.

C. transmarina V. Krecz., l. c. 130.

C. orthocaula V. Krecz., l. c. 131.

C. camptotropa V. Krecz., l. c. 131.

C. jucunda V. Krecz., l. c. 136.

C. setina (Christ) V. Krecz., l. c. 136; id. in Fl. SSSR III, 192.

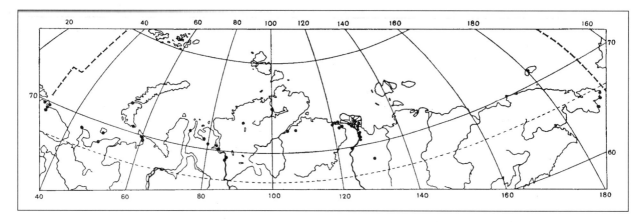

Map III–18 Distribution of *Carex maritima* Gunn.

C. psychroluta V. Krecz., l. c. 138.

Ill.: Fl. SSSR III, pl. XIV, fig. 10, 11; Fl. Murm. II, pl. XX; Polunin, l. c. 81.

Circumpolar arctic species reaching arctic districts at comparatively high latitudes. Mainly associated with sea shores, but sometimes occurring at a considerable distance from the sea in mountains and river valleys.

Growing in small clonal groups at sandy or gravelly sites on sea shores, also ascending maritime slopes; sometimes occurring in sedge-moss tundra adjacent to the shore, and in Arctic Yakutia on fresh alluvial deposits (of sand or gravel) on the terraces of large rivers and slightly penetrating the zone of open forests. In the Verkhoyansk Range the plant has been found at a considerable distance from the Arctic near a gigantic ice sheet on the River Ulakhan-Sakkyryr (south of 68°N) together with a series of other arctic species (*Phippsia algida*, *Saxifraga hyperborea*, *S. tenuis*, *Ranunculus hyperboreus*, etc.).

V.I. Krechetovich (Izv. Bot. sada XXX) separated from *C. maritima* seven further species, which he himself subsequently (Fl. SSSR III) reduced to synonymy with it. Almost all the species separated by V.I. Krechetovich (with the exception of *C. setina*) are merely scarcely differentiated infraspecific populations. The characters used by V.I. Krechetovich to characterize these species do not exceed the limits of variation in *C. maritima*, and can be observed in specimens from throughout its range.

The only one of the species related to *C. maritima* described by V.I. Krechetovich which has a definite taxonomic significance is *C. setina*. But in consideration of the fact that the differences between the latter and *C. maritima* s. str. are too insignificant and blurred, we think it more correct to treat plants of the *C. setina* type as the subspecies *C. maritima* ssp. *setina* (Christ) Egor. *Carex maritima* s. str. described from Norway is distributed in the Kola Peninsula, Scandinavia, Scotland, Iceland, Greenland and North America. In the expanse between the White and Bering Seas, *C. maritima* ssp. *maritima* is replaced by subspecies *setina*. The two subspecies differ in the size of the inflorescence and perigynia. In *C. maritima* s. str. the capitula of the inflorescence are 0.8–1.5 cm long and 0.7–1.3 cm wide, and the perigynia 4–5(5.5) mm long and (1.7)2–2.5(3) mm wide; in *C. maritima* ssp. *setina* the inflorescence is almost globose, no more than 1 cm in diameter, and the perigynia 3.7–4 mm long and 1.4–1.7 mm wide.

Plants referred to ssp. *setina* were originally described as a variety, var. β *setina* Christ (from the lower reaches of the Yenisey), to which V.I. Krechetovich (Izv. Bot. sada XXX; Fl. SSSR III) subsequently accorded the rank of species.

Carex maritima ssp. *setina* (*C. setina* V. Krecz.) is distributed not only in the territory of the Soviet Arctic as assumed by V.I. Krechetovich, but also in Alaska, the Canadian Arctic Archipelago, Greenland, and even in Scandinavia. In some of these districts the typical form of *C. maritima* and forms intermediate between the two subspecies also occur.

Carex juncifolia All., a high mountain plant from the Alps, is almost completely indistinguishable from *C. maritima* ssp. *setina*, and apparently represents one of the subspecies of *C. maritima* s. l.

Also very doubtfully distinct are the North American species *C. incurviformis* Mackenz. and *C. perglobosa* Mackenz.

The East Siberian species *C. sajanensis* V. Krecz. and *C. reptabunda* V. Krecz. very closely approach *C. maritima* s. l.

Soviet Arctic. *C. maritima* ssp. *maritima*: Murman (Pechenga district, Rybachiy Peninsula, Ara-Guba, Kola Bay). *C. maritima* ssp. *setina*: Northern Kanin; Malozemelskaya Tundra; Kolguyev Island; south shore of Yugorskiy Shar, Vaygach Island; South Island of Novaya Zemlya; Gydanskaya Tundra; lower reaches of Yenisey; Taymyr (rare); River Khatanga; right bank of the River Popigay; lower reaches of the Olenek and Olenek Bay; lower reaches and delta of the Lena; Ayon Island; Wrangel Island; Chukotka Peninsula (Uelen, Lawrence Bay, Litke Bay, Arakamchechen Island). (Map III–18).

Due to inadequate herbarium material we cannot give precise details of the distribution of each subspecies outside the Soviet Union, all the more so because relevant information is lacking in the literature due to foreign authors interpreting *C. maritima* in a wide sense. Therefore for the foreign Arctic and districts outside the Arctic we give the distribution of *C. maritima* s. l.

Foreign Arctic. West and north coasts of Alaska (probably predominantly forms referable to ssp. *setina*); arctic coast of Canada; Labrador; Canadian Arctic Archipelago (apparently only ssp. *setina*); coasts of Greenland all the way to its northern extremity; Iceland; Arctic Norway (predominantly ssp. *maritima*).

Outside the Arctic. Ssp. *maritima*: Scotland and extreme north of England; Faroe Islands; coast and mountains of Norway; west coast of Sweden; Northern Denmark; Kola Peninsula (Kuzomen village). Ssp. *setina*: Yakutia (the north of the Verkhoyansk Range); coast and mountains of Alaska; Hudson Bay, Labrador and Newfoundland; disjunctly in the Rocky Mountains. Recently *C. maritima* was discovered by Roivainen [Ann. Bot. Soc. Vanamo 28, 2 (1954), 198] in Tierra del Fuego.

16. *Carex leporina* L., Sp. pl. (1753), 973; Treviranus in Ledebour, Fl. ross. IV, 278; Kükenthal, Cyper. Caricoid. 210, pro parte; id., Cyper. Sibir. 62; Krylov, Fl. Zap. Sib. III, 450; Krechetovich in Fl. SSSR III, 161; Hultén, Atlas, map 343; Kuzeneva in Fl. Murm. II, 58, map 11.

Ill.: Fl. SSSR III, pl. XI, fig. 5; Fl. Murm. II, pl. XII, fig. 2.

One of the most frequently encountered plants in moist meadows, forest edges and glades, distributed mainly in the forest zone of Europe and West Siberia and in the forest and subalpine zones of the Caucasus. Within the Arctic occurring only on the Kola Peninsula.

Soviet Arctic. Murman [district of Pechenga, Kildin Island, the mouth of the Voroney (Fl. Murm. II), Ara-Guba].

Foreign Arctic. Not occurring.

Outside the Arctic. Western Europe (almost everywhere south of 65°N, in Norway reaching 68°N, absent from Albania and Greece); European part of the USSR (except southern districts); Caucasus; West Siberia (south of 58°N); the south of East Siberia west of Baykal; North America (from Newfoundland to New York State, introduced); New Zealand (possibly introduced).

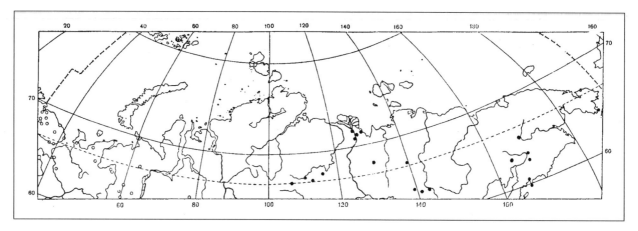

Map III–19 Distributions of *Carex dioica* L. (open circles) and *C. gynocrates* Wormsk. (black circles).

17. **Carex dioica** L., Sp. pl. (1753), 972; Treviranus in Ledebour, Fl. ross. IV, 264 (excl. var.); Ostenfeld, Fl. arct. 1, 60 (quoad α typ.); Kükenthal, Cyper. Caricoid. 78; id., Cyper. Sibir. 33; Gorodkov in Tr. Bot. muz. XX, 206 (incl. ssp. *asiatica*); Krylov, Fl. Zap. Sib. III, 433; Perfilev, Fl. Sev. I, 107; Krechetovich in Fl. SSSR III, 164; Leskov, Fl. Malozem. tundry 36; Hultén, Atlas, map 316; id., Circump. pl. I, 170, map 161; Kuzeneva in Fl. Murm. II, 59, map 12.

   Ill.: Fl. SSSR III, pl. XIV, fig. 3; Fl. Murm. II, pl. XIII, fig. 1.

   Boreal, predominantly European species occurring more or less frequently in temperate districts of the Arctic.

   Growing in mossy (usually sphagnous) bogs, marshy moss-covered meadows, moist meadows, and on the shores of rivers and streams.

   Soviet Arctic. Murman; Kanin; Timanskaya Tundra (River Indiga); SE part of Bolshezemelskaya Tundra (Sivaya Maska); Kolguyev Island (Hultén, Circump. pl. I). (Map III–19).

   Foreign Arctic. Iceland, Arctic Scandinavia (frequent).

   Outside the Arctic. Fennoscandia; Atlantic, Central and parts of Southern Europe; northern half of European part of USSR; Urals; West Siberia south of 63°N; Yenisey Basin south of Yeniseysk; isolated discoveries in Prebaikalia.

18. **Carex gynocrates** Wormsk. apud Drejer, Revis. Car. bor. (1841), 16; Kükenthal, Cyper. Caricoid. 79; id., Cyper. Sibir. 33; Gorodkov in Tr. Bot. muz. XX, 206; Komarov, Fl. Kamch. I, 223; Hultén, Fl. Kamtch. I, 172; id., Fl. Al. II, 302, map 227; Krechetovich in Fl. SSSR III, 165; A.E. Porsild, Ill. Fl. Arct. Arch. 48, map 73; Böcher & al., Grønl. Fl. 250; Karavayev, Konsp. fl. Yak. 64; Polunin, Circump. arct. fl. 76, 102.

   C. dioica γ Trev. in Ledebour, Fl. ross. IV (1852), 264.

   C. dioica β *gynocrates* Ostenf., Fl. arct. 1 (1902), 60.

   C. dioica ssp. *gynocrates* (Wormsk.) Hult., Circump. pl. I (1962), 170, map 161.

   Ill.: Fl. SSSR III, pl. XIV, fig. 5; Polunin, l. c. 77.

   East Siberian and North American, predominantly boreal species penetrating certain districts of the Arctic Region along rivers. Growing in moist *Dryas*-sedge-moss and sedge-cottongrass tundras, in small-sedge bogs with *Sphagnum* or *Hypnum* (including flarks), and in moist open larch woods.

   Carex gynocrates is a morphologically well differentiated race replacing *C. dioica* east of Baykal. Although *C. gynocrates* is closely related to *C. dioica*, it should not be considered a subspecies of the latter. The differential characters of the two species are constant. In addition to morphological differences, *C. dioica* and *C.*

gynocrates possess different chromosome numbers: 2n=52 in the former, 2n=48 in the latter. There are no transitional forms between the two species.

Specimens with androgynous spikelets also turn up in Asia (contrary to the opinion of B.N. Gorodkov), not only in North America.

Soviet Arctic. Lower reaches of Lena; Tiksi Bay; Chukotka (River Kuvet, Chaplino Hot Springs); Anadyr Basin (mouth of River Yablonovaya); Penzhina Basin; Bay of Korf. (Map III–19).

Foreign Arctic. Arctic Alaska; lower reaches of Mackenzie River; Melville Peninsula (one locality); Southern Baffin Island (one locality); West Greenland south of the 72nd parallel.

Outside the Arctic. Central Siberian Plateau (Rivers Olenek and Muna, upper reaches of River Markha); Central Yakutia; Verkhoyansk Range; basins of the upper courses of the Indigirka and Kolyma; Transbaikalia; Zeya Basin and the sources of the Bureya; Northern Sikhote-Alin; coasts of Sea of Okhotsk; Malyy Shantar Island; Kamchatka; Commander and Kurile Islands; isolated localities on Sakhalin and the islands of Hokkaido and Honshu; Alaska including the Aleutian Islands; all Canada; northern districts of the United States and disjunctly in the Rocky Mountains.

19. **Carex Redowskiana** C.A. Mey. in Mém. Acad. Sci. St.-Pétersb. Sav. Étr. I (1831), 207; Treviranus in Ledebour, Fl. ross. IV, 265; Kükenthal, Cyper. Caricoid. 77; id., Cyper. Sibir. 32; Tolmatchev, Contr. Fl. Vaig. 144, pro parte; Krylov, Fl. Zap. Sib. III, 431; Gorodkov in Tr. Bot. muz. XX, 203; Krechetovich in Fl. SSSR III, 166; Perfilev, Fl. Sev. I, 107; Karavayev, Konsp. fl. Yak. 65; Yegorova in Sp. rast. Gerb. Fl. SSSR XV, 22; Hultén, Circump. pl. I, 170, map 161; Polunin, Circump. arct. fl. 77, 104.

C. Davalliana auct. non Smith — Scheutz, Pl. jeniss. 173.

C. dioica δ *Redowskiana* Ostenf., Fl. arct. 1 (1902), 60.

Ill.: C.A. Meyer, l. c., tab. IV; Fl. SSSR III, pl. XIV, fig. 7.

Boreal, mainly Siberian species also distributed in temperate parts of the Arctic. Mainly associated with the forest-tundra zone. Rather common in certain districts, but everywhere occurring in small quantity. Growing in mossy bogs, on mounds in dwarf birch-moss tundra, in moist *Dryas*-sedge-moss tundras, in grassy openings in shrubby willow tundra, in moist mossy open larch woods, as well as in moist peaty meadows on river shores.

In the Siberian portion of its range *C. Redowskiana* shows only insignificant variation in the size of the perigynia and in the size and colour of the scales (from light ferruginous to castaneous). But in the Polar Ural (as rightly noted by B.N. Gorodkov), as well as in the Karskaya and Bolshezemelskaya Tundras, typical specimens of *C. Redowskiana* occur along with specimens intermediate between this species and the European arctic species *C. parallela*. These differ from *C. Redowskiana* in having longer scales and shorter beaks on the perigynia. Typical *C. parallela* does not occur in the Polar Ural and the Karskaya Tundra; it appears in the Bolshezemelskaya Tundra (River Korotaikha) and thence ranges westwards as far as Greenland. Within the range of *C. parallela* as far as Scandinavia (inclusively), forms indistinguishable from typical *C. Redowskiana* occur occasionally, as well as intermediate forms approaching one or other of these species. Only in Greenland and Spitsbergen are forms resembling *C. Redowskiana* absent.

To all appearances *C. parallela* and *C. Redowskiana* are subspecies of a single species, but for more definite judgement of the taxonomic status of these forms it is necessary to acquire considerably more material of *C. parallela* than we have available at the present time.

Soviet Arctic. Malozemelskaya Tundra (between Pechora Bay and Kolokolkova Bay); Kolguyev Island; Bolshezemelskaya Tundra (not uncommon); Karskaya Tundra; Polar Ural (rather common); Vaygach Island (Varnek Bay); lower reaches of the Ob

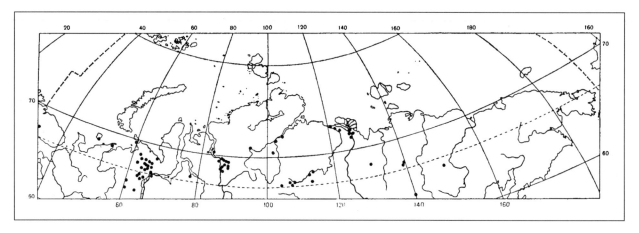

MAP III–20 Distribution of *Carex Redowskiana* C.A. Mey.

(Salekhard, Poluy Valley); lower reaches of the Yenisey (rather frequent); Khatanga Basin; lower reaches of the Olenek and Lena. (Map III–20).

Foreign Arctic. Arctic Scandinavia (isolated localities).

Outside the Arctic. Kola Peninsula (Khibins Mountains); Prepolar, Northern and Middle Urals; West Siberia; Altay; East Siberia (rare east of the Lena and Baykal).

20. **Carex parallela** (Laest.) Sommerf., Suppl. fl. lapp. (1826), 39; Kükenthal, Cyper. Caricoid. 78; Lynge, Vasc. pl. N.Z. 95; Perfilev, Fl. Sev. I, 107; Krechetovich in Fl. SSSR III, 167; Leskov, Fl. Malozem. tundry 39; Tolmachev, Mat. fl. Mat. Shar 286; Kuzeneva in Fl. Murm. II, 60, map 13; Hultén, Atlas, map 317; id., Amphi-atl. pl. 82, map 63; Böcher & al., Grønl. Fl. 251; Polunin, Circump. arct. fl. 77, 104.

C. dioica β *parallela* Laest. in Kongl. Vet. Acad. Nya Handl. (1822), 338.

C. dioica β Trev. in Ledebour, Fl. ross. IV (1852), 263.

C. dioica γ *parallela* Ostenfeld, Fl. arct. 1 (1902), 60.

Ill.: Fl. SSSR III, pl. XIV, fig. 6; Polunin, l. c. 77.

Almost exclusively arctic species but of very rare occurrence. Growing mainly in mossy bogs.

Soviet Arctic. Murman (Pechenga district, Rybachiy Peninsula, Cape Svyatoy Nos and Cape Orlov); Malozemelskaya Tundra (Korovinskiy Range); Bolshezemelskaya Tundra (western part and basin of River Korotaikha); ? Karskaya Tundra; Novaya Zemlya (South Island and the south of the North Island). (Map III–21).

Foreign Arctic. East Greenland (between 70° and 75°N), Spitsbergen; Arctic Scandinavia.

Outside the Arctic. Mountains of Sweden and Norway (rare); Khibins Mountains.

21. **Carex Kreczetoviczii** Egor. in Bot. mat. Gerb. Bot. inst. AN SSSR XIX (1959), 62.

C. laeviculmis auct. non Meinsh. — Kükenthal, Cyper. Caricoid. 232, pro parte; Krechetovich in Fl. SSSR III, 145.

Ill.: Yegorova, l. c. 62.

Growing in moist or marshy meadows and grassy marshes; very rare species.

Soviet Arctic. Anadyr Basin (lowland tundra below Ilmuve, shore of a channel).

Foreign Arctic. Not occurring.

Outside the Arctic. East Siberia (Barguzinskiy Range); coasts of Sea of Okhotsk, Kamchatka.

22. **Carex brunnescens** (Pers.) Poir., Encycl. Suppl. III (1813), 286; Ostenfeld, Fl. arct. 1, 56; Kükenthal, Cyper. Caricoid. 219; id., Cyper. Sibir. 68; Krylov, Fl. Zap. Sib. III, 454;

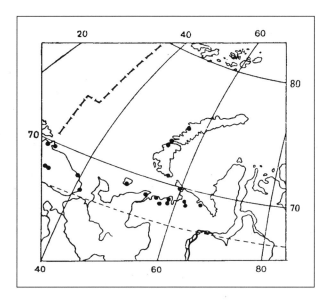

Map III–21 Distribution of *Carex parallela* (Laest.) Sommerf.

Perfilev, Fl. Sev. I, 114; Krechetovich in Fl. SSSR III, 179; Leskov, Fl. Malozem. tundry 35; Hultén, Fl. Al. II, 326, map 261; id., Atlas, map 348; id., Circump. pl. I, 40, map 34; Kuzeneva in Fl. Murm. II, 66, map 16; Böcher & al., Grønl. Fl. 253; Polunin, Circump. arct. fl. 84, 102.

C. vitilis Fries, Nov. Fl. Suec. Mant. III (1842), 137 (incl. var.); Krechetovich, l. c. 180 (quoad var. α).

C. canescens β *alpestris* Trev. in Ledebour, Fl. ross. IV (1852), 283.

Ill.: Fl. SSSR III, pl. XII, fig. 10; Fl. Murm. II, pl. XV, fig. 2; Polunin, l. c. 82.

Boreal species occurring not uncommonly in temperate districts of the European sector of the Arctic. Most characteristic of dry peat bogs and tundras with peat mounds; growing also in moss-sedge-mixed herb tundras with dwarf willow or dwarf birch, in streamside peaty meadows, and along trails and roads.

Soviet Arctic. Murman (everywhere, not uncommon); Kanin (mainly in the northern part, rather common); Timanskaya Tundra; Malozemelskaya Tundra (not uncommon); Kolguyev Island; lower reaches of the Pechora (River Yushina); Bolshezemelskaya Tundra (frequent); Polar Ural; lower reaches of Ob (Salekhard); Obsko-Tazovskiy Peninsula; lower reaches of Yenisey. (Map III–22).

Foreign Arctic. Arctic Alaska (one locality); Great Bear Lake; west coast of Greenland south of 72° N and east coast south of the Arctic Circle (not uncommon); Iceland; Arctic Scandinavia (common).

Outside the Arctic. Fennoscandia; mountains of Central Europe; Poland; the north of the European part of the USSR (south to Smolensk and Moscow; absent from the Lithuanian and Byelorussian Republics); all the Urals; West Siberia; Altay; Saur Range; Yenisey Basin south of Turukhansk; Prebaikalia; basin of the River Chara; coasts of the Sea of Okhotsk; Sakhalin; Kamchatka; Kurile Islands; Korean Peninsula; islands of Hokkaido and Honshu; Alaska; almost all Canada; in the United States in the Rocky Mountains, the Great Lakes district and Appalachia.

23. *Carex Mackenziei* V. Krecz. in Fl. SSSR III (1935), 183; Hultén, Fl. Al. II, 323, map 257; id., Atlas, map 347; id., Amphi-atl. pl. 292, map 274; Kuzeneva in Fl. Murm. II, 70, map 19; Böcher & al., Grønl. Fl. 254; Polunin, Circump. arct. fl. 83, 103.

C. norvegica Willd. ex Schkuhr, Riedgr. I (1801), 50 and II (1806), 17; Treviranus in Ledebour, Fl. ross. IV, 280; Ostenfeld, Fl. arct. 1, 56; Kükenthal, Cyper. Caricoid.

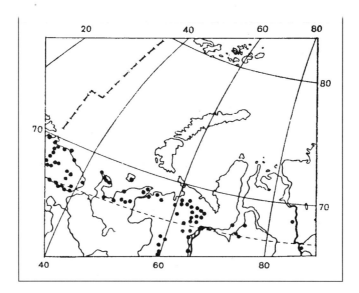

MAP III–22 Distribution of *Carex brunnescens* (Pers.) Poir.

216; id., Cyper. Sibir. 65; Perfilev, Fl. Sev. I, 111; Leskov, Fl. Malozem. tundry 39; non Retz. (1779).

C. pribylovensis auct. non J.M. Macoun — Krechetovich, l. c. 181.

Ill.: Fl. SSSR III, pl. XII, fig. 16; Fl. Murm. II, pl. XVII, fig. 1; Polunin, l. c. 82.

Plant of northern sea shores. Growing in saline moist or marshy meadows (sometimes subject to tidal flooding), on coastal rocks, and on sand and pebble beaches; locally occurring in large quantity, becoming a "background" plant.

Soviet Arctic. Murman; Kanin; Malozemelskaya Tundra (one locality near Kuznetskiy Nos); Anadyr Basin (Konchalanskiy Lagoon and the River Volchikha, shore of Anadyr Bay); Koryakia (the mouth of the Penzhina and the Bay of Korf). (Map III–23).

Foreign Arctic. West coast of Alaska; southern extremity of Greenland; Iceland; Arctic Scandinavia.

Outside the Arctic. Coast of Fennoscandia, Gulf of Finland, coasts and islands of the White Sea; Pskov Oblast (Ryushskiy saltflat); Shantar Islands; Kamchatka; Sakhalin; Hokkaido; Alaska; Hudson Bay; Atlantic coast of Canada and the State of Maine (USA).

24. ***Carex canescens*** L., Sp. pl. (1753), 974; Treviranus in Ledebour, Fl. ross. IV, 280 (excl. var.); Ostenfeld, Fl. arct. 1, 54; Kükenthal, Cyper. Caricoid. 216, pro parte; id., Cyper. Sibir. 66, pro parte; Komarov, Fl. Kamch. I, 230; Hultén, Fl. Kamtch. I, 181, map 184; Krylov, Fl. Zap. Sib. III, 452; Andreyev, Mat. fl. Kanina 161; Perfilev, Fl. Sev. I, 113; Krechetovich in Fl. SSSR III, 176; Leskov, Fl. Malozem. tundry 36, pro parte; Hultén, Fl. Al. II, 324, map 258; id., Atlas, map 349; id., Circump. pl. I, 90, map 82; Kuzeneva in Fl. Murm. II, 64, map 15; Polunin, Circump. arct. fl. 84, 102.

C. curta Good. in Trans. Linn. Soc. II (1794), 145; Böcher & al., Grønl. Fl. 254.

Ill.: Fl. SSSR III, pl. XII, fig. 8, 9; Fl. Murm. II, pl. XV, fig. 1; Polunin, l. c. 85.

Common plant of the forest zone of Eurasia and North America, also occurring in the Southern Hemisphere. Widespread in temperate districts of the European Arctic; penetrating the Asian and American Arctic only to a limited extent. Growing in grass-sedge fens and marshy meadows, on the margins of sphagnum bogs, in dwarf birch-moss tundra, on slightly marshy shores of waterbodies, in flarks, and in shrubbery.

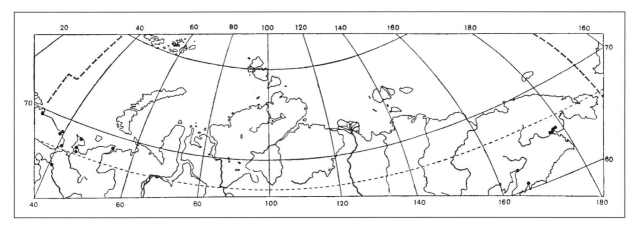

Map III–23 Distribution of *Carex Mackenziei* V. Krecz.

Soviet Arctic. Murman (everywhere, but mainly in the western part); Northern Kanin; Northern Timan; Kolguyev Island (Perfilev, l. c.); Malozemelskaya Tundra (not uncommon); Bolshezemelskaya Tundra; Polar Ural; lower reaches of Ob (rare); River Nyda; lower reaches of Yenisey, vicinity of Norilsk; district of Chaun Bay (River Umku-Veyem); Anadyr Basin (Vayegi settlement). (Map III–24).

Foreign Arctic. Arctic Alaska (isolated discoveries); district of Great Bear Lake; Labrador; West Greenland south of 70°N, SE Greenland; Iceland; Arctic Scandinavia.

Outside the Arctic. Fennoscandia; Northern and Central Europe; European part of the USSR (except southern districts); Urals; all West Siberia; Altay; Tien Shan; Yenisey Basin south of the Podkammenaya Tunguska; Prebaikalia; basins of the Vilyuy and Aldan; Preamuria; Primorskiy Kray; coasts of the Sea of Okhotsk; Kamchatka; Sakhalin; islands of Hokkaido and Honshu; the north of the Korean Peninsula; NE China; Northern Mongolia; Kashmir; in North America throughout Canada and in the Western United States as far as California and Arizona (inclusively), in the east north of 38°N; Southern South America; Australia.

25. *Carex lapponica* O.F. Lang in Linnaea XXIV (1851), 539; Krechetovich in Fl. SSSR III, 179; Hultén, Fl. Al. II, 325; id., Atlas, map 350; Kuzeneva in Fl. Murm. II, 66, map 15; Karavayev, Konsp. fl. Yak. 65.

C. canescens var. *subloliacea* Laest. in Nov. Acta Regiae Soc. Sci. Upsal. XI (1839), 282; Kükenthal, Cyper. Caricoid. 217; Leskov, Fl. Malozem. tundry 36 (under *C. canescens*).

C. Cajanderi auct. non Kük. — Andreyev, Mat. fl. Kanina 161; Perfilev, Fl. Sev. I, 113.

Ill.: Fl. Murm. II, pl. XV, fig. 3.

Carex lapponica is very close to *C. bonanzensis* Britt. These species differ mainly in the colour of the scales: in *C. lapponica* these are whitish green with very weak brownish yellow tint and ⅔-¾ (rarely only half) as long as the perigynia, in *C. bonanzensis* they are light to dark brown and usually ½-⅔ as long as the perigynia. In addition, the spikelets and perigynia in *C. bonanzensis* are somewhat smaller than in *C. lapponica*.

Growing in depressions in sedge-shrub tundra, in flarks, in hummocky bogs, on marshy lake shores, and in shrubbery in stream valleys.

Soviet Arctic. Murman (Drozdovka); Northern Kanin; Timanskaya and Malozemelskaya Tundras (River Indiga and between Pechora Bay and Kolokolkova Bay); Bolshezemelskaya Tundra (western part and eastern edge); Polar Ural (Krechetovich, l. c.); lower reaches of the Ob (vicinity of Salekhard and further

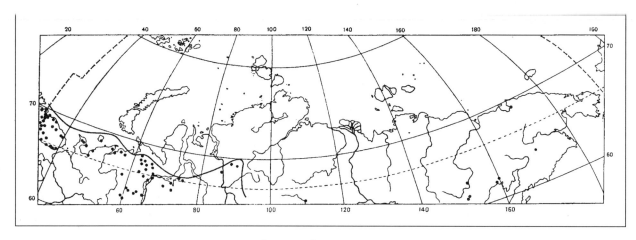

MAP III–24 Distribution of *Carex canescens* L.

north); lower reaches of River Nyda; Anadyr Basin (eastern spurs of the Pekulney Range and the vicinity of Ust-Belaya); Penzhina basin (River Oklan and vicinity of Penzhino settlement). (Map III–25).

Foreign Arctic. Arctic Scandinavia (extremely rare).

Outside the Arctic. Fennoscandia (isolated discoveries in the northern half); West Siberia (isolated localities in the basins of the rivers Severnaya Sosva, Pelym and Poluy); upper course of Yenisey; upper reaches of the Angara and Lena (very rare); occasionally in Transbaikalia and the Amur Basin; Yakutia (Khodak in the Vilyuy Basin, Zhigansk, between Yakutsk and the mouth of the Aldan); coasts of the Sea of Okhotsk (Ola district); Southern Kamchatka; Alaska; Yukon.

26. **Carex bonanzensis** Britt. in Bull. N.Z. Bot. gard. II, 6 (1901), 160; Krechetovich in Fl. SSSR III, 175; Hultén, Fl. Al. II, 325.

C. Cajanderi Kük. in Öfver. Finska Vet. Soc. Förhandl. XLV, 8 (1902–1903), 3; id., Cyper. Caricoid. 219; id., Cyper. Sibir. 67; Krylov, Fl. Zap. Sib. III, 453.

Ill.: Kükenthal, Cyper. Sibir., fig. 51 (under *C. Cajanderi*); Fl. SSSR III, pl. XII, fig., 12.

Growing in moist or marshy meadows, in floodplain shrubbery, and on marshy shores of waterbodies.

Soviet Arctic. Lower course of Yana; River Omolon; the basins of the Anadyr, of the Belaya (Bitcho Range), and of the middle course of the Penzhina (west slope of the Palmatkina Mountains).

Foreign Arctic. Not occurring.

Outside the Arctic. West Siberia (basin of the Severnaya Sosva); isolated localities in the basin of the middle course of the Yenisey and the Lower Tunguska; Prebaikalia; Aldan and Amur Basins; Kolyma Basin (Srednekolymsk, Tuastakh); north coast of the Sea of Okhotsk (Ola district); North America (Yukon Basin).

27. **Carex glareosa** Wahlb. in Kongl. Vet. Akad. Nya Handl. XXIV (1803), 146; Treviranus in Ledebour, Fl. ross. IV, 284; Ostenfeld, Fl. arct. 1, 58; Kükenthal, Cyper. Caricoid. 215; id., Cyper. Sibir. 65; Tolmachev, Fl. Kolg. 16; Andreyev, Mat. fl. Kanina 161; Perfilev, Fl. Sev. I, 111; Krechetovich in Fl. SSSR III, 182; Leskov, Fl. Malozem. tundry 37; Hultén, Fl. Al. II, 323, map 256; id., Atlas, map 345; id., Circump. pl. I, 194, map 183; Kuzeneva in Fl. Murm. II, 68, map 18; Böcher & al., Grønl. Fl. 252; Polunin, Circump. arct. fl. 83, 102.

C. marina Dew. in Amer. Journ. Sci. XXIX (1836), 247; Krechetovich, l. c. 182; Kuzeneva, l. c. 70, map 18; Karavayev, Konsp. fl. Yak. 65.

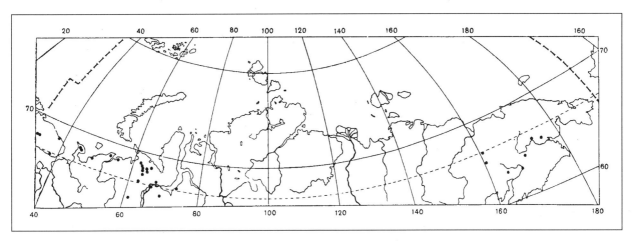

Map III–25 Distribution of *Carex lapponica* O.F. Lang.

C. glareosa var. *amphigena* Fern. in Rhodora 7 (1906), 47; A.E. Porsild, Ill. Fl. Arct. Arch. 52, map 82.

Ill.: Fl. SSSR III, pl. XII, fig. 14; Fl. Murm. II, pl. XVI, fig. 2 and pl. XVII, fig. 2; Polunin, l. c. 82.

Plant of northern sea shores. Very characteristic of maritime saline meadows (sometimes flooded by tides), as well as of sand and pebble beaches; occurring on rocky slopes on sea shores and on coastal cliffs; sometimes forming pure stands.

Usually *C. glareosa* is considered to include mainly European plants, while Siberian and North American plants are considered to belong to *C. marina* Dew. described from the arctic coast of Canada. Study of herbarium material has shown that the majority of European plants do not show any differences when compared with plants from Siberia and North America. These plants possess ovoid or obovoid perigynia more or less abruptly tapering to the beak. On the Kola Peninsula and in Karelia (Rybachiy Peninsula, Yarnyshnaya Bay, Iokanga, Kandalaksha, Solovetskiye Islands, Kem), as well as in certain districts of Scandinavia and in the Eastern United States, there occurs a form with elongate-ovoid or lanceolate perigynia which taper gradually to the beak (without incurvation). The ecology of plants of both types, so far as can be judged from herbarium labels, is completely identical. In addition to plants of the extreme types, specimens intermediate between them in perigynial structure also occur. In all probability the more rarely occurring form with elongate perigynia should be ranked as a variety. We reduce *C. marina* Dew. to synonymy with *C. glareosa* since it was described later than the latter.

Soviet Arctic. Murman (everywhere); Kanin; River Pesha (Perfilev, l. c.); Malozemelskaya Tundra (district of Pechora Bay and between it and Kolokolkova Bay); Kolguyev Island; Baydaratsk Coast (lower reaches of the River Igoy-Yaga), Litke Islands, Vaygach (Perfilev, l. c.), Malyy Storozhevoy; Novaya Zemlya (on the South Island at Belushya Bay and Malyye Karmakuly); Olenek Bay; Buorkhaya Bay; Chukotka (district of Chaun Bay, Vankarem, Kolyuchin Island, and the Beringian shore of the Chukotka Peninsula from Uelen to Provideniye Bay); Ratmanov Island; lower reaches of the Anadyr; Bay of Korf. (Map III–26).

Foreign Arctic. West and north coast of Alaska; western part of the arctic coast of Canada; Labrador; Canadian Arctic Archipelago (Devon, Ellesmere and Baffin Islands); west and east coasts of Greenland south of 78°N; Iceland; Spitsbergen; Arctic Scandinavia.

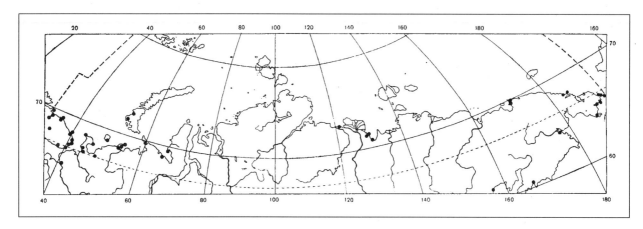

MAP III–26 Distribution of *Carex glareosa* Wahlb.

Outside the Arctic. Coasts of Fennoscandia and the Gulf of Finland; coasts of the White Sea; coasts of the Sea of Okhotsk; Kamchatka; Commander Islands; Sakhalin; south coast of Alaska; Hudson Bay; Newfoundland.

28. ***Carex ursina*** Dew. in Amer. Journ. Sci. XXVII (1835), 240; Ostenfeld, Fl. arct. 1, 59; Kükenthal, Cyper. Caricoid. 74; id., Cyper. Sibir. 31; Kjellman & Lundström, Phanerogam. N.Z., Waig. 155; Krylov, Fl. Zap. Sib. III, 431; Krechetovich in Fl. SSSR III, 183; Tolmachev, Obz. fl. N.Z. 150; A.E. Porsild, Ill. Fl. Arct. Arch. 49, map 77; Böcher & al., Grønl. Fl. 254; Hultén, Circump. pl. I, 192, map 181; Polunin, Circump. arct. fl. 77, 105.

    Ill.: Fl. SSSR III, pl. XII, fig. 15; A.E. Porsild, l. c., fig. 16, 7 and 18, e; Polunin, l. c. 77.

    Plant of arctic sea coasts, reaching high arctic districts for which it is rather characteristic. Occurring comparatively rarely within the Soviet Arctic; a considerably greater number of localities are known from the American sector of the Arctic. Growing on low-lying shores of marine lagoons (flooded by high tides) in low-herb *Puccinellia*-sedge communities of maritime saltflats.

    Soviet Arctic. Vaygach Island (Varnek Bay); Novaya Zemlya (South Island and the south of the North Island); Franz Josef Land; Belyy Island; Krestovskiy Island in Yenisey Bay; Preobrazheniye Island; Olenek Bay (vicinity of Stannakh-Khocho settlement); Buorkhaya Bay (Tiksi Bay, Bulunkan; Mostakh Island); Belkovskiy Island; district of Chaun Bay (Pevek and Cape Shelagskiy); Wrangel Island; Chukotka Peninsula (Uelen). (Map III–27).

    Foreign Arctic. North coast of Alaska; arctic coast of Canada; Labrador; Canadian Arctic Archipelago; west coast of Greenland (between the Arctic Circle and 71°N) and east coast (between 70° and 75° N); Spitsbergen.

    Outside the Arctic. Hudson Bay, Northern Labrador and Newfoundland.

29. ***Carex loliacea*** L., Sp. pl. (1753), 974; Treviranus in Ledebour, Fl. ross. IV, 281, pro parte; Kükenthal, Cyper. Caricoid. 225; id., Cyper. Sibir. 71; Krylov, Fl. Zap. Sib. III, 457; Perfilev, Fl. Sev. I, 112; Krechetovich in Fl. SSSR III, 173; Hultén, Fl. Al. II, 328, map 264; id., Atlas, map 351; id., Circump. pl. I, 78, map 69; Kuzeneva in Fl. Murm. II, 60, map 14; Karavayev, Konsp. fl. Yak. 65.

    Ill.: Fl. SSSR III, pl. XII, fig. 6; Fl. Murm. II, pl. XIV, fig. 1.

    Plant of the taiga zone of Eurasia and North America, just penetrating arctic limits. Growing in the forest-tundra zone in sphagnum bogs or more rarely in sedge-grass fens in forest, on the banks of forest streams, and in marshy shrubbery.

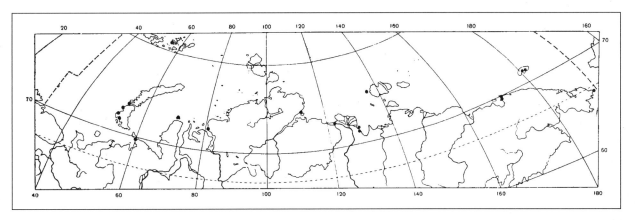

MAP III–27 Distribution of *Carex ursina* Dew.

Soviet Arctic. Murman [Pechenga district and the lower reaches of the Kola (Hultén, Atlas), Ponoy], very rare; lower reaches of the Yenisey (Plakhino); Bay of Korf.
Foreign Arctic. Arctic Alaska (single locality in the upper reaches of the River Noatak); Arctic Scandinavia.
Outside the Arctic. Fennoscandia; Poland; northern half of the European part of the USSR; Urals; western and southern part of West Siberia; Altay; Kazakhstan (Karkaralinsk); basin of the upper course of the Yenisey; isolated localities in the basins of the Lower Tunguska and Podkamennaya Tunguska Rivers; Prebaikalia; Transbaikalia; basins of the Aldan and of upper right-bank tributaries of the Kolyma and Amur; Primorskiy Kray; coasts of the Sea of Okhotsk; Kamchatka; Sakhalin; Northern Mongolia; the north of the Korean Peninsula; Hokkaido; Alaska; taiga zone of Canada.

30. ***Carex tenuiflora*** Wahlb. in Kongl. Vet. Akad. Nya Handl. XXIV (1803), 147; Treviranus in Ledebour, Fl. ross. IV, 282; Scheutz, Pl. jeniss. 176; Kükenthal, Cyper. Caricoid. 224; id., Cyper. Sibir. 70; Krylov, Fl. Zap. Sib. III, 456; Perfilev, Fl. Sev. I, 113; Krechetovich in Fl. SSSR III, 174; Hultén, Fl. Al. II, 328, map 263; id., Atlas, map 353; id., Circump. pl. I, 82, map 74; Kuzeneva in Fl. Murm. II, 62, map 14; Karavayev, Konsp. fl. Yak. 65; Polunin, Circump. arct. fl. 83, 105.
C. macilentha Fries in Bot. Notis. (1884) 23; Perfilev, l. c. 113.
Ill.: Fl. SSSR III, pl. XII, fig. 7; Fl. Murm. II, pl. XIV, fig. 2.

Boreal species occasionally occurring in a few districts of the Arctic.
Growing in moss-sedge tundra with dwarf birch, in *Hypnum* fens, in moist hollows in open larch forests with mosses and lichens, and in flarks.

Soviet Arctic. Murman (lower reaches of the rivers Kola and Voroney); SE edge of the Bolshezemelskaya Tundra (Sivaya Maska); Polar Ural (basin of the River Sob); lower reaches of Ob (Salekhard); lower reaches of Yenisey (Dudinka); Kheta Basin (Volochanka); lower reaches of Kolyma; Anadyr Basin (very rare); Penzhina Basin (upper course of the Oklan). (Map III–28).
Foreign Arctic. Arctic Alaska (isolated discoveries); lower reaches of the Mackenzie River and the district of Great Bear Lake; Arctic Scandinavia (extremely rare).
Outside the Arctic. Fennoscandia; the north of the European part of the USSR; Precaucasus; West Siberia (southern part and the basins of the rivers Severnaya Sosva and Poluy); Altay (Terektinskiy Belok); basin of the upper course of the Yenisey (Rivers Azas and Syda), basin of the Lower Tunguska (66°N at Lake Amnundakan); northern edge of the Central Siberian Plateau; upper course of the Vilyuy; lower course of the Lena; the south of the Verkhoyansk Range; middle course of the Indigirka; basins of the Aldan and of upper right-bank tributaries of

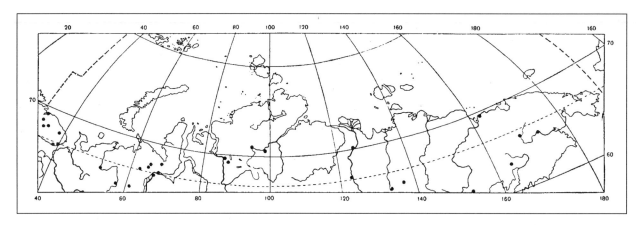

Map III-28 Distribution of *Carex tenuiflora* Wahlb.

the Kolyma; Prebaikalia; Amur Basin; Primorskiy Kray; Kamchatka; Kurile Islands; Sakhalin; the north of the Korean Peninsula; islands of Hokkaido and Honshu; Northern Mongolia; NE China; Alaska; Canada; Great Lakes district.

31. **Carex heleonastes** Ehrh. in L. fil., Suppl. (1781), 414; id., Beitr. I, 186; Treviranus in Ledebour, Fl. ross. IV, 279; Kükenthal, Cyper. Caricoid. 214, pro parte; id., Cyper. Sibir. 64; Krylov, Fl. Zap. Sib. III, 452; Perfilev, Fl. Sev. I, 113; Krechetovich in Fl. SSSR III, 184; Hultén, Fl. Al. II, 321, map 253; id., Atlas, map 346; id., Circump. pl. I, 68, map 59; Kuzeneva in Fl. Murm. II, 72; Polunin, Circump. arct. fl. 83, 102, pro parte.

C. aa Kom., Fl. Kamch. I (1927), 234; Krechetovich, l. c. 185.

Ill.: Fl. SSSR III, pl. XII, fig. 7.

Boreal species just penetrating the limits of the Arctic, where it grows in mossy bogs, marshy meadows and shrubbery, and on the shores of rivers and streams.

Carex aa Kom., here reduced to synonymy with *C. heleonastes*, was described from Eastern Kamchatka where only two localities for it are known (Lake Kronotskoye and Kirganik village). The plants upon which the description of *C. aa* was based have proved to be identical to *C. heleonastes* Ehrh. despite the considerable distance of the Kamchatka localities from the main range of the latter. The looser tufts and pale castaneous scales ostensibly characterizing *C. aa* sometimes also occur in *C. heleonastes*. In general the degree of looseness of the tuft and the intensity of scale coloration depends on the ecological conditions under which the species grows. The presence of brown nerves on the perigynia is also not a distinguishing character of *C. aa*, since nerves are also present on the perigynia of *C. heleonastes* and are always conspicuous when they turn brown.

Soviet Arctic. Murman (Kildin Island according to Hultén, Atlas); SE part of Bolshezemelskaya Tundra (Sivaya Maska).

Foreign Arctic. Arctic Norway (isolated discoveries); Iceland.

Outside the Arctic. Fennoscandia; the north and mountains of Central Europe; Bulgaria; northern half of the European part of the USSR (south to Moscow Oblast, southwest to Smolensk and Klintsy); Precaucasus; West Siberia (certain localities in the basins of the rivers Severnaya Sosva, Upper Pur, Ishim, Tara and Chulym); Altay (Biysk district); East Siberia (the mouth of the Kureyka and isolated discoveries south from the Podkamennaya Tunguska and east to Baykal inclusively); Kamchatka; Southern Alaska (one locality); Canada (provinces of Alberta, Saskatchewan and Manitoba).

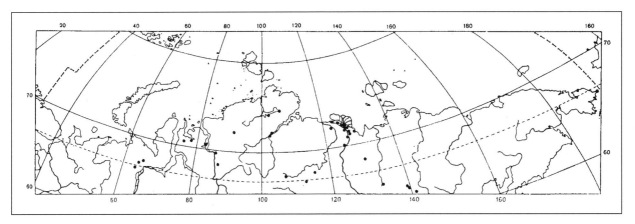

MAP III–29 Distribution of *Carex amblyorhyncha* V. Krecz.

32. **Carex amblyorhyncha** V. Krecz. in Fl. SSSR III (1935), 184 and 595 (descr.); id. in Areal I, map 26; Böcher in Acta Arctica 5, 22; Böcher & al., Grønl. Fl. 253; A.E. Porsild, Ill. Fl. Arct. Arch., map 80; Hultén, Circump. pl. I, 68, map 59; Karavayev, Konsp. fl. Yak. 65; Derviz-Sokolova in Bot. mat. Gerb. Bot. inst. AN SSSR XXI, 69.

C. heleonastes auct. non Ehrh. — Scheutz. Pl. jeniss. 175; Ostenfeld, Fl. arct. 1, 58; Kükenthal, Cyper. Caricoid. 215, pro minima parte; id., Cyper. Sibir. 64, pro minima parte; Polunin, Circump. arct. fl. 83, 102, pro parte.

Ill.: Fl. SSSR III, pl. XII, fig. 18; Böcher in Acta Arctica, fig. 7, 9, 10; Böcher & al., Grønl. Fl., fig. 44, d.

Siberian and North American subarctic species, also distributed in the mountains of Siberia. Rather rare plant. Growing in moist mossy tundras and open forests, in marshy hollows, in moss-cottongrass-small sedge marshes on river terraces, in polygonal marshes, and on river shores (at sites well covered by snow in winter). Apparently most common in the zone of subarctic open forests and in the more southern districts of the tundra zone.

Soviet Arctic. Polar Ural; Gydanskaya Tundra; lower reaches of Yenisey; basins of Lake Taymyr and the Khatanga; lower reaches of Olenek; lower reaches and delta of the Lena (numerous localities); Buorkhaya Bay; Chukotka Peninsula (district of Cape Dezhnev). (Map III–29).

Foreign Arctic. Arctic part of Alaska; arctic coast of Canada (lower reaches of Mackenzie River, district of Bathurst Inlet); Arctic Labrador (one locality); Victoria, Southampton and Baffin Islands; SW and East Greenland between the Arctic Circle and 75°N; Spitsbergen.

Outside the Arctic. Prepolar Ural (Voykar Basin); Altay (one locality on the upper course of the Katun); Eastern Sayan; NE part of Central Siberian Plateau; Lena (Sektyakh); Verkhoyansk Range; Alaska; Hudson Bay.

33. **Carex tripartita** All., Fl. Pedem. II (1785), 265, tab. 92, fig. 51; Krechetovich in Fl. SSSR III, 181; Kuzeneva in Fl. Murm. II, 67, map 17; Karavayev, Konsp. fl. Yak. 65.

C. Lachenalii Schkuhr, Riedgr. I (1801), 51; Krylov, Fl. Zap. Sib. III, 450; Andreyev, Mat. fl. Kanina 161; Leskov, Fl. Malozem. tundry 38; Hultén, Fl. Al. II, 320, map 252; id., Atlas, map 344; id., Circump. pl. I, 50, map 44; A.E. Porsild, Ill. Fl. Arct. Arch. 51, map 80; Böcher & al., Grønl. Fl. 252; Polunin, Circump. arct. fl. 83, 102.

C. lagopina Wahlb. in Kongl. Vet. Akad. Nya Handl. XXIV (1803), 145; Treviranus in Ledebour, Fl. ross. IV, 279; Scheutz, Pl. jeniss. 175; Ostenfeld, Fl. arct. 1, 58; Kükenthal, Cyper. Caricoid. 213, pro parte; id., Cyper. Sibir. 63; Lynge, Vasc. pl. N.Z. 94; Tolmachev, Mat. fl. Mat. Shar 286; id., Fl. Taym. I, 103; id., Obz. fl. N.Z. 150.

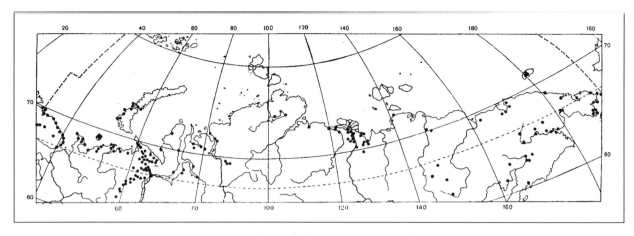

MAP III–30. Distribution of *Carex tripartita* All.

Ill.: Fl. SSSR III, pl. XII, fig. 13; Fl. Murm. II, pl. XVI, fig. 1; A.E. Porsild, l. c., fig. 19, b, c.

Arctic-alpine circumpolar species extremely characteristic of the Arctic Region. Within the Soviet Arctic occurring almost everywhere and common in many districts, although not ranging particularly far to the north. An eutrophic plant associated with richer soils. It prefers adequately or even strongly moistened sites (but always with flow), normally where snow accumulates in winter.

Growing in moist or slightly marshy meadows, in shrubby tundras with herbs and mosses, in bogs with small hummocks, in riverine meadows and on stream shores, on vegetated sandy shoreline slopes, in moist rock crevices, and sometimes on moist rocky slopes and along roads. It often grows at sites where snow long persists (in nival meadows and in herb-moss communities with layer of *Salix polaris*); associated mainly with mountainous districts of the Arctic.

Soviet Arctic. Murman (everywhere, rather common); Northern Kanin (frequent); Timanskaya and Malozemelskaya Tundras (more or less common); Kolguyev Island (very frequent); lower reaches of the Pechora and the Bolshezemelskaya Tundra (frequent); Karskaya Tundra; Polar Ural (frequent); Yugorskiy Peninsula (Pay-Khoy and south coast of Yugorskiy Shar); Vaygach Island; Novaya Zemlya; Yamal; lower reaches of Ob; Obsko-Tazovskiy Peninsula; Gydanskaya Tundra; lower reaches of Yenisey (not uncommon); Taymyr (basin of Lake Taymyr); lower reaches of the Anabar and Olenek; lower reaches and delta of the Lena (common); Buorkhaya Bay (rather common in the Tiksi Bay district, Mostakh Island); middle course of the River Kharaulakh, Nyaybinskiy Range; Svyatoy Nos Peninsula; lower reaches of the Indigirka and Kolyma; Chukotka (district of Chaun Bay, Cape Schmidt, Kolyuchin Bay, River Chegitun, and the Beringian coast of the Chukotka Peninsula from Uelen to its SE extremity); Wrangel, Ratmanov and Arakamchechen Islands; Anadyr Basin (numerous localities); Koryakia (Penzhina Basin, Bay of Korf). (Map III–30).

Foreign Arctic. Arctic Alaska (frequent); lower reaches of Mackenzie River; north shore of Great Bear Lake; Melville Peninsula; Labrador; Southern Baffin Island; Southampton Island; SW and East Greenland; Iceland; Jan Mayen Island, Spitsbergen; Arctic Scandinavia.

Outside the Arctic. Mountains of Scotland; Norway, North and West Sweden; Northern Finland; mountains of Central and Southern Europe (south to the Sierra Nevada in Spain and the Serra da Estrela in Portugal); Kola Peninsula (Khibins Mountains); Prepolar, Northern and Middle Urals; West Siberia (basin of the Severnaya Sosva); Altay; northern edge of the Central Siberian Plateau; Eastern Sayan; Transbaikalia; Suntar-Khayata Range; the sources of the Bureya; basin of the upper course of the Kolyma; coasts of the Sea of Okhotsk; Kamchatka; Kurile Islands; islands of Hokkaido and Honshu; Korean Peninsula; India (Khasia); Alaska; Rocky Mountains and subarctic districts of Eastern Canada; New Zealand (South Island).

34. *Carex pribylovensis* J.M. Macoun in Jordan, Fur Seals N Pacif. Ocean III (1899), 572; Mackenzie in N Amer. Fl. XVIII, 2, 91; Hultén, Fl. Al. II, 322, map 255; Yegorova in Bot. mat. Gerb. Bot. inst. AN SSSR XIX, 65.

C. lagopina var. *pribylovensis* (J.M. Macoun) Kük., Cyper. Caricoid. (1909) 214.

Very rare plant, reliably known so far only from islands adjacent to Alaska. In the "Flora of the USSR" (III, 181) *C. pribylovensis* was recorded for the Anadyr Basin, but the sole herbarium specimen from that district identified as *C. pribylovensis* by V.I. Krechetovich has proved to belong in fact to *C. Mackenziei*. But there is another herbarium specimen also collected in the Anadyr Basin which agrees with the description of *C. pribylovensis* (type not seen by us). The data are: shore of right channel of River Tanyurer, opposite settlement, river floodplain, willow carr with grasses and mixed forbs, 10 IX 1950, A. Reut. This specimen is deposted in the herbarium of Moscow State University.

Soviet Arctic. Anadyr Basin (River Tanyurer).

Foreign Arctic. Not occurring.

Outside the Arctic. Aleutian Islands; Pribilof Islands (St. Paul Island); St. Lawrence Island.

35. *Carex disperma* Dew. in Amer. Journ. Sci. VIII (1824), 266; Krechetovich in Fl. SSSR III, 185; Hultén, Fl. Al. II, 327, map 262; id., Atlas, map 352; id., Circump. pl. I, 80, map 72; Kuzeneva in Fl. Murm. II, 72, map 19; Karavayev, Konsp. fl. Yak. 65.

C. tenella Schkuhr, Riedgr. I (1801), 23; Kükenthal, Cyper. Caricoid. 223; id., Cyper. Sibir. 69; Komarov, Fl. Kamch. I, 231; Hultén, Fl. Kamtch. I, 183; Krylov, Fl. Zap. Sib. III, 456; Perfilev, Fl. Sev. I, 111.

C. loliacea auct. non L. — Treviranus in Ledebour, Fl. ross. IV, 281, pro parte.

Ill.: Fl. SSSR III, pl. XII, fig. 19; Fl. Murm. II, pl. XVIII.

Species of the forest zone of Eurasia and North America with circumpolar distribution. In the Arctic only isolated localities situated close to the northern boundary of forest are known.

Growing in moist or marshy mossy coniferous forests and shrubbery, and in the valleys of forest streams and rivulets.

Soviet Arctic. Murman (Ponoy); lower reaches of Ob (near Cape Orniol).

Foreign Arctic. Arctic Alaska (one locality); lower reaches of River Mackenzie; the north of Great Bear Lake.

Outside the Arctic. Norway (isolated in the southeast); Sweden; Finland; NE Poland; northern half of the European part of the USSR; Middle and Southern Urals; West and East Siberia south of 62°N; Amur Basin, Primorskiy Kray; Kamchatka (basin of the Kamchatka River); Sakhalin; Korean Peninsula; Hokkaido; forest zone of North America.

36. *Carex acuta* L., Sp. pl. (1753) 978 (excl. var. α); Treviranus in Ledebour, Fl. ross. IV, 313, pro parte; Perfilev, Fl. Sev. I, 117; Kuzeneva in Fl. Murm. II, 82.

C. gracilis Curt., Fl. Londin. IV (1777–1787), 282; Kükenthal, Cyper. Caricoid. 319; id., Cyper. Sibir. 86; Krylov, Fl. Zap. Sib. III, 465; Krechetovich in Fl. SSSR III, 210; Leskov, Fl. Malozem. tundry 37; Hultén, Atlas, map 360; Polunin, Circump. arct. fl. 90.

C. fusco-vaginata auct. non Kük. — Krechetovich, l. c. 211, pro maxima parte; Karavayev, Konsp. fl. Yak. 66.

Ill.: Kükenthal, Cyper. Caricoid., fig. 49; id., Cyper. Sibir., fig. 67.

Species of the temperate forest zone of Eurasia, penetrating the Arctic only in the northwest of its range. Growing on the shores of waterbodies, in marshes and marshy places, and in floodplain meadows.

Soviet Arctic. Murman; Kanin; Malozemelskaya Tundra (rare); Bolshezemelskaya Tundra; Polar Ural; lower reaches of Ob.

Foreign Arctic. Not occurring.

Outside the Arctic. All Europe; Precaucasus; West Siberia (mainly southern half); Altay; southern part of East Siberia (approximately to the latitude of Yakutsk); Eastern and Northern Kazakhstan; Northern Mongolia; ? North Africa.

37. ***Carex nigra*** (L.) Reichard, Fl. moeno-francof. II (1778), 96; Kuzeneva in Fl. Murm. II, 80, map 22; Böcher & al., Grønl. Fl. 260; Polunin, Circump. arct. fl. 90, 113.

C. acuta L., Sp. pl. (1753) 978 (var. α *nigra*); Krechetovich in Fl. SSSR III, 207; Mackenzie in N Amer. Fl. XVIII, 6, 388.

C. fusca Bell. ex All., Fl. Pedem. II (1785), 269.

C. Goodenoughii Gay in Ann. Sci. Nat., 2 sér., IX (1839), 191; Kükenthal, Cyper. Caricoid. 313; id., Cyper. Sibir. 84; Krylov, Fl. Zap. Sib. III, 470; Hultén, Atlas, map 361.

C. vulgaris Fries, Nov. Fl. Suec. Mant. III (1842), 153; Treviranus in Ledebour, Fl. ross. IV, 311; Perfilev, Fl. Sev. I, 146.

C. vulgaris **juncella* Fries in Bot. Notis. (1843), 105.

C. juncella (Fries) Th. Fries in Bot. Notis. (1857), 207; Krechetovich, l. c. 208.

Ill.: Fl. Murm. II, pl. XXII.

One of the commonest sedges of the forest zone of Eurasia and the far east of North America. Often growing in large masses. Just penetrating arctic limits.

Growing in grassy or sphagnous marshes, at moist or marshy sites, and on shores of waterbodies.

An extremely variable species, which has occasioned the description of numerous varieties and forms. One of the forms, possessing tall slender culms and relatively longer leaves, was described under the name *C. juncella* (Fries) Th. Fries and has been accepted as a species by certain authors. In our opinion this form should not be considered a separate species because, apart from these differences in habit, it possesses no other distinguishing features when compared with the typical form of *C. nigra*. The tall and short forms of *C. nigra* are connected by an endless host of intermediates and constitute ecological forms.

The record of the discovery of this species at Dudinka (Scheutz, Pl. jeniss.) is clearly erroneous and refers, in all probability, to *C. wiluica* Meinsh.

Soviet Arctic. Murman (western part); SE Bolshezemelskaya Tundra.

Foreign Arctic. Southern Greenland; ? Iceland; Arctic Scandinavia.

Outside the Arctic. Europe, West Siberia (to 66°15′N in the lower reaches of the Ob); extreme eastern part of North America north of 42°N.

38. ***Carex wiluica*** Meinsh. in Maack, Vilyuysk. okr. Yakutsk. obl. II (1886), 308; Kükenthal, Cyper. Caricoid. 325; id., Cyper. Sibir. 86; Andreyev, Mat. fl. Kanina 162; Perfilev, Fl. Sev. I, 118; Krechetovich in Fl. SSSR III, 208; Leskov, Fl. Malozem. tundry 41; Karavayev, Konsp. fl. Yak. 66; Polunin, Circump. arct. fl. 90.

C. vulgaris **zonata* Nyl., Spicil. pl. Fenn. (1843), 19.

C. caespitosa ssp. *wiluica* Kryl., Fl. Zap. Sib. III (1929), 468.

C. wiluica ssp. *europaea* Gorodk. in Zhurn. Russk. bot. obshch. 7 (1922), 230.

C. juncella auct. non Th. Fries — Kuzeneva in Fl. Murm. II, 80, pro parte.

Boreal species distributed mainly in the European part of the USSR and in Siberia west of the Lena. It occurs with moderate frequency, mainly in southern districts of the Arctic, in the forest-tundra zone. Growing in moss-sedge fens, in

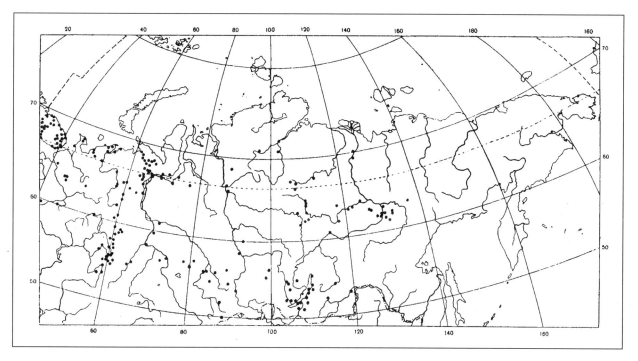

MAP III–31 Distribution of *Carex wiluica* Meinsh.

marshy willow carr and sparsely treed tundra, and on the shores of rivers and lakes. Forming large tussocks.

Many arctic specimens of *C. wiluica*, especially from the European part of the Arctic, differ from the typical form in having broader leaves [2.5–3(3.5) mm wide instead of 1–2 mm], as well as sometimes by the presence of phyllopodia. At the base of the shoots of these plants there are only a small number of short scalelike reddish-brown sheaths, while the remaining (similarly coloured) sheaths are prolonged as leaf blades. The proportion of sheaths of the two types varies, so a sharp distinction cannot be drawn between phyllopodic and aphyllopodic specimens. Nevertheless a tendency towards phyllopody in individuals of *C. wiluica* from the arctic part of the range is evident. B.N. Gorodkov (l. c.) separated phyllopodic forms of *C. wiluica* as ssp. *europaea*, which "differt a typo culmis phyllopodis, vaginis foliorum vix fibrillosis." There is a certain link between phyllopody and relatively broad leaves. But not all broad-leaved forms are phyllopodic. We can speak of the predominant occurrence of such forms in the Arctic, but should refrain from treating them as a subspecies.

On account of frequent confusion of *C. wiluica* with *C. nigra* (L.) Reichard [= *C. juncella* (Fries) Th. Fries] the distribution of *C. wiluica* in Scandinavia remains unclear, although there are reliable records of its presence in Northern Sweden.

Soviet Arctic. Murman (everywhere); Kanin (Shomokhovskiye Hills and River Khalesed-Yaga); Timanskaya Tundra (River Belaya); Malozemelskaya Tundra (Bolvanskaya Hill and River Sula); lower reaches of the Pechora (River Yushina); Bolshezemelskaya Tundra; middle course of the Kara; Polar Ural (frequent); lower reaches of the Ob (Salekhard and islands in the Ob Delta); Nyda; lower reaches of Yenisey (occasional); River Kheta (at 70 km below the Volochanka). (Map III–31).

Foreign Arctic. ? Arctic Scandinavia.

Outside the Arctic. Swedish Lappland; European part of the USSR (south to the latitude of Kuybyshev, mainly in the northern and eastern parts); throughout the

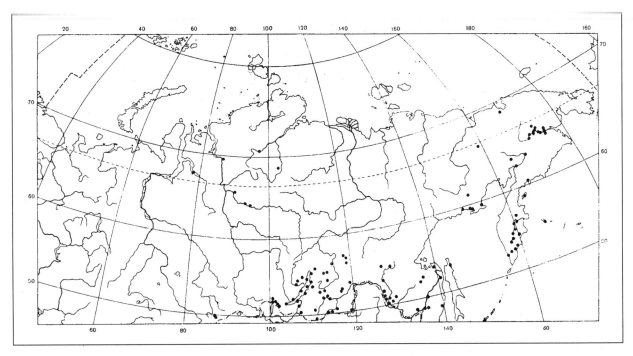

Map III–32 Distribution of *Carex appendiculata* (Trautv. et C.A. Mey.) Kük.

Urals; Altay; Siberia west of the Lena; Transbaikalia; occasionally occurring between the Lena and the Amga.

39. *Carex appendiculata* (Trautv. et C.A. Mey.) Kük. in Bull. Herb. Boiss., 2 sér., IV (1904), 54; id., Cyper. Caricoid. 338; id., Cyper. Sibir. 91; Krechetovich in Fl. SSSR III, 214; Karavayev, Konsp. fl. Yak. 66.

C. acuta var. *appendiculata* Trautv. et C.A. Mey., Flor. ochot. phaen. (1856), 100.
C. gracilis auct. non Curt. — Ostenfeld, Fl. arct. 1, 72, pro parte.
Ill.: Russk. bot. zhurn. 3/6 (1911), fig. 72.

Boreal species widespread in the forest zone of East Siberia and the Far East. Penetrating the temperate part of the Arctic to a limited extent, where it occurs in river valleys. Growing among floodplain shrubbery (often in willow thickets), and on moist or marshy shores of rivers and lakes; also penetrating lowland tundra. Sometimes forming large tussocks.

Arctic specimens of *C. appendiculata* differ from typical specimens in their shorter growth, smaller pistillate spikelets (sometimes no more than 1 cm long), and darker scales of the pistillate spikelets (sometimes completely black, without pale stripe along the midrib).

Carex appendiculata replaces *C. wiluica* east of the Lena, and shows very close relationship to that species. Typical specimens of both species are well distinguished from one another by a series of rather constant characters, but in districts where the species grow together (in Prebaikalia and between the Lena and the Amga) there occur intermediate specimens that combine the characters of the two species in various ways. These specimens possibly represent forms of hybrid origin.

Soviet Arctic. Mouth of River Taz; lower reaches of Yenisey (Dudinka); right bank of the Kheta (at 10 km below Boyarka); lower reaches of the Kolyma (mouth of the Oloy); district of Chaun Bay (Baranikha); Anadyr Basin (Rivers Belaya, Tanyurer

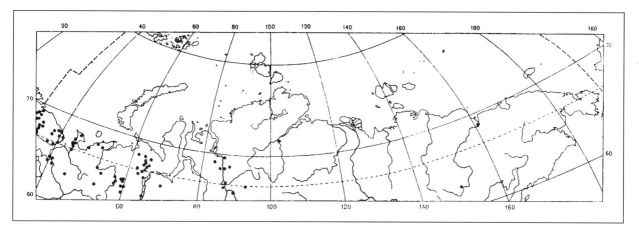

MAP III–33 Distribution of *Carex caespitosa* L.

and Volchikha) and Penzhina Basin (upper course of the Oklan, near Penzhino village, and 12 km above Kamenskoye village); Bay of Korf. (Map III–32).

Foreign Arctic. Not occurring.

Outside the Arctic. Siberia, mainly east of the Lena, west of the Lena only in the basin of the Lower Tunguska and in Altay (Kosh-Agach); Amur Basin; Primorskiy Kray; Sakhalin; Kamchatka; Kurile and Commander Islands; Northern Mongolia; NE China.

40. ***Carex caespitosa*** L., Sp. pl. (1753), 978; Treviranus in Ledebour, Fl. ross. IV, 310; Kükenthal, Cyper. Caricoid. 328 (excl. var. *minuta*); Krylov, Fl. Zap. Sib. III, 467, pro parte; Perfilev, Fl. Sev. I, 117; Krechetovich in Fl. SSSR III, 217; id. in Mat. ist. fl. rastit. I, 161, map 11; Leskov, Fl. Malozem. tundry 35; Hultén, Atlas, map 359; Kuzeneva in Fl. Murm. II, 84, map 25; Polunin, Circump. arct. fl. 90, 102.
   C. rubra Lévl. et Vaniot in Bull. Acad. Intern. Géogr. Bot. XIX (1909), 33; Krechetovich, l. c. 218; Karavayev, Konsp. fl. Yak. 66.
   Ill.: Fl. SSSR III, pl. XV, fig. 6.

   Rather common plant of the forest zone of Eurasia, also occurring (but infrequent) in temperate districts of the Arctic. Growing in river valleys in riparian marshy shrubbery, in willow tundras, and in marshy open forests. Creating tussock formations, which sometimes occupy rather extensive areas.

   Soviet Arctic. Murman; Kanin; Malozemelskaya Tundra (mainly in the southern part); Eastern Bolshezemelskaya Tundra; Polar Ural; lower reaches of the Ob and Yenisey. (Map III–33).

   Foreign Arctic. Arctic Scandinavia.

   Outside the Arctic. Western Europe; European part of the USSR (except the most southern districts); Northern Caucasus and Western Transcaucasus; West Siberia; western part of Central Siberian Plateau; Central Yakutia (north of Olekminsk); Sayans; Prebaikalia; Amur Basin; Primorskiy Kray; coasts of Sea of Okhotsk; Kazakhstan (northern and eastern parts); Northern Mongolia (very rare); Sakhalin; Japan; NE China.

41. ***Carex Schmidtii*** Meinsh. in Baer & Helmersen, Beitr. zur Kenntn. russ. Reich. XXVI (1871), 224; id. in Acta Horti Petropol. XVIII, 340; Kükenthal, Cyper. Caricoid. 326; id., Cyper. Sibir. 88; Krechetovich in Fl. SSSR III, 223; Karavayev, Konsp. fl. Yak. 66.
    C. Maximowiczii Fr. Schmidt, Reis. Amurl. ins. Sachal. (1868) 71; non Miq. (1866).
    C. lineolata Cham. ex V. Krecz. in Fl. SSSR III, 223, 598; Meinshausen in Acta Horti Petropol. XVIII, 228 (as synonym under *C. Schmidtii*).

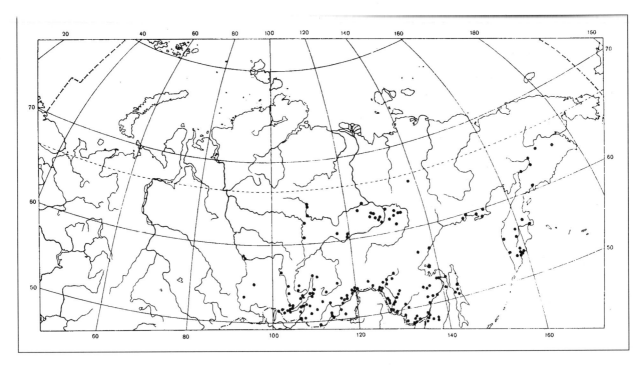

Map III–34 Distribution of *Carex Schmidtii* Meinsh.

Ill.: Kükenthal, Cyper. Caricoid., fig. 50; id., Cyper. Sibir., fig. 70.

East Siberian boreal species just penetrating arctic limits in the northeast of its range. Growing in moist or marshy meadows, in marshes, in floodplain shrubbery in river valleys, and on the edges of oxbows.

According to V.I. Krechetovich (l. c.), *C. Schmidtii* does not occur in the Arctic. For the Anadyr Basin he reported another species, *C. lineolata* Cham. ex V. Krecz., whose range also includes Kamchatka.

Chamisso denoted with the name *C. lineolata* ("lined") plants collected by him in Kamchatka. Their perigynia are brown spotted, something occurring not at all uncommonly in *C. Schmidtii*. Study of extensive material of *C. Schmidtii* and of the specimens determined by V.I. Krechetovich as *C. lineolata* has shown that both species are identical. Contrary to V.I. Krechetovich's statement, the perigynia in *C. lineolata* are just as inflated as in *C. Schmidtii*. Elliptical or ovoid perigynia, as allegedly distinguishing *C. lineolata* from *C. Schmidtii*, are observed throughout the range of the latter, though more rarely than globose-ovoid perigynia. In the Anadyr Basin plants with elliptical or ovoid perigynia are just as rare as in other parts of the range of *C. Schmidtii*. Emarginate beaks of the perigynia are also not peculiar to Anadyr plants; the form of the beak varies throughout the range of *C. Schmidtii* from almost entire to shallowly bidentate, with plants possessing perigynia with emarginate or bidentate beaks occurring considerably more frequently than those with entire beaks.

Soviet Arctic. Anadyr Basin; Penzhina Basin and Bay of Korf. (Map III–34).
Foreign Arctic. Absent.
Outside the Arctic. Widespread in East Siberia south of the Arctic Circle, Preamuria, Primorskiy Kray, Northern Kamchatka, the Kurile Islands, Sakhalin, Hokkaido, the north of the Korean Peninsula, Northern Mongolia, and NE China.

42. ***Carex aquatilis*** Wahlb. in Kongl. Vet. Akad. Nya Handl. XXIV (1803), 165; Treviranus in Ledebour, Fl. ross. IV, 312 (excl. var.); Ostenfeld, Fl. arct. 1, 70, pro parte;

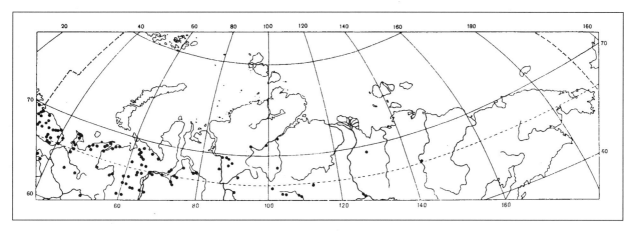

MAP III–35 Distribution of *Carex aquatilis* Wahlb.

Kükenthal, Cyper. Caricoid. 309, pro parte; id., Cyper. Sibir. 83; Krylov, Fl. Zap. Sib. III, 463, pro parte; Perfilev, Fl. Sev. I, 115, pro parte; Mackenzie in N Amer. Fl. XVIII, 7; Krechetovich in Fl. SSSR III, 229; Leskov, Fl. Malozem. tundry 35, pro parte; Hultén, Fl. Al. II, 338, pro parte; id., Atlas, map 366, pro parte; Kuzeneva in Fl. Murm. II, 86, map 27; Karavayev, Konsp. fl. Yak. 67; Polunin, Circump. arct. fl. 90, 100, pro parte.

Ill.: Fl. Murm. II, pl. XXIV.

Species characteristic of the taiga zone of Eurasia and North America, but also penetrating the forest-tundra and tundra, where it occurs in more favourable protected habitats.

Growing in watermeadows on floodplains, on shores of waterbodies (sometimes in the water), in marshy places and fens; often forming pure beds.

Soviet Arctic. Murman (throughout); Kanin; Timanskaya and Malozemelskaya Tundras; lower reaches and mouth of the Pechora; Bolshezemelskaya Tundra; Karskaya Tundra (River Nenzi-Yaga); Polar Ural; lower reaches and delta of the Ob; Southern Yamal (Nakhodka Bay, River Yada); Obsko-Tazovskiy Peninsula (Nyda settlement, River Yepoko, and near the mouth of the River Taz); lower reaches of the Yenisey (Dudinka and River Solenaya); forest-tundra on the southern boundary of the Taymyr Peninsula (Volochanka and the mouth of the River Medvezhya); Khatanga. (Map III–35).

Foreign Arctic. Iceland; Arctic Norway (Tromsö); probably also occurring in other districts of Arctic Scandinavia.

Outside the Arctic. Scandinavia; Finland; Ireland; England; the north of the European part of the USSR (approximately to the latitude of Kazan and Zlatoust); West Siberia; Central Siberian Plateau; the sources of the Yenisey; Baykal; Central Yakutia; northern part of the Verkhoyansk Range; Southern Canada and the Atlantic states of the USA.

43. **Carex stans** Drej., Revis. Car. bor. (1841), 40; Andreyev, Mat. fl. Kanina 162; Krechetovich in Fl. SSSR III, 230; Kuzeneva in Fl. Murm. II, 88, map 28; A.E. Porsild, Ill. Fl. Arct. Arch. 53, map 86; Karavayev, Konsp. fl. Yak. 67; Böcher & al., Grønl. Fl. 259.

C. aquatilis var. β et γ Trev. in Ledebour, Fl. ross. IV (1852), 312, pro parte.

C. aquatilis var. *stans* (Drej.) Boott, Illustr. Carex IV (1867), 163; Ostenfeld, Fl. arct. 1, 70; Kükenthal, Cyper. Caricoid. 311; id., Cyper. Sibir. 84; Perfilev, Fl. Sev. I, 116; Hultén, Fl. Al. II, 339; Tolmachev, Mat. fl. Mat. Shar 286; id., Fl. Taym. I, 102.

C. aquatilis auct. non Wahlb. — Krylov, Fl. Zap. Sib. III, 464, pro parte; Perfilev, Fl. Sev.

I, 116, Mackenzie in N Amer. Fl. XVIII, 7, 396, pro parte; Leskov, Fl. Malozem. tundry 35; Polunin, Circump. arct. fl. 90, 100, pro parte.

C. subspathacea auct. non Wormsk. — Krylov, l. c. 474.

Ill.: Fl. Murm. II, pl. XXIV.

Predominantly arctic species replacing the lowland marsh species *C. aquatilis* Wahlb. in the majority of arctic districts, as well as in the Verkhoyansk-Kolymsk mountain country and in the barrens of the Urals and Transbaikalia. In the arctic region *C. stans* is characteristic of districts with more temperate climate, although it ranges north to the arctic tundra subzone (inclusively). Normally absent from truly high arctic floras. In marshy lowland tundras of the arctic coast of Eurasia this is one of the most abundant plants, which cloaks the landscape and also possesses considerable importance as fodder for grazing animals.

Carex stans is most typical of tundra *Hypnum* fens at low or intermediate elevation (containing sedges or sedges and cotton grasses, together with *Calliergon* and *Drepanocladus* spp., etc.), but it is also very common in moist sedge-moss or moss-sedge tundras (with *Tomenthypnum nitens* and *Aulacomnium palustre*) developed on the slopes of poorly drained loamy areas, where it shows less vigour and forms a sparse cover. The species is not characteristic of oligotrophic conditions.

Carex stans is a typical rhizomatous plant; its long rhizomes grow sympodially and branch where they break through to the soil surface, forming there so-called tillering nodes; after prolonged tillering in the flarks of polygonal fens, *C. stans* forms small tussocks. The rhizomatous clones of this species attain large size; reproduction by seed is of subordinate importance, and can be absent in more northern districts. In lowland marshes the species is very often dominant in the herb cover, forming a closed layer; no less frequently *C. stans* participates in mixed herb stands with rhizomatous species of cotton grass (*Eriophorum angustifolium*, *E. medium* and *E. russeolum*, more rarely *E. Scheuchzeri*; at the bottom of gullies dominance often passes to *E. angustifolium*). The tips of the leaves of *C. stans* in the herb stands of lowland marshes often become completely dry and whitened (on account of spring and early summer frosts?).

Carex stans is normally associated with peaty or peaty-clayey soils of tundra fens, also with tundra clay soils (with weak peat horizon) developing on loam, alluvial sandy loam or sand with surplus moisture. It is sensitive to the degree of soil moisture, but not to moisture flow. Growing where there is adequate winter snow cover.

In the more southern subzones of the tundra zone, beds of *C. stans* frequently contain dwarf birch and low shrub willows (*Salix phylicifolia*, *S. pulchra*, *S. glauca*, *S. lanata*, *S. reptans*).

Carex stans is extremely closely related to *C. aquatilis*. The morphological differences between these species are purely quantitative and not of an absolute nature. *Carex stans* differs from *C. aquatilis* mainly in the smaller size of the plant as a whole, shorter inflorescences (due to aggregation of the spikelets), shorter spikelets, and darker scale colour. Additionally, in *C. stans* the leaves are more or less yellowish and sometimes folded lengthwise, distinctly shorter than the culm, and the perigynia normally brownish distally; while in *C. aquatilis* the leaves are green or greyish green, flat, as long as or slightly longer than the culm, and the perigynia always pale. Typical specimens (those most frequently encountered) of *C. stans* sharply differ in habit from typical *C. aquatilis*. The former possess inflorescences mostly 7–12 cm long and pistillate spikelets mostly 3–3.5 cm long; the latter inflorescences mostly 22–27 cm long and pistillate spikelets mostly 4–6 cm long. But specimens of *C. aquatilis* sometimes occur outside the Arctic with short inflorescences (14–15 cm or even only 10 cm long). Specimens of *C. aquatilis* with long inflorescences may also possess small pistillate spikelets (2–3 cm long). On

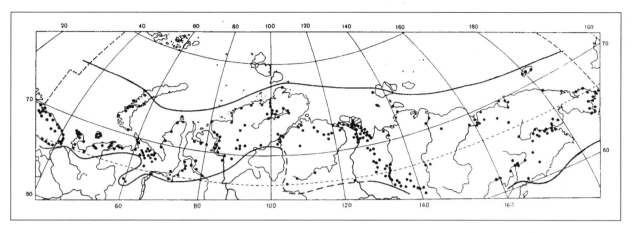

MAP III–36 Distribution of *Carex stans* Drej.

the other hand, long pistillate spikelets (up to 5 cm long) are observed in some specimens of *C. stans*.

In certain arctic districts of the European part of the USSR, West Siberia and parts of East Siberia (as far as the Lena), both *C. stans* and *C. aquatilis* are found. Transitional forms occur as well as typical specimens of both species. We have noted such transitional forms from Murman, Northern Kanin, the Malozemelskaya, Timanskaya and Bolshezemelskaya Tundras, the islands of Kolguyev and Vaygach, the Polar Ural, and the lower reaches of the Yenisey (Dudinka) and of the Lena (Bulun district). Sometimes transitional specimens tend more towards one or other of the stated species, but in the majority of cases it is impossible to determine to which species such specimens belong.

Because of the absence of a clear morphological boundary between *C. aquatilis* and *C. stans*, some authors have not recognized the latter as a separate species but have either treated it as a variety of *C. aquatilis* or not distinguished it at all. But study of an enormous quantity of herbarium material from such a large territory as the arctic districts of the USSR has convinced us that *C. stans* exists as a distinct arctic race, whose range is to a considerable extent different from the range of *C. aquatilis*. The existence of transitional forms and the presence of only quantitative differences between *C. aquatilis* and *C. stans* attest that these are subspecies of a single species (*C. aquatilis* s. l.). *Carex stans* is readily distinguished from *C. Bigelowii*, with which it is sometimes confused, by having the bract of the lowest-but-one spikelet equalling or only just shorter than the inflorescence; in *C. Bigelowii* only the lowest bract occasionally equals the inflorescence, while the bract of the lowest-but-one spikelet only just equals its spikelet. Furthermore, the beak of the perigynia is thickened like a callus in *C. stans*. *Carex stans* differs from *C. subspathacea* in lacking a mucro on the scales of the staminate spikelets, in having perigynia which abruptly (not gradually) taper to a beak, and in having more numerous leaves on the culms (up to 15 or more; in *C. subspathacea* there are no more than 6–8).

Soviet Arctic. Murman (everywhere); Kanin; Timanskaya and Malozemelskaya Tundras; Kolguyev Island; lower reaches of the Pechora, Bolshezemelskaya Tundra; Polar Ural; Yugorskiy Peninsula; Vaygach Island; South Island and the southern part of the North Island of Novaya Zemlya; Yamal; Belyy Island; Obsko-Tazovskiy Peninsula; Gydanskaya Tundra; lower reaches of the Yenisey (from the forest-tundra to its mouth), coast and islands of Yenisey Bay; Taymyr; Khatanga Basin; basins of the Anabar and Olenek; lower reaches and delta of the Lena; Buorkhaya Bay; Cape Svyatoy Nos; lower reaches of the Yana, Indigirka and

Kolyma, New Siberian Islands, Wrangel Island, Chukotka (from Ayon Island and Chaun Bay to Kolyuchin Bay and the Bay of Krest); Arakamchechen and Ratmanov Islands; Anadyr and Penzhina Basins. (Map III–36).

Foreign Arctic. Arctic part of Alaska; arctic coast of Canada; Labrador; Canadian Arctic Archipelago; west and east coasts of Greenland between the Arctic Circle and 80°N; Arctic Scandinavia.

Outside the Arctic. Northern Scandinavia; Kola Peninsula; Solovetskiye Islands; Karelian Coast (Shuya); Prepolar and Northern Urals; basin of the Upper Angara; Transbaikalia; Verkhoyansk Range; Aldan Mountains; coast of the Sea of Okhotsk north of Ola; Commander and Aleutian Islands; Hudson Bay and the NE part of Canada.

44. *Carex Bigelowii* Torr. ex Schwein in Ann. Lyc. Nat. hist. N.Y. I (1824), 67; Hultén, Fl. Al. II, 334, pro parte; id., Atlas, map 364; Kuzeneva in Fl. Murm. II, 85; A.E. Porsild, Ill. Fl. Arct. Arch. 52, map 84; Polunin, Circump. arct. fl. 90, 101; Böcher & al., Grønl. Fl. 259.

? *C. concolor* R. Br. in Suppl. App. Parry's Voyage XI (1824), 283; Mackenzie in N Amer. Fl. XVIII, 6, 378.

? *C. rigida* Good. in Trans. Linn. Soc. II (1794), 193; Ostenfeld, Fl. arct. 1, 77, pro parte; Kükenthal, Cyper. Caricoid. 299, pro parte; Hultén, Circump. pl. I, 50 (pro parte), map 43; non Schrank (1789).

C. rigida var. *inferalpina* Laest. in Nov. Acta Regiae Soc. Sci. Upsal. XI (1839), 287.

C. hyperborea Drej., Revis. Car. bor. (1841), 43; Krechetovich in Fl. SSSR III, 226.

C. rigida var. *concolor* (R. Br.) Kük., l. c. 301, pro parte.

C. rigida ssp. *inferalpina* (Laest.) Gorodk. in Zhurn. Russk. bot. obshch. XV (1930), 181, pro parte (quoad typ.).

Ill.: Fl. Murm. II, pl. XXIII.

Amphiatlantic arctic species. Growing in mountain and lowland tundras and in forest-tundra; extremely common in dwarf shrub-lichen tundras and in glades in birch woods (*Betula tortuosa*), also growing on moist shores of rivers and lakes and in shrub thickets.

Occurring within the Soviet Arctic only in Murman; east of the Kola Peninsula replaced by *C. ensifolia* ssp. *arctisibirica*. Plants approaching *C. ensifolia* ssp. *arctisibirica* in several characters already occur among material from Murman (Teriberka, Iokanga village, Kildin Island), along with typical specimens of *C. Bigelowii* which are completely identical with specimens from Eastern North America (whence this species was described). Such plants have the upper spikelets closely aggregated, completely black spikelet scales (without pale midrib), and perigynia with a sharply differentiated short beak. Specimens tending towards ssp. *arctisibirica* also occur in Iceland, Scotland and Scandinavia. The identification of *C. Bigelowii* with *C. concolor* R. Br. described from Melville Island is apparently incorrect.

Carex Bigelowii varies in the size of the plant as a whole, the length of the lowest bract, the degree of separation of the spikelets, the number of pistillate spikelets, and the length of the pedicel of the lowest pistillate spikelet; the scales are on average equal to the perigynia in length and width and mostly rounded apically, but may be slightly longer or shorter than the perigynia.

Soviet Arctic. Murman (everywhere).

Foreign Arctic. Eastern part of the arctic coast of Canada and of the Canadian Arctic Archipelago; Labrador; Greenland (almost to 80°N); Iceland; Spitsbergen; Faroe Islands; Arctic Scandinavia.

Outside the Arctic. Distributed in Canada (Quebec, Manitoba, Newfoundland), the mountains of the NE United States, Great Britain (Scotland and Wales), Ireland, the northern part of Scandinavia, and the Kola Peninsula.

45. *Carex ensifolia* (Turcz. ex Gorodk.) V. Krecz. ssp. ***arctisibirica*** Jurtz. in Nov. sist. vyssh. rast. (1965) 308.

C. rigida auct. non Good. — Ostenfeld, Fl. arct. 1, 77, pro parte; Scheutz, Fl. jeniss. 180.

C. rigida var. *concolor* (R. Br.) Kük., Cyper. Caricoid. (1909) 301 (respecting Siberian plants); Krylov, Fl. Zap. Sib. III, 460.

C. rigida f. *infuscata* auct. non Drej. — Kükenthal, Cyper. Sibir. 82, pro parte.

C. hyperborea auct. non Drej. — Leskov, Fl. Malozem. tundry 37; Krechetovich in Fl. SSSR III, 226, pro parte; Karavayev, Konsp. fl. Yak. 66.

C. rigida var. *typica* Kryl., l. c. 462; Andreyev, Mat. fl. Kanina 161.

C. rigida ssp. *inferalpina* (Laest.) Gorodk. in Zhurn. Russk. bot. obshch. XV (1930), 181, pro parte (excl. typ.); Perfilev, Fl. Sev. I, 115; Tolmachev, Fl. Taym. I, 103; id., Mat. fl. Mat. Shar 287.

One of the plants cloaking the landscape in Arctic Siberia, also playing a major role in the vegetational cover of subarctic barrens; in the High Arctic it loses its importance, being absent from many of the Siberian polar islands or occurring very rarely. The region where *C. ensifolia* ssp. *arctisibirica* is widespread on the watersheds lies between the tundras of Kanin and those of the Indigirka Basin; further east this race is replaced by the very closely related species *C. lugens* Holm.

The southern typical race of *C. ensifolia* (ssp. *ensifolia*) is a common and abundant plant on the barrens of Southern Siberia (Transbaikalia, Khamar-Daban, Sayans, Tuva, Altay) and Northern Mongolia.

In the subalpine zone of mountains of the Verkhoyansk-Kolymsk country the predominantly arctic race of *C. ensifolia* (ssp. *arctisibirica*) is characteristic of eutrophic moss-sedge fens at the bottom of mountain valleys: these commonly develop a low shrub layer of dwarf birch (*Betula exilis*), small-leaved rhododendron and willows, sometimes with the presence of *Cassiope tetragona*, *Salix reticulata*, *Tofieldia coccinea*, etc.; the most abundant mosses are *Tomenthypnum nitens*, *Aulacomnium* spp., etc.

In the Arctic *C. ensifolia* ssp. *arctisibirica* spreads all over the watersheds. So-called mossy tundras are very often vegetated by it to some degree. [In the majority of geobotanical works it is referred to as *C. rigida* Good., *C. rigida* ssp. *inferalpina* (Laest.) Gorodk. or *C. hyperborea* Drej.]. This sedge is rather characteristic for northern variants of placorn tundras with mosses and cotton grass tussocks in the lower reaches of the Olenek, Lena, Omoloy, etc.; east of the Indigirka (under analogous conditions) it is replaced by *C. lugens*, which forms dense tufts.

Sometimes *C. ensifolia* ssp. *arctisibirica* occurs in association with fruticose lichens (in hummocky alpine *Cetraria* tundras with *C. islandica*, *C. crispa*, *C. nivalis* and *C. cucullata*, and *Alectoria* tundras with *Alectoria ochroleuca*, *Cornicularia divergens*, etc.) or in association with mesophilic mosses (*Hylocomium splendens* var. *alascanum*, *Ptilidium ciliare*, etc.). But it is most typically associated with hygrophilic eutrophic mosses (*Tomenthypnum nitens*, *Aulacomnium turgidum*, *Au. palustre*, *Drepanocladus uncinatus*) in moist patchy tundras composed of mosses and sedges or dwarf willow with mosses and sedges. The snow cover here in wintertime may be deep or extremely thin in response to microrelief caused by frost and solifluction (in the form of convex bare patches of various shapes separated by vegetated hollows, or in the form of solifluction terraces). Fragments of moss-sedge turf (with *C. ensifolia* ssp. *arctisibirica*) are confined in these tundras to depressions of the microrelief; the more convex, better drained areas of microrelief (edges of bare patches, upper slopes of hummocks, etc.) are occupied by mats of *Dryas punctata*, etc. Often embedded in the moss-sedge turf are mats of prostrate willows (most frequently *Salix reticulata*), arctic-alpine mixed herbage, and certain other sedge species (*Carex misandra*, *C. atrofusca*, *C. fuscidula*, *C. vaginata*, etc.); in the southern tundra

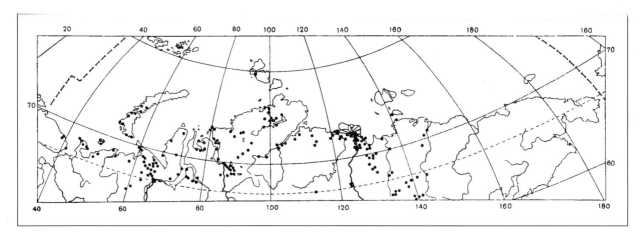

Map III–37 Distribution of *Carex ensifolia* (Turcz. ex Gorodk.) V. Krecz. ssp. *arctisibirica* Jurtz.

subzone a sparse layer of low shrub willows and dwarf birch is sometimes developed here.

In more windswept, better drained places with patchy *Dryas*-moss tundras this species grows less vigorously. It is rather characteristic of fissured *Dryas* tundras on fine (usually sandy) soil on the south-facing slopes of valleys or mounds, where good drainage and deep summer thawing of the ground is combined with adequate winter snow cover. Sometimes the plant can also be found in moister variants of *Cassiope*-moss tundra (the *Cassiope tetragona-Tomenthypnum nitens* association). In a less typical form (sometimes transitional to *C. rigidioides*) it also occurs in subarctic open forests with dwarf shrubs and mosses in East Siberia, especially in districts of surface exposure of calcareous rocks.

Carex ensifolia ssp. *arctisibirica* almost always grows as extensive rhizomatous clones with a rather dense arrangement of aerial shoots.

The variability of this species is very great. In particular, plants from more southern tundra (and forest-tundra) sometimes possess relatively broad, more or less flat leaves and larger, more widely separated pistillate spikelets with paler coloured perigynia.

Carex ensifolia ssp. *arctisibirica* is morphologically very weakly differentiated from *C. ensifolia* ssp. *ensifolia*; the differences between them are purely quantitative. The former differs in having narrower leaves, more slender culms, a sessile staminate spikelet or with pedicel not more than 4 mm long (in ssp. *ensifolia* the pedicel length fluctuates within a range of 0–9 mm), and smaller perigynia.

Plants from Central Europe referred by many authors to *C. rigida* Good. are indistinguishable from the Siberian ssp. *ensifolia*; since the name *C. rigida* Good. is unavailable as a subsequent homonym, the Central European *C. rigida* should be named *C. ensifolia* ssp. *ensifolia*.

Thus, *C. ensifolia* s. l. is an arctic-alpine species, the southern part of whose range is rather highly fragmented.

Carex rigida ssp. *altaica* Gorodk. is in our opinion very close to ssp. *ensifolia*.

Soviet Arctic. ? Murman (Iokanga); Kanin; Timanskaya and Malozemelskaya Tundras; lower reaches of Pechora; Bolshezemelskaya Tundra; Polar Ural; Pay-Khoy; Vaygach Island; Novaya Zemlya (South Island and southern part of North Island); Yamal; lower reaches of Ob; Obsko-Tazovskiy Peninsula; Gydanskaya Tundra; lower reaches of Yenisey; Taymyr; Severnaya Zemlya; Khatanga Basin; lower reaches of the Anabar and Olenek; lower reaches and delta of the Lena; Buorkhaya Bay; lower reaches of the Yana and Indigirka. (Map III–37).

Foreign Arctic. Occasionally occurring in a not entirely typical form in Arctic Scandinavia and in Iceland (see commentary on *C. Bigelowii*).

Outside the Arctic. Upper reaches of Pechora; Urals; NW edge of Central Siberian Plateau; Verkhoyansk Range; Cherskiy Range.

46. ***Carex rigidioides*** (Gorodk.) V. Krecz. in Fl. SSSR III (1935), 228; Karavayev, Konsp. fl. Yak. 66.

   C. rigida Good. ssp. *rigidioides* Gorodk. in Zhurn. Russk. bot. obshch. XV (1930), 182.

   East Siberian species of suboceanic barrens, just penetrating the limits of the Arctic where it occurs only in the district of the northern extremity of the Verkhoyansk Range and adjacent mountain districts to the west. Most characteristic of dry stony or rocky mountain slopes and the lower summits; here it grows most frequently in the company of fruticose lichens (species of *Cladonia*, *Cetraria*, *Alectoria* and *Cornicularia*) or among isolated patches of flowering plants scattered among boulders and rock debris. Under these conditions *C. rigidioides* forms small tufts with a very short, densely branching rhizome and dense array of aerial shoots. Less characteristic of alpine *Dryas* tundras.

   Additionally, *C. rigidioides* occurs in mountain sphagnum bogs and in open forests with sphagnum or sphagnum and green mosses; here the rhizomes become longer and the aerial shoots are more loosely arranged. Sometimes *C. rigidioides* grows in dwarf birch carr in valleys.

   In the tundra zone proper, *C. rigidioides* normally gravitates towards dwarf shrub-*Cladonia* or dwarf shrub-*Alectoria* communities (with blueberry, cranberry, crowberry, alpine bearberry, dwarf birch) on well drained slopes with adequate winter snow cover. In the typical tundra subzone this species can sometimes also be found in *Cassiope-Cladonia* tundras; thus, on the Olenek slope of Chekanovskiy Ridge the species occurring in *Cassiope* tundras is almost exclusively *C. rigidioides*, while *C. ensifolia* ssp. *arctisibirica* grows here in areas with surplus moisture in patchy moss-sedge tundras.

   Carex rigidioides is closely related to *C. ensifolia* ssp. *arctisibirica*. At a series of sites (in the Olenek Basin, the lower reaches of the Lena, and the mountain system of the Verkhoyansk Range south to the Suntar-Khayata Range), the two species occur sympatrically but occupy different habitats. In these districts individuals of a transitional character, tending towards one of the stated species, also occur.

   Within the remainder of its range, *C. rigidioides* shows little variation.

   Soviet Arctic. Lower reaches of the Olenek and Lena; Buorkhaya Bay (Tiksi Bay).

   Foreign Arctic. Not occurring.

   Outside the Arctic. Eastern part of Central Siberian Plateau (basin of the River Markha); Verkhoyansk Range; Cherskiy Range; the sources of the Kolyma; Dzhugdzhur Range; Aldan Mountains (basin of the River Timpton); the system of the Stanovoy Range (all the way to Northern Preamuria and Primorskiy Kray).

47. ***Carex lugens*** H.T. Holm in Amer. Journ. Sci., ser. IV, X (1900), 269, fig. A; Hultén, Fl. Al. II, 335, pro parte (excl. *C. Soczavaeana* Gorodk.), map 271, 272; A.E. Porsild, Ill. Fl. Arct. Arch. 53, map 85; Polunin, Circump. arct. fl. 90, 103.

   C. rigida Good. f. *infuscata* auct. non Drej. — Kükenthal in Russk. Bot. zhurn. 3–6 (1911), 82, pro parte.

   C. Soczavaeana auct. non Gorodk. — Krechetovich in Fl. SSSR III (1935), 219 (respecting American plants).

   Beringian arctic species.

   Growing in various types of tundras (sedge-cottongrass, sedge-moss, lichen-moss, *Dryas*, etc.), as well as on moist shores of rivers and lakes.

   Carex lugens is very close to *C. ensifolia* ssp. *arctisibirica*. The most substantial difference between them lies in the nature of their tuft structure. *Carex lugens* is a

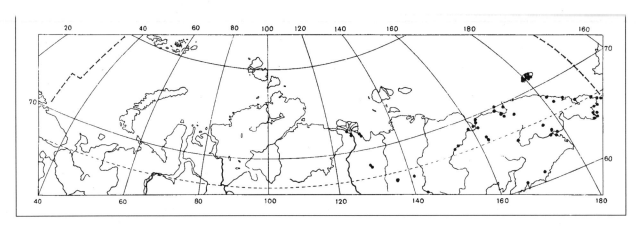

MAP III–38 Distribution of *Carex lugens* H.T. Holm.

densely tufted plant with short but distinct (in well collected specimens) horizontal portions of the shoots (i.e., rhizomes). *Carex ensifolia* ssp. *arctisibirica* is a loosely tufted plant with the horizontal portions of the shoots (which constitute the rhizomes) usually elongated. Additionally, *C. lugens* possesses gradually acuminate narrower leaves, longer pistillate spikelets, and a staminate spikelet sometimes borne on a pedicel up to 5 mm long (rarely longer). The scales and perigynia of these species do not differ (see also the comments under *C. Soczavaeana*).

Carex lugens could in fact be considered a Beringian (Chukotkan-Alaskan) subspecies of *C. ensifolia* s. l., but after taking into account the stability of the differences in tuft structure throughout the range of the two races, we retain for the present the treatment of *C. lugens* as a separate species.

Soviet Arctic. Lower reaches of Lena; Buorkhaya Bay (Tiksi Bay); lower reaches of Kolyma; district of Chaun Bay; Wrangel Island; the whole of the Chukotka Peninsula (west to Vankarem, south to Provideniye Bay); Ratmanov and Arakamchechen Islands; Anadyr Basin; Bay of Korf. (Map III–38).

Foreign Arctic. Arctic Alaska; Victoria and Banks Islands.

Outside the Arctic. Northern part of the Verkhoyansk arc of mountains; basins of the upper course of the Indigirka and of the Rivers Tompo and Menkyule; Alaska; Yukon; Mackenzie River Basin.

48. *Carex Soczavaeana* Gorodk. in Zhurn. Russk. bot. obshch. XV (1930), 185; Krechetovich in Fl. SSSR III, 219, pro parte.

Ill.: Zhurn. Russk. bot. obshch. XV, 184, fig. 1.

Asian Beringian subarctic species.

Widespread in tundra districts near the Pacific Ocean both in placorn tundras and in tussocky (cottongrass-sedge, *Hypnum* or *Hypnum-Sphagnum*) tundras in valleys, where this species and *Eriophorum vaginatum* are the basic tussock-formers.

Also growing in hummocky sphagnum bogs and in marshy meadows, entering the subzones of cedar-pine stlanik and open subarctic larch forests in habitats with excess moisture.

V.I. Krechetovich (herbarium annotation) and several other authors following him (Hultén, M.N. Karavayev, etc.) reduced *C. Soczavaeana* to synonymy with *C. lugens*, which in our opinion is erroneous.

Although difficult to distinguish from *C. lugens* and *C. ensifolia* s. l. by characters of the generative organs, *C. Soczavaeana* is rather sharply distinguished from them by its ability to form tussocks. Additionally, the shoots of *C. Soczavaeana* are

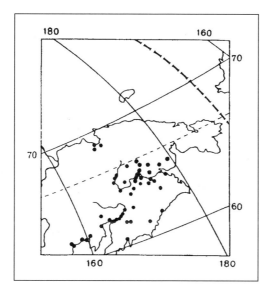

MAP III–39 Distribution of *Carex Soczavaeana* Gorodk.

usually surrounded at their bases by more or less numerous dull brownish-yellow long scalelike sheaths (i.e., the plant is aphyllopodic); in *C. lugens* and *C. ensifolia* the bases of the shoots are clothed with brownish sheaths which are prolonged as leaf blades (i.e., the plant is phyllopodic); in these species scalelike sheaths, usually short, are present only at the extreme base of the shoots and on the rhizomes. Specimens with inadequately excavated subterranean parts collected in the Anadyr Basin or in the district of the Bay of Korf, where both *C. lugens* and *C. Soczavaeana* occur, are practically indistinguishable.

B.N. Gorodkov did not distinguish this species from specimens of *C. lugens* H.T. Holm, of whose description he was unaware. The herbarium contains specimens of *C. lugens* identified by him as *C. Soczavaeana*. He described the range of *C. Soczavaeana* as follows: the mouth of the Lena, the mouth of the Kolyma, the Chukotka Peninsula, the Anadyr Basin, and the northern part of the coast of the Sea of Okhotsk.

B.N. Gorodkov did not designate a type of *C. Soczavaeana*. Specimens from the Anadyr Basin (mouth of the Belaya River) selected as lectotype by V.I. Krechetovich are not entirely typical and deviate somewhat towards *C. lugens*. Therefore we think it expedient, in order to prevent muddling and confusion of *C. Soczavaeana* and *C. lugens*, to select a new lectotype agreeing better with the description from the plants annotated by B.N. Gorodkov [Lectotypus: Tauysk Bay, Nagayevo Bay, *Hypnum* fen on mountain slope in zone of cedar-pine stlanik, 5 VII 1932, Gorodkov and Tikhomirov. Sinus Nagaevo maris Ochotensis; in palude hypnoso in declivio montis, 5 VII 1932, Gorodkov et Tichomirov. In Herb. Inst. Bot. Ac. Sci. URSS(LE) conservatur].

Specimens of *C. Soczavaeana* from the Sea of Okhotsk differ from Anadyr specimens in the larger size of the plant as a whole and of its separate parts.

Soviet Arctic. District of Chaun Bay; Anadyr Basin; Koryakia (Penzhina Basin, Bay of Korf). (Map III–39).

Foreign Arctic. Not occurring.

Outside the Arctic. On the coast of the Sea of Okhotsk from Ayan to Gizhiga Bay; in the upper reaches of the Kolyma.

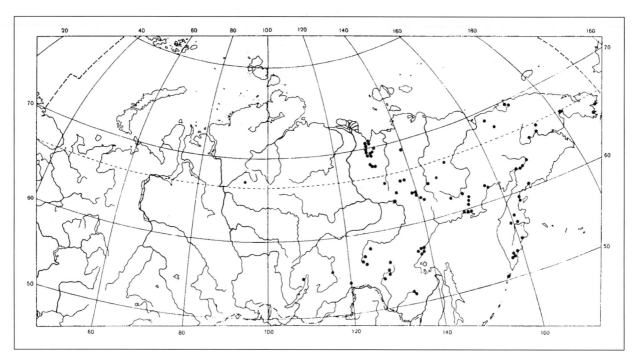

Map III–40 Distribution of *Carex eleusinoides* Turcz. ex Kunth.

49. Carex eleusinoides Turcz. ex Kunth, Enum. pl. II (1837), 407; Turczaninow, Fl. baic.-dahur. I, 391; Kükenthal, Cyper. Caricoid. 299; id., Cyper. Sibir. 80; Krechetovich in Fl. SSSR III, 231; Ohwi, Cyper. Japon. 290; Karavayev, Konsp. fl. Yak. 67; Gjaerevoll, Bot. inv. centr. Alaska 50, map (Fig. 14).

C. caespitosa β Trev. in Ledebour, Fl. ross. IV (1852), 311.

? *C. kokrinensis* Porsild in Rhodora 41 (1939), 206.

Ill.: Kükenthal, Cyper. Sibir., fig. 62.

Characteristic plant of the subalpine and lower alpine (barrens) zone of the mountains of East Siberia. Growing on moist sandbars and gravelbars on the shores of rivers and streams, on the margins of spring-fed marshes, and near gigantic ice sheets, usually at poorly vegetated sites. Penetrating the Arctic at both northern ends of the Lena-Chaun mountain arc; extremely sporadically distributed.

Soviet Arctic. Buorkhaya Bay (Nyayba settlement) and middle course of the River Kharaulakh; Northern Anyuyskiy Range (River Tymkaveyem); district of Chaun Bay; Chukotka Range; Chukotka Peninsula; Bay of Krest; Anadyr Basin (Otvorotnaya and Belaya Rivers); Penzhina Basin. (Map III–40).

Foreign Arctic. Known only from Alaska.

Outside the Arctic. Central Siberian Plateau (Lake Nyakshinda, 67°N); lower course of the Lena; Verkhoyansk Range; Cherskiy Range; upper reaches of the Kolyma; Dzhugdzhur Range; coasts of the Sea of Okhotsk; Kamchatka; Kurile Islands; Hokkaido; barrens in the basins of the Aldan and of northern tributaries of the Amur; mountains of Transbaikalia and Northern Mongolia; Sayans; Eastern Altay.

50. Carex bicolor Bell. ex All., Fl. Pedem. II (1785), 267; Treviranus in Ledebour, Fl. ross. IV, 285; Ostenfeld, Fl. arct. 1, 79; Kükenthal, Cyper. Caricoid. 297; id., Cyper. Sibir. 79; Krylov, Fl. Zap. Sib. III, 460; Krechetovich in Fl. SSSR III, 264; Leskov, Fl. Malozem. tundry 35; Hultén, Fl. Al. II, 332, map 268; id., Atlas, map 396; A.E.

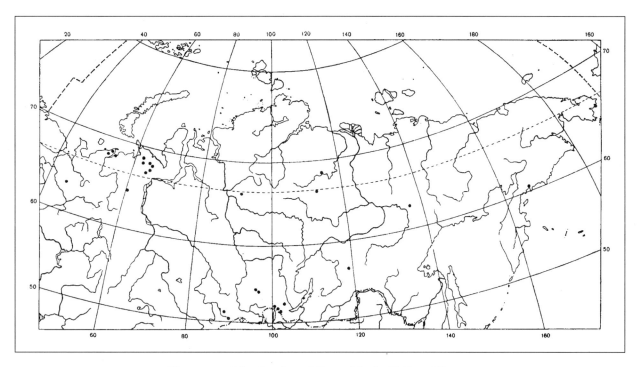

Map III–41 Distribution of *Carex bicolor* Bell. ex All.

Porsild, Ill. Fl. Arct. Arch. 62, map 83; Böcher & al., Grønl. Fl. 257; Karavayev, Konsp. fl. Yak. 67; Polunin, Circump. arct. fl. 84, 101.

Ill.: Kükenthal, Cyper. Caricoid., fig. 45; Polunin, l. c. 85.

Arctic-alpine species with almost circumpolar but extremely fragmented distribution. Very rare everywhere. Growing on sandbars and gravelbars, in moist meadows on the shores of rivers and lakes, in marshes, and sometimes at springs.

Absent from continental districts of Arctic Siberia, although occurring in somewhat more southern districts on the Central Siberian Plateau.

Soviet Arctic. Malozemelskaya Tundra (Tulovo Hill); lower reaches of the Pechora (River Yushina); eastern part of Bolshezemelskaya Tundra; Polar Ural (River Yelets); Lawrence Bay; Bay of Korf. (Map III–41).

Foreign Arctic. Arctic Alaska; arctic coast of Canada (mainly western part); Southern Baffin Island; arctic part of Labrador; SW and East Greenland; ? Iceland; Arctic Scandinavia.

Outside the Arctic. Mountains of Sweden and Norway; Alps; basin of the Severnaya Dvina (River Pinega); Prepolar Ural (basin of River Lyapin); Altay; Central and Eastern Sayans; Prebaikalia; lower reaches of the Lower Tunguska (66°N); basins of the Markha, Olenek, Aldan (River Khandyga) and Chara; Aleutian Islands (Unimak Island); Alaska and Canada.

51. *Carex paleacea* Wahlb. in Kongl. Vet. Akad. Nya Handl. XXIV (1803), 164; Mackenzie in N Amer. Fl. XVIII, 7, 414; Hultén, Atlas, map 369; Kuzeneva in Fl. Murm. II, 93, map 30; Hultén, Amphi-atl. pl. 284, map 265; Polunin, Circump. arct. fl. 88, 104.

C. maritima O.F. Muell. in Fl. Dan. IV, 12 (1777), 5; Treviranus in Ledebour, Fl. ross. IV, 313; Kükenthal, Cyper. Caricoid. 359; Perfilev, Fl. Sev. I, 119; non Gunner (1772).

C. paralia V. Krecz. in Fl. SSSR III (1935), 237.

Ill.: Fl. Murm. II, pl. XXVII.

Amphiatlantic species growing on sea coasts.

Soviet Arctic. Murman (only Korabelnaya Bay on Kola Bay).
Foreign Arctic. Arctic Scandinavia.
Outside the Arctic. Coasts of Norway and Sweden; east coast of the Gulf of Bothnia; coast of the White Sea (at the mouth of the Severnaya Dvina); east coast of North America between 42° and 60°N.

52. ***Carex cryptocarpa*** C.A. Mey. in Mém. Ac. Sci. St.-Pétersb. Sav. Étr. I (1831), 226, tab. 14; Treviranus in Ledebour, Fl. ross. IV, 313; Krechetovich in Fl. SSSR III, 239.
? *C. Lyngbyei* Hornem. in Fl. Dan. XI, 32 (1827), 6; Ostenfeld, Fl. arct. 1, 75; Kükenthal, Cyper. Caricoid. 363; id., Cyper. Sibir. 96; Mackenzie in N Amer. Fl. XVIII, 7, 415; Böcher & al., Grønl. Fl. 259; Polunin, Circump. arct. fl. 88, 103; Hultén, Amphi-atl. pl. 292, map 273.
C. Lyngbyei ssp. *cryptocarpa* (C.A. Mey.) Hult., Fl. Al. II (1942), 343; id., Amphi-atl. pl. 292.
C. suifunensis Kom. in Acta Horti Petropol. XVIII (1901), 445; Krechetovich, l. c. 239.
C. pedunculifera Kom. in Feddes Repert. sp. nov. XIII (1914), 163; Krechetovich, l. c. 240.
C. Riabushinskii Kom., l. c. 163; Krechetovich, l. c. 239.

Growing mostly in maritime marshes and marshy meadows; sometimes penetrating moist tundra. In some districts occurring at a more or less considerable distance from the sea.

Carex cryptocarpa is very close to and possibly identical with *C. Lyngbyei*. The former was described from Unalaska, the latter from the Faroe Islands. Certain authors, such as Hultén and Mackenzie, are of the opinion that *C. Lyngbyei* grows on the Atlantic coast of Canada (Labrador, Gaspé, Anticosti Island), Greenland, Iceland and the Faroe Islands. They consider the Pacific coast of North America and temperate Asia to be the region of distribution of *C. cryptocarpa*, which many authors have not distinguished from *C. Lyngbyei*. Hultén considers *C. cryptocarpa* to be a subspecies of *C. Lyngbyei* on the basis of the insignificance of the differences between the Atlantic and Pacific races and the impossibility of distinguishing them in certain cases. *Carex Lyngbyei* s. str. is absent from the herbarium of the Botanical Institute. Study of herbarium material received from Canada has shown that Icelandic plants are completely indistinguishable from *C. cryptocarpa*. Plants from NE Canada (Gaspé, Anticosti Island) identified as *C. Lyngbyei* proved to belong to *C. paleacea* or *C. recta*. We have not seen specimens of *C. Lyngbyei* from Greenland or the Faroe Islands, nor the type of this species. Therefore at the present time we cannot confirm the identity of *C. cryptocarpa* and *C. Lyngbyei* with complete reliability.

Carex cryptocarpa varies throughout its range in the length of the scales, which either considerably exceed the length of the perigynia (by 1.5 times) or are only slightly longer (or rarely of about equal length), as well as in the size of the pistillate spikelets and the length of their pedicels.

Carex Riabushinskii Kom. and *C. pedunculifera* Kom., described from Kamchatka, and *C. suifunensis* Kom. from the south of Primorskiy Kray represent forms within the individual variation of *C. cryptocarpa* occurring throughout the range of the species.

Soviet Arctic. Eastern part of the Chukotka Peninsula (Lorino, Chaplino and Senyavin Hot Springs); Anadyr Basin (River Tanyurer and the middle course of the Anadyr, Geka Land); Penzhina Basin and district of Penzhina Bay; Bay of Korf.
Foreign Arctic. Arctic Alaska (Kotzebue Sound); southern extremity and SW part of Greenland (south of 62°N); Iceland.
Outside the Arctic. The whole of the Far East of the USSR; Japan; Aleutian Islands; Pacific coast of North America from Alaska to Northern California; Faroe Islands.

53. ***Carex recta*** Boott in Hook., Fl. bor.-amer. II (1839), 220, tab. 222; Mackenzie in N Amer. Fl. XVIII, 7, 418; Kuzeneva in Fl. Murm. II, 92, map 30; Hultén, Amphi-atl. pl. 282, map 264.
? *C. kattegatensis* Fries in Ind. sem. Hort. Upsal. (1857), nomen; Krechetovich in Fl. SSSR III, 237.
C. salina var. *kattegatensis* Almq. in Hartman, Handb. Scand. Fl., ed. II (1879), 466; Kükenthal, Cyper. Caricoid. 363.
C. salina s. l. — Ostenfeld, Fl. arct. 1, 73; Böcher & al., Grønl. Fl. 259; Polunin, Circump. arct. fl. 89, 104, pro parte.
C. Lyngbyei auct. non Hornem. — Perfilev, Fl. Sev. I, 118.
Ill.: Fl. Murm. II, pl. XXVI, fig. 2.

Amphiatlantic plant just penetrating arctic limits. Growing in maritime meadows and marshes.

Soviet Arctic. Murman (shore of Kola Bay and Iokanga).

Foreign Arctic. Southern extremity of Greenland; Iceland; Arctic Scandinavia.

Outside the Arctic. Scotland; Faroe Islands; entire coast of Norway; the northeast of the Gulf of Bothnia; coast and islands of the White Sea as far as the mouth of the Severnaya Dvina; Atlantic coast of North America north of 42°N.

54. ***Carex salina*** Wahlb. in Kongl. Vet. Akad. Nya Handl. XXIV (1803), 165; Ostenfeld, Fl. arct. 1, 73, pro parte; Kükenthal, Cyper. Caricoid. 361, pro parte; Hultén, Atlas, map 368, pro parte; id., Amphi-atl. pl. 282, map 263; ? Böcher & al., Grønl. Fl. 259, pro parte; Polunin, Circump. arct. fl. 89, 104, pro parte.
? *C. lanceata* Dew. in Amer. Journ. Sci. XXIX (1836), 249; Kuzeneva in Fl. Murm. II, 90, map 29.
C. discolor Nyl., Spicil. pl. fenn. (1846) 12; Krechetovich in Fl. SSSR III, 236.
C. trinervis auct. non Degl. — Perfilev, Fl. Sev. I, 116.
Ill.: Fl. Murm. II, pl. XXVI, fig. 1.

Amphiatlantic species growing on sandbars or siltbars on sea coasts. A very rarely encountered plant occupying a morphologically intermediate position between *C. subspathacea* and *C. recta*. From the former species *C. salina* differs in having longer pistillate spikelets, usually 3 (not 2) in number, and all usually borne on short pedicels (in *C. subspathacea* only the lower spikelet is pedicellate, the upper sessile). Additionally, the scales of the pistillate spikelets in *C. salina* are more or less awned, and the perigynia taper abruptly to a short beak. In *C. subspathacea* the scales are normally obtusish (more rarely acutish), sometimes with a very short mucro apically; the perigynia are gradually extended (without incurvation) to the beak. Plants of *C. salina* are larger than those of *C. subspathacea*. *Carex salina* differs from *C. recta* in the smaller size of the plants, the shorter pedicels of the spikelets, the more weakly awned scales, and the broader beaks of the perigynia (usually twice as wide as in *C. recta*). No sharp boundary exists between *C. salina*, *C. subspathacea* and *C. recta*, and some plants cannot be precisely identified. The ranges of *C. salina* and *C. recta* almost completely coincide. We have the impression that *C. salina* may be the hybrid between *C. subspathacea* and *C. recta*.

Soviet Arctic. Murman (lower reaches of the Kola and the shore of Kola Bay, Kildin Island, Lumbovka, Kharlovka, Ponoy).

Foreign Arctic. ? Southern extremity of Greenland; Iceland; Arctic Scandinavia. The distribution of *C. salina* is not known precisely, because many authors interpret this species in a wide sense inclusive of *C. recta*. In particular, it is unclear which of these species occurs in Greenland.

Outside the Arctic. Coast of Norway; Kola Peninsula; coast and islands of the White Sea as far as the mouth of the Severnaya Dvina; Hudson Bay; Labrador and

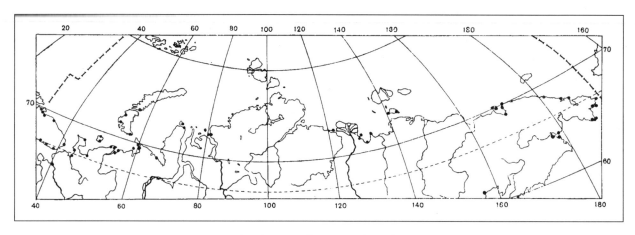

MAP III–42 Distribution of *Carex subspathacea* Wormsk. ex Hornem.

Newfoundland. The record of the occurrence of this species on the lower course of the Yenisey (Scheutz, Pl. jeniss. 180) is erroneous.

55. *Carex subspathacea* Wormsk. ex Hornem. in Fl. Dan. IX, 26 (1820), 6, tab. 1530; Treviranus in Ledebour, Fl. ross. IV, 304; Kükenthal, Cyper. Caricoid. 361; id., Cyper. Sibir. 95; Andreyev, Mat. fl. Kanina 162; Tolmachev, Mat. fl. Mat. Shar 287; Perfilev, Fl. Sev. I, 118; Mackenzie in N Amer. Fl. XVIII, 7, 417; Krechetovich in Fl. SSSR III, 235; Tolmachev, Obz. rast. o. Sibiryakova 215; Leskov, Fl. Malozem. tundry 41; Hultén, Fl. Al. II, 341; id., Atlas, map 367; Kuzeneva in Fl. Murm. II, 90, map 29; A.E. Porsild, Ill. Fl. Arct. Arch. 54, map 87; Böcher & al., Grønl. Fl. 258; Karavayev, Konsp. fl. Yak. 67; Polunin, Circump. arct. fl. 89, 105; Derviz-Sokolova in Bot. mat. Gerb. Bot. inst. AN SSSR XXI, 73.

C. salina var. *subspathacea* (Wormsk.) Ostenf., Fl. arct. 1 (1902), 75.

Ill.: Fl. Murm. II, pl. XXV.

Predominantly arctic plant with circumpolar distribution. Very characteristic of low-herb maritime salt "meadows" and estuarine saltmarshes, where it sometimes occurs in great quantity, forming dense low beds. Often growing in the zone flooded by marine tides. On the Beringian coast of the Chukotka Peninsula, *C. subspathacea* can be found flooded for up to 7 days. It also penetrates lowland tundra, where it occurs on low shores of lakes united with the sea during high tides.

Soviet Arctic. Murman (western part as far as Teriberka and in the district of the mouth of the Ponoy); Kanin; Timanskaya Tundra (River Pesha); Malozemelskaya Tundra (eastern part); Bolshezemelskaya Tundra; Kolguyev Island; district of Baydaratskaya Bay; Vaygach Island; Yugorskiy Peninsula; Novaya Zemlya (South Island at Kostin Shar and Malyye Karmakuly, North Island in the district of Matochkin Shar); coasts and islands of Yenisey Bay; lower reaches of Olenek; Buorkhaya Bay; Bolshoy Lyakhovskiy Island; Chukotka (Ayon Island, district of Chaun Bay, Cape Schmidt, Vankarem, Kolyuchin Island, and the Beringian coast from Uelen to Provideniye Bay); Anadyr Basin (Volchikha and Velikaya Rivers, Geka Land). (Map III–42).

Foreign Arctic. Arctic coast of Alaska and Canada; southern islands of Canadian Arctic Archipelago; Labrador; SW and East Greenland; Arctic Scandinavia; ? Spitsbergen.

Outside the Arctic. Coast of Norway; coasts and islands of the White Sea; coasts of Sea of Okhotsk; Sakhalin; Kamchatka; Kurile Islands; Hokkaido; west coast of Alaska; Hudson Bay and Labrador.

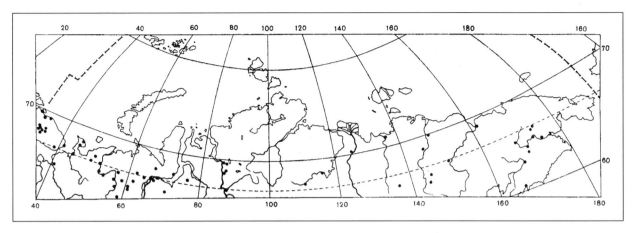

MAP III–43 Distribution of *Carex limosa* L.

56. **Carex limosa** L., Sp. pl. (1753) 977; Treviranus in Ledebour, Fl. ross. IV, 307; Ostenfeld, Fl. arct. 1, 68; Kükenthal, Cyper. Caricoid. 504; Komarov, Fl. Kamch. I, 257; Hultén, Fl. Kamtch. I, 201; Krylov, Fl. Zap. Sib. III, 499; Perfilev, Fl. Sev. I, 123; Krechetovich in Fl. SSSR III, 244; id. in Mat. ist. fl. rastit. SSSR I, 153, map 3; Leskov, Fl. Malozem. tundry 38; Hultén, Fl. Al. II, 369, map 305; id., Atlas, map 386; id., Circump. pl. I, 92, map 84; Kuzeneva in Fl. Murm. II, 98, map 32; Karavayev, Konsp. fl. Yak. 67; Tyrtikov in Byull. MOIP, otd. biol. LX, 5, 139; Polunin, Circump. arct. fl. 98.
   C. limosa var. *fusco-cuprea* Kük., l. c. 505.
   C. fusco-cuprea (Kük.) V. Krecz., l. c. 244.
   Ill.: Fl. Murm. II, pl. XXVIII.

   Circumpolar species distributed in association with sphagnum bogs in the forest zone; rarely occurring in the Arctic. Growing in sedge-moss tundra, sphagnum bogs, flarks, vegetated lake beds, and on peaty river shores.

   Ostenfeld's report that *C. limosa* grows on Kolguyev Island was erroneous; apparently he mistook *C. rariflora* Wahlb., which is often encountered there, for *C. limosa*.

   In individuals of *C. limosa* the pistillate spikelets sometimes bear staminate florets distally.

   Soviet Arctic. Murman; Kanin; Malozemelskaya Tundra (Rivers Shchuchya and Sula); Southern Bolshezemelskaya Tundra (River Kolva, Sivaya Maska); Polar Ural; lower reaches of Ob; lower reaches of the Yenisey, Yana (70°40′N according to Tyrtikov's data), Indigirka (River Malaya) and Kolyma; Anadyr and Penzhina Basins; Bay of Korf. (Map III–43).

   Foreign Arctic. Arctic Alaska; district of the mouth of the Mackenzie River and Great Bear Lake; Iceland; Arctic Scandinavia.

   Outside the Arctic. Northern and Central Europe; Fennoscandia; European part of the USSR (except southern districts); Western Transcaucasus; Urals; West Siberia; sporadically in East Siberia (mainly in the southern half); the whole of the Far East of the USSR; the north of the Korean Peninsula; Japan; Alaska; Canada; eastern states of the USA (approximately north of 36°N), as well as Oregon and Montana and occasionally in central states.

57. **Carex magellanica** Lam., Encycl. III (1789), 385; Ostenfeld, Fl. arct. 1, 66; Kükenthal, Cyper. Caricoid. 505; id., Cyper. Sibir. 142; Komarov, Fl. Kamch. I, 258; Hultén, Fl. Kamtch. 202; Krylov, Fl. Zap. Sib. III, 500; Leskov, Fl. Malozem. tundry 38; Hultén, Fl. Al. 370, map 306; id., Atlas, map 385; Böcher & al., Grønl. Fl. 262; Polunin, Circump. arct. fl. 98, 103.

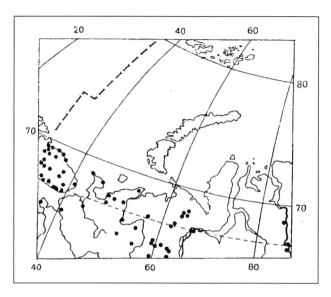

Map III-44 Distribution of *Carex magellanica* Lam.

C. paupercula Michx., Fl. bor.-amer. II (1803), 172; Mackenzie in N Amer. Fl. XVIII, 6, 351.
C. limosa β *irrigua* Wahlb. in Kongl. Vet. Akad. Nya Handl. XXIV (1803), 162.
C. irrigua (Wahlb.) Smith ex Hoppe, Caric. germ. (1826), 72; Treviranus in Ledebour, Fl. ross. IV, 307; Perfilev, Fl. Sev. I, 123; Krechetovich in Fl. SSSR III, 245; id. in Mat. ist. fl. rastit. SSSR I, map 8; Kuzeneva in Fl. Murm. II, 100, map 33.
C. magellanica ssp. *irrigua* (Wahlb.) Hult., Circump. pl. I (1962), 90, map 81.
Ill.: Fl. Murm. II, pl. XXIX.

Bipolar species, with circumpolar distribution in the Northern Hemisphere and also occurring at the southern extremity of South America (whence it was described).

South American specimens of this sedge preserved in the herbarium of the Botanical Institute in Leningrad are completely indistinguishable from Northern Hemisphere plants known under the name *C. irrigua*. Roivainen in his study of the bogs of Tierra del Fuego [in Ann. Bot. Soc. Vanamo 28, 2 (1954), 198] suggested that *C. magellanica* and *C. irrigua* are distinct races. He based this opinion on the fact that in specimens of *C. magellanica* collected by him the uppermost and lateral spikelets were gynaecandrous in distinction from *C. irrigua*, and the perigynia narrower (3.5–4 mm long). However, gynaecandrous spikelets, as supposedly characterizing *C. magellanica*, occur not uncommonly in individuals of *C. irrigua* throughout its range. Usually only the uppermost spikelet is gynaecandrous, somewhat more rarely also the lateral. As for differences in the width of the perigynia, this is evidently within the range of individual variation.

Carex magellanica is a characteristic plant of mossy bogs of the forest zone and the forest-tundra. It penetrates arctic limits mainly in Northern Europe. Growing in sphagnum peat bogs, flarks, sedge-moss tundra, and open boggy forests.

Soviet Arctic. Murman (throughout); northern part of Kanin (Nattey and Krutaya Rivers); Timanskaya, Malozemelskaya and Bolshezemelskaya Tundras; Polar Ural; lower reaches of the Ob (Salekhard). (Map III–44).

Foreign Arctic. Arctic Alaska; lower reaches of Mackenzie River; Greenland south of 63°N (isolated localities); Iceland; Arctic Scandinavia.

Outside the Arctic. Northern and Central Europe; northern half of the European part of the USSR; Caucasus; Urals; West Siberia; Altay; Yenisey Basin south of 65°N;

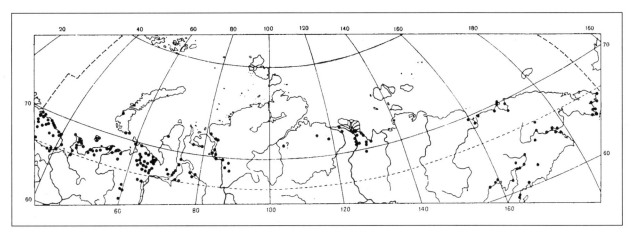

MAP III-45 Distribution of *Carex rariflora* (Wahlb.) Smith.

Prebaikalia; Transbaikalia; Lake Toko; Dusse-Alin; coast of Sea of Okhotsk (Ola district); Kamchatka; Kurile Islands; Honshu; Alaska; Canada; eastern states of the USA (approximately north of 42°N) and isolated occurrences in the central states; extreme south of South America.

58. *Carex rariflora* (Wahlb.) Smith, Engl. Bot. (1813), tab. 2516; Treviranus in Ledebour, Fl. ross. IV, 297; Scheutz, Pl. jeniss. 178; Ostenfeld, Fl. arct. 1, 67; Kükenthal, Cyper. Caricoid. 502; id., Cyper. Sibir. 141; Lynge, Vasc. pl. N.Z. 96; Komarov, Fl. Kamch. I, 257; Hultén, Fl. Kamtch. I, 200; id., Fl. Al. II, 367, map 303; id., Atlas, map 387; id., Circump. pl. I, 14, map 8; Tolmatchev, Contr. Fl. Vaig. 128; Tolmachev, Fl. Kolg. 16; id., Obz. fl. N.Z. 150; Krylov, Fl. Zap. Sib. III, 498; Andreyev, Mat. fl. Kanina 161; Perfilev, Fl. Sev. I, 122; Krechetovich in Fl. SSSR III, 242; Leskov, Fl. Malozem. tundry 39; Kuzeneva in Fl. Murm. II, 94, map 31; A.E. Porsild, Ill. Fl. Arct. Arch. 56, map 92; Böcher & al., Grønl. Fl. 262; Karavayev, Konsp. fl. Yak. 67; Polunin, Circump. arct. fl. 98, 104.

C. limosa γ *rariflora* Wahlb. in Kongl. Vet. Akad. Nya Handl. XXIV (1803), 162.

Ill.: Fl. Murm. II, pl. XXVIII.

Circumpolar tundra species, a characteristic inhabitant of lowland tundra bogs in relatively temperate districts of the Arctic. Outside the tundra zone it is most widespread along the Pacific coast of Asia (as far as Sakhalin and the Kurile islands); in Northern Yakutia it only just penetrates the subarctic zone of open forests, and is even absent from high subarctic mountains of the Verkhoyansk-Kolymsk mountain country.

Carex rariflora is very characteristic of *Sphagnum*-sedge bogs or *Hypnum*-sedge fens on tundra at low or intermediate elevation, and of peaty areas with little or no water flow, normally adequately protected by snow in winter. In particular, this is one of the basic plants of the wide saucer-shaped closed depressions in polygonal bogs, whether these are persistently flooded with water or more or less drained; here *C. rariflora* spreads by rhizomes and forms extensive beds; its companions are *Carex chordorrhiza*, *C. stans* and *Hierochloë pauciflora*. In the complex communities of dwarf willow-sedge-moss tundra with flat hummocks and cottongrass-sedge-moss boggy tundras (dominated by *Carex stans*, *Eriophorum angustifolium*, *Salix reptans* and *S. pulchra*), *C. rariflora* is more common in the wetter depressed areas including flarks. In more southern parts of its range (including the zone of subarctic open forests) it sometimes also occurs in mesotrophic *Sphagnum* or *Hypnum-Sphagnum* fens with dwarf shrubs, cotton grasses and sedges (with *Eriophorum vaginatum* and *E. angustifolioum*).

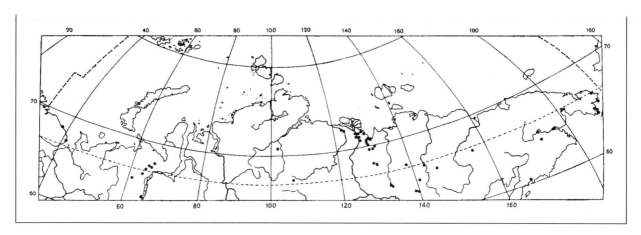

MAP III–46 Distribution of *Carex atrofusca* Schkuhr.

Soviet Arctic. Murman (everywhere); Kanin; Kolguyev Island; Timanskaya and Malozemelskaya Tundras; lower reaches of Pechora; Bolshezemelskaya Tundra; Polar Ural; district of Baydaratskaya Bay; Vaygach Island; South Island of Novaya Zemlya; Yamal (southern and northwestern parts); Obsko-Tazovskiy Peninsula; Gydanskaya Tundra; lower reaches of Yenisey; Yenisey Bay; lower reaches of the Anabar and Olenek; lower reaches and delta of the Lena; Buorkhaya Bay; lower reaches of Kolyma; basin of Chaun Bay and Ayon Island; eastern part of Chukotka Peninsula from Cape Dezhnev to Provideniye Bay; Arakamchechen Island; Anadyr and Penzhina Basins; Bay of Korf. (Map III–45).

Foreign Arctic. Arctic Alaska; arctic coast of Canada; the south of the Canadian Arctic Archipelago (Victoria Island and Southern Baffin Island); Labrador; SW and East Greenland; Iceland; Arctic Scandinavia.

Outside the Arctic. Scotland; Norway; the north of Sweden and Finland; Kola Peninsula; coasts of the White Sea; Northern Ural; basin of the River Kotuy; coasts of the Sea of Okhotsk; Kamchatka; Sakhalin; Kurile and Commander Islands; Alaska; northern and eastern Canada.

59. ***Carex atrofusca*** Schkuhr, Riedgrass. I (1801), 106, tab. V, fig. 82; Kükenthal, Cyper. Caricoid. 553, pro parte; id., Cyper. Sibir. 149; Krylov, Fl. Zap. Sib. III, 506 (excl. var. *fusca* Kryl.); Krechetovich in Fl. SSSR III, 282; Hultén, Fl. Al. II, 373, map 310; id., Atlas, map 398; Kuzeneva in Fl. Murm. II, 110, map 36; A.E. Porsild, Ill. Fl. Arct. Arch. 56, map 94; Böcher & al., Grønl. Fl. 263; Karavayev, Konsp. fl. Yak. 68; Polunin, Circump. arct. fl. 93, 101; Hultén, Circump. pl. I, 46, map 39.

C. ustulata Wahlb. in Kongl. Vet. Akad. Nya Handl. XXIV (1803), 156; Ostenfeld, Fl. arct. 1, 90; Perfilev, Fl. Sev. I, 124.

C. stilbophaea V. Krecz. in Fl. SSSR III (1935), 283 and 605.

C. oxyleuca V. Krecz., l. c. 284 and 605.

Ill.: Fl. Murm. II, pl. XXXIV, fig. 2; Polunin, l. c. 93; A.E. Porsild, l. c., fig. 21, d.

Arctic-alpine species with fragmented distribution; more common in mountain districts with surface exposures of carbonate rocks. Normally associated with areas of good or even surplus (but always flowing) moisture, with adequate snow cover in winter, with strongly mineralized hard groundwater and a well developed moss layer consisting of *Tomenthypnum nitens, Aulacomnium* spp., etc. (banks of mountain torrents, outflows of springs, mountain moss or sedge fens, patchy moist moss-sedge, moss-*Kobresia* or moss tundras on aprons of slopes and in runoff channels). Occurring more rarely in drier areas, such as *Cassiope*-moss tundras (with *Tomenthypnum nitens* and *Carex ensifolia* ssp. *arctisibirica*) or even

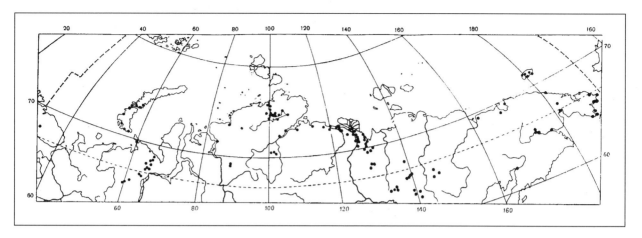

MAP III–47 Distribution of *Carex misandra* R. Br.

Dryas-sedge, Dryas-Kobresia-sedge or Kobresia-sedge turfy tundras with mesoxerophilic mosses (*Rhytidium rugosum, Thuidium abietinum*), lichens (*Cetraria nivalis, C. chrysantha*) and numerous species of arctic-alpine herbs.

In the district of the mouth of the Anadyr, specimens have been noted in which the uppermost spikelet is gynaecandrous. Specimens with pale brown scales occur very rarely; such plants have been collected especially in the lower reaches of the Lena.

Soviet Arctic. Murman (Cape Orlov and, according to Hultén, in the Pechenga district and between the settlements of Gremikha and Varzino); Polar Ural (basin of the River Sob and the upper reaches of the River Nemyr-Yugan); lower reaches of the Olenek and Lena; Buorkhaya Bay (Tiksi Bay); middle course of River Kharaulakh; Chukotka Peninsula (from the River Chegitun to the SE extremity of the Bay of Krest); middle course and mouth of the Anadyr; Bay of Korf. (Map III–46).

Foreign Arctic. Arctic Alaska; northern coast of Canada; arctic part of Labrador; Canadian Arctic Archipelago; Greenland (north of the Arctic Circle); Arctic Scandinavia.

Outside the Arctic. Scandinavia; mountains of Scotland; ? Pyrenees and mountains of Central Europe; Prepolar Ural and Northern Ural (Denezhkin Rock); Altay; northern part of Central Siberian Plateau; Central and Eastern Sayans; Southern Muyskiy Range; Verkhoyansk Range; Suntar-Khayata Range and Cherskiy Range (Momskiy district); Kolyma Basin (River Berezovka); coast of Sea of Okhotsk (Ayan); Central Asia (mountainous districts); Afghanistan; Kashmir; Alaska; northern part of Canada.

60. ***Carex misandra*** R. Br. in Suppl. App. Parry's Voyage XI (1824), 283; Ostenfeld, Fl. arct. 1, 88; Komarov, Fl. Kamch. I, 262; Hultén, Fl. Kamtch. I, 204; Tolmatchev, Contr. Fl. Vaig. 127; Krylov, Fl. Zap. Sib. III, 507; Perfilev, Fl. Sev. I, 114; Krechetovich in Fl. SSSR III, 292; Tolmachev, Fl. Taym. I, 103; id., Obz. fl. N.Z. 150; Hultén, Fl. Al. II, 373, map 311; id., Atlas, map 397; id., Circump. pl. I, 20, map 14; Kuzeneva in Fl. Murm. II, 112, map 36; A.E. Porsild, Ill. Fl. Arct. Arch. 56, map 95; Böcher & al., Grønl. Fl. 263; Karavayev, Konsp. fl. Yak. 68; Polunin, Circump. arct. fl. 84, 103.

C. frigida var. β Trev. in Ledebour, Fl. ross. IV (1852), 294.

C. fuliginosa var. β *misandra* (R. Br.) O.F. Lang in Linnaea XXIV (1851), 597; Kükenthal, Cyper. Caricoid. 557; id., Cyper. Sibir. 152.

Ill.: Fl. SSSR III, pl. XVIII, fig. 1; Fl. Murm. II, pl. XXXIV.

Circumpolar species, most widely distributed in the Arctic and in high subarctic mountains; scarcely penetrating more southern high alpine districts.

Growing in the Arctic in areas adequately protected by snow in winter, with abundant (but flowing) or moderate moisture and comparatively rich soils. Most characteristic of patchy moss-sedge tundras (with *Tomenthypnum nitens* and *Carex ensifolia* ssp. *arctisibirica*), patchy small-sedge tundras (with *C. atrofusca, C. fuscidula, C. vaginata, C. melanocarpa*, etc.), patchy moss tundras, mossy springfed fens, and moist gravelbars. Also very common in drier tundras (*Dryas*, *Dryas*-sedge, *Kobresia*, *Cassiope*-moss and certain others).

Soviet Arctic. Polar Ural; Vaygach Island; Novaya Zemlya (South Island and the south of the North Island); lower reaches of the Yenisey and the district of Yenisey Bay; Taymyr; Khatanga Basin; lower reaches of the Anabar and Olenek; lower reaches and delta of the Lena; Buorkhaya Bay; middle course of the Kharaulakh; Northern Anyuyskiy Range (valley of the River Sukharnaya); district of Chaun Bay (Pevek, Elveney); Cape Schmidt; Wrangel Island; eastern and southern parts of the Chukotka Peninsula; Anadyr Basin (Ust-Belaya settlement and near Anadyr Bay); lower reaches of the Penzhina (Kamenskoye village). (Map III–47).

Foreign Arctic. Arctic part of Alaska; northern coast of Canada; Labrador; Canadian Arctic Archipelago; Greenland north of 65°N; Spitsbergen; Arctic Scandinavia.

Outside the Arctic. Mountains of Norway; Khibins Mountains; Prepolar and Northern Urals (south to Denezhkin Rock); northern edge of Central Siberian Plateau; Verkhoyansk Range; Suntar-Khayata Range; upper reaches of Aldan; Kamchatka; Kurile Islands; Alaska; Canada (mainly the eastern part); isolated localities in the states of Utah and Colorado.

61. ***Carex macrogyna*** Turcz. ex Steud., Syn. Cyp. (1855) 236; Turcz. ex Bess. in Flora XVII, I Beibl. (1834), 27 (nomen); Turcz. in Bull. Soc. Nat. Mosc. XI, 1, 104 (nomen); Kükenthal, Cyper. Caricoid. 560; id., Cyper. Sibir. 150; Krechetovich in Fl. SSSR III, 293; Karavayev, Konsp. fl. Yak. 68.

C. ferruginea var. β Trev. in Ledebour, Fl. ross. IV (1852), 294, pro parte.

C. ustulata var. γ *macrogyna* Regel in Acta Horti Petropol. VII (1880), 571.

C. tristis auct. non M.B. — Ostenfeld, Fl. arct. 1, 88, pro parte.

Ill.: Kükenthal, Cyper. Sibir., fig. 125; Fl. SSSR III, pl. XVIII, fig. 6.

Species of the barrens of continental Siberia, with restricted distribution also in the high mountains of Central Asia and Mongolia and just penetrating the Arctic; particularly characteristic of the lower alpine (barrens) zone and the subalpine zone. Normally associated with carbonate rocks.

This species is an active stabilizer of slopes and sometimes dominates in the herb cover of dry stony tundras and dry glades in Dahurian larch forest (together with *Carex Trautvetteriana, C. ensifolia* ssp. *arctisibirica, C. rupestris, Kobresia Bellardii, K. simpliciuscula* and *Dryas* spp.); in moister dwarf willow-moss-sedge and moss-*Kobresia* tundras it occurs as an admixture; also growing in herb communities of the forest zone (along streams). In the lower reaches of the Lena (Tuora-Sis Range), *C. macrogyna* is a characteristic companion of *Caragana jubata*.

Carex macrogyna is extremely close to, if not identical with, *C. petricosa* Dew. distributed in Alaska and in the mountains of Western Canada. *Carex petricosa* also occurs in arctic districts of those countries, where it grows in the same habitats as *C. macrogyna*.

Carex macrogyna (and also *C. petricosa*) varies in the structure of the upper spikelets of the inflorescence. The most widespread plants are those in which three of the upper spikelets are entirely staminate and the remainder pistillate, but specimens also occur not uncommonly in which the uppermost staminate spikelet possesses 1–4 or more pistillate florets at its base. The next two spikelets below it are either entirely staminate or androgynous.

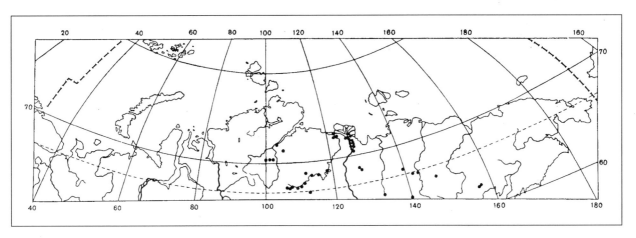

Map III-48 Distribution of *Carex macrogyna* Turcz. ex Steud.

Soviet Arctic. Lower reaches of the Khatanga (near the mouth of the River Zhdanikha); lower reaches of the Olenek and Lena. (Map III–48).
Foreign Arctic. Not occurring (see text).
Outside the Arctic. Northern part of Central Siberian Plateau; Narymskiy Range; Central and Eastern Sayans; Northern Mongolia; Southern Transbaikalia; basins of the Muya and Aldan; Verkhoyansk Range and Cherskiy Range; coast of Sea of Okhotsk (basin of River Lantar).

62. ***Carex Ktausipalii*** Meinsh. in Acta Horti Petropol. XVIII, 3 (1901), 359; Krechetovich in Fl. SSSR III, 286.
    C. stenantha auct. non Franch. et Sav. — Kükenthal, Cyper. Caricoid. 565, pro parte; id., Cyper. Sibir. 153; Hultén, Fl. Kamtch. I, 204.
    C. stenantha var. *taisetsuensis* Akiyama in Journ. Coll. Sci. Hokk. Univ., ser. V, I (1931), 60.
    C. Gorodkovii V. Krecz. in Fl. SSSR III (1935), 287 and 606.
        Growing on rocky slopes.
Soviet Arctic. Penzhina Bay (near the shore of the bay at Lovat village).
Foreign Arctic. Not occurring.
Outside the Arctic. Coast of Sea of Okhotsk (Tauysk Bay); Sakhalin; Kamchatka; Kurile Islands; Hokkaido.

The sole herbarium specimen of this species known from the Arctic was referred by V.I. Krechetovich to *C. Gorodkovii*, described by him from the coast of the Sea of Okhotsk (Nagayevo Bay). However, study of the types of the two species has shown that *C. Gorodkovii* is identical with *C. Ktausipalii*. The type specimens of *C. Gorodkovii* were collected in the flowering stage and consequently the size of the plant as a whole is somewhat less than in *C. Ktausipalii*. V.I. Krechetovich's statement that the perigynia in *C. Gorodkovii* possess elongate pedicels was not confirmed by study of the type material.

63. ***Carex atrata*** L., Sp. pl. (1753) 976; Ostenfeld, Fl. arct. 1, 64, pro parte; Treviranus in Ledebour, Fl. ross. IV, 287, pro parte; Kükenthal, Cyper. Caricoid. 396, pro parte; id., Cyper. Sibir. 107, pro parte; Krylov, Fl. Zap. Sib. III, 480 (respecting Uralian plants); Krechetovich in Fl. SSSR III, 252; Hultén, Atlas, map 395; Kuzeneva in Fl. Murm. II, 100, map 34; Böcher & al., Grønl. Fl. 261; Polunin, Circump. arct. fl. 87, 100, pro parte.
    Ill.: Fl. SSSR III, pl. XVII, fig. 6; Fl. Murm. II, pl. XXX.

Arctic-alpine species. Growing in moist and marshy places, on the shores of rivers and streams, in alpine and lowland tundra.

Soviet Arctic. Murman (mainly the western part, very rare in the east); Polar Ural (upper reaches of River Yelets).

Foreign Arctic. Greenland south of 66°N; Iceland; Arctic Scandinavia.

Outside the Arctic. Scandinavia; the north of England; mountains of Western and Eastern Europe (Carpathians, Tatra); Kola Peninsula; Northern Ural.

64. ***Carex perfusca*** V. Krecz. in Fl. SSSR III (1935), 600; Karavayev, Konsp. fl. Yak. 67.
? *C. aterrima* Hoppe, Caric. germ. (1826), 51.
C. atrata auct. non L. — Treviranus in Ledebour, Fl. ross. IV, 287, pro parte; Scheutz, Pl. jeniss. 176; Krylov, Fl. Zap. Sib. III, 480 (excluding Uralian plants).
? *C. atrata* var. *aterrima* (Hoppe) Hartm. in Svensk och Norsk Excursionfl. (1846), 131; Kükenthal, Cyper. Caricoid. 398, pro parte; id., Cyper. Sibir. 108; Krylov, l. c. 480.
C. atrata ssp. *perfusca* (V. Krecz.) T. Koyama in Journ. Jap. Bot. 30 (1955), 312.
C. caucasica ssp. *perfusca* (V. Krecz.) T. Koyama in Journ. Fac. Sci. Univ. Tokyo III, 8 (4) (1962), 198.

Ill.: Fl. SSSR III, pl. XVII, fig. 7.

Growing on river shores, and in moist meadows and marshes.

Carex perfusca is extremely close to *C. atrata*, from which it differs mainly in having dark purplish brown ripe perigynia. The difference in the colour of the perigynia is extremely constant.

Soviet Arctic. Lower reaches of the Yenisey (Lukovaya Channel, Dudinka, Khantayka, Norilsk Mountains).

Foreign Arctic. Not occurring.

Outside the Arctic. Southern Ural (Yaman-Tau); Tien Shan; Tarbagatay; Altay; Kuznetskiy Alatau; Sayans; basin of the upper course of the Yenisey; Prebaikalia; Transbaikalia; Northern Mongolia; northern part of Central Siberian Plateau; basins of the Zeya, of the Aldan and of upper right-bank tributaries of the Kolyma; coasts of Sea of Okhotsk; Southern Sikhote-Alin; Sakhalin; Honshu; Korean Peninsula.

Among the numerous herbarium specimens of *C. perfusca* we have not noticed plants with ferruginous yellow perigynia, like in *C. atrata*. However, within the range of *C. atrata* (both in Western Europe and Scandinavia) specimens whose perigynia are coloured like in *C. perfusca* are occasionally found. Similar specimens from the Alps were described by Hoppe (l. c.) as a separate species, *C. aterrima*.

We have not seen plants from the Alps with dark purplish brown perigynia. Specimens with such perigynia collected in the Eastern Carpathians and Western Tatra are practically indistinguishable from *C. perfusca*. Nevertheless, we cannot yet reliably assert that *C. aterrima* and *C. perfusca* are identical. The question of the relationships between these two species and between them and other closely related species, especially the Caucasian *C. Medwedewii* Lesk. and certain Western North American species, requires special investigation.

65. ***Carex Buxbaumii*** Wahlb. in Kongl. Vet. Akad. Nya Handl. XXIV (1803), 163; Treviranus in Ledebour, Fl. ross. IV, 285; Kükenthal, Cyper. Caricoid. 393, pro parte; id., Cyper. Sibir. 105, pro parte; Krylov, Fl. Zap. Sib. III, 479 (excl. var. *alpica*); Perfilev, Fl. Sev. I, 114; Krechetovich in Fl. SSSR III, 276; Hultén, Fl. Al. II, 350, map 281; id., Atlas, map 390; Kuzeneva in Fl. Murm. II, 106, map 36; Böcher & al., Grønl. Fl. 262; Hultén, Amphi-atl. pl. 272, map 254; Karavayev, Konsp. fl. Yak. 68; Polunin, Circump. arct. fl. 87, 101, pro parte
C. fusca auct. non All. — Ostenfeld, Fl. arct. 1, 65.
C. polygama Schkuhr, Riedgrass. I (1801), 84; non J.F. Gmel. (1791).

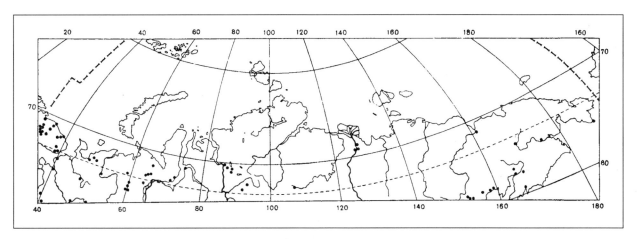

MAP III-49 Distribution of *Carex angarae* Steud.

Ill.: Russk. bot. zhurn. 3/6 (1911), fig. 86; Fl. Murm. II, pl. XXXII.
   Circumpolar boreal species just penetrating the Arctic. Growing in marshes, marshy meadows, and on shores of rivers and lakes.

Soviet Arctic. Murman [Taybola settlement in the valley of the Kola River, also according to O.I. Kuzeneva (l. c.) the lower reaches of the River Iokanga and Cape Svyatoy Nos].

Foreign Arctic. Greenland (extreme south); Arctic Scandinavia.

Outside the Arctic. Western Europe; Scandinavia; northern and central parts of the European part of the USSR; Precaucasus; the south of West Siberia; Kazakhstan; basin of the upper reaches of the Yenisey; Prebaikalia; Transbaikalia (lower reaches of River Selenga); Yakutia (Lena Valley near the mouth of the River Biryuk); northern half of North America; North Africa (extreme north of Algeria); ? Australia.

66. **Carex adelostoma** V. Krecz. in Fl. SSSR III (1935), 275 and 603; Hultén, Atlas, map 391; id., Amphi-atl. pl. 36, map 18; Kuzeneva in Fl. Murm. II, 105, map 35.

   C. Buxbaumii var. *alpicola* Hartm., Handb. Scand. Fl., ed. 1 (1820), 41; Kükenthal, Cyper. Caricoid. 394; Krylov, Fl. Zap. Sib. III, 479.

   Ill.: Fl. Murm. II, pl. XXXII.
       Growing in moss-sedge fens, in marshy tundras, and on rocky slopes.

   Soviet Arctic. Murman (district of Kola Bay, Teriberka, River Iokanga, Cape Svyatoy Nos).

   Foreign Arctic. Iceland (single locality in NW part); Arctic Scandinavia.

   Outside the Arctic. Kola Peninsula; Karelia; the Baltic; Prepolar Ural (basin of the River Lyapin); sporadically in East Siberia (district of Achinsk, basin of the River Daldyn a tributary of the Markha, Barguzinskiy Range); in North America at Great Bear Lake, at isolated localities near the west side of Hudson Bay, and in the northern part of Labrador and Newfoundland.

67. **Carex angarae** Steud., Syn. Cyper. (1855), 190; Krechetovich in Fl. SSSR III, 171; Kalela in Ann. Bot. Soc. Vanamo 19, 3, 1, taf. VII; Hultén, Atlas, map 394; Kuzeneva in Fl. Murm. II, 105, map 35; Karavayev, Konsp. fl. Yak. 68.

   C. alpina β *inferalpina* Wahlb., Fl. Lapp. (1812), 241; Kükenthal, Cyper. Caricoid. 386; id., Cyper. Sibir. 101; Krylov, Fl. Zap. Sib. III, 475 (pro f. *inferalpina* Wahlb.); Perfilev, Fl. Sev. I, 114 (pro var. *inferalpina* Wahlb.).

   ? *C. media* R. Br. in Richardson, Bot. App. Franklin Journ. (1823) 750; Hultén, Circump. pl. I, 36, map 29.

   C. brachylepis Turcz. ex Bess. in Flora XVII, 1 (1834), 26 (nomen nudum).

C. alpina auct. non Sw. — Kükenthal, Cyper. Caricoid. 304, pro parte; Krylov, l. c. 475, pro parte; Leskov, Fl. Malozem. tundry 34.

C. norvegica ssp. *inferalpina* (Wahlb.) Hult., Fl. Al. II (1942), 348, map 280.

Ill.: Fl. Murm. II, pl. XXXI.

Species predominantly of the forest zone, rarely occurring in the subalpine zone and still more rarely in the alpine zone of mountains; penetrating the Arctic in river valleys.

Growing on the shores of rivers and streams (in shrub–mixed herb communities, alder carr, grass–mixed herb meadows), and on margins of marshes.

Soviet Arctic. Murman; Malozemelskaya Tundra (Kharisova Viska and the River Sula near the mouth of the River Shchuchya, upper reaches of the River Shchuchya); Polar Ural; district of Ob Sound; lower reaches of the Yenisey, Lena and Kolyma; Anadyr Basin (Pekulney Range and mouth of the Bolshoy Peledon River); Penzhina Basin; Bay of Korf. (Map III–49).

Foreign Arctic. Arctic Alaska; lower reaches of Mackenzie River; ? Arctic Scandinavia.

Outside the Arctic. Scandinavia; Tirolean Alps; Kola Peninsula; the northeast of the European part of the USSR; Northern and Middle Urals; Altay; southern half of Siberia, north to the Verkhoyansk Mountains; basins of upper right-bank tributaries of the Kolyma; Primorskiy Kray; coasts of the Sea of Okhotsk; Northern Mongolia; the north of the Korean Peninsula; Alaska; Canada and the northern United States (approximately north of 46°N).

Carex angarae is in all probability identical with *C. media* R. Br., described earlier from the northwestern forested part of Canada. The specimen from the Hooker herbarium considered to be the type of *C. media*, which we received from Kew, is completely identical with *C. angarae*. This specimen is annotated in the hand of an old worker (signature illegible): "Probably *C. media* R. Br. in Richardson suppos..." In addition the name "D-r Richardson," the author of the Botanical Appendix in which the description of *C. media* was published, has been written on it by someone unknown. But there are no author's labels on this specimen. Nor also is the place of collection indicated. Consequently, we do not have complete confidence that the stated specimen is in fact the type of *C. media*.

Carex angarae varies throughout its range in the degree of granulosity of the perigynia and in the relative length of the scales and perigynia (see also the comments under *C. norvegica* Retz.).

68. **Carex norvegica** Retz., Fl. Scand. (1779) 179; Kalela in Ann. Bot. Soc. Vanamo 19, 3, 12, taf. VII; A.E. Porsild, Ill. Fl. Arct. Arch. 54, map 88; Böcher & al., Grønl. Fl. 261; Hultén, Amphi-atl. pl. 92, map 74; Polunin, Circump. arct. fl. 87, 103.

C. Halleri Gunn., Fl. Norv. (1772) 106, pro parte (excl. typ.).

C. alpina Sw. in Liljeblad, Svensk Fl., ed. 2 (1798), 26; Treviranus in Ledebour, Fl. ross. IV, 286; Ostenfeld, Fl. arct. 1, 62, pro parte; Kükenthal, Cyper. Caricoid. 384, pro parte; id., Cyper. Sibir. 100, pro parte; Krylov, Fl. Zap. Sib. III, 475, pro parte; Perfilev, Fl. Sev. I, 114, pro parte; non Schrank (1789), nec Honck. (1792).

C. mimula V. Krecz. in Fl. SSSR III (1935), 266 and 603.

C. Halleri auct. non Gunn. — Krechetovich in Fl. SSSR III, 265; id. in Areal, map 31, pro parte; Hultén, Atlas, map 393; Kuzeneva in Fl. Murm. II, 104, map 34; Karavayev, Konsp. fl. Yak. 68.

Ill.: Fl. Murm. II, pl. XXXI; Polunin, l. c. 86.

Arctic-alpine species. Growing on (often rocky) slopes of mountains and valleys, on shoreline cliffs, in moist or marshy places in shrub-moss tundra, on river shores, and on the margins of mossy fens.

Soviet Arctic. Murman; Kanin; Timanskaya Tundra (Kharisova Viska on the Velikaya River); eastern edge of the Bolshezemelskaya Tundra; Polar Ural; lower reaches of the Yenisey (Dudinka); lower reaches of the Lena (River Ayakit); lower reaches of

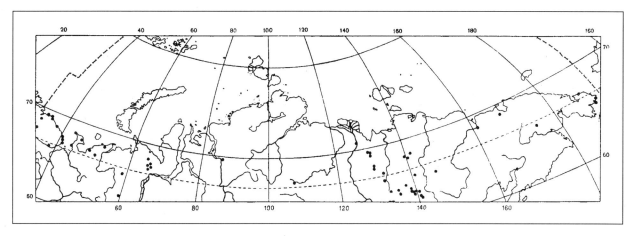

Map III–50 Distribution of *Carex norvegica* Retz.

the Kolyma (vicinity of Panteleikha settlement); district of Chaun Bay (Baranikha, Pevek); Chukotka Peninsula (Lawrence and Litke Bays); near Chuvanskoye settlement in the valley of the Belaya River (tributary of the Anadyr); lower reaches of Penzhina. (Map III–50).

Foreign Arctic. Southern part of Baffin Island; Labrador; SW and East Greenland; Iceland; Arctic Scandinavia.

Outside the Arctic. Extreme north of Scotland; Swiss Alps; Northern Scandinavia; Kola Peninsula; Northern and Middle Urals; Altay; basin of the upper course of the Yenisey; Sayans (Tunkinskiy district); Verkhoyansk Range; Suntar-Khayata Range; Yablonovyy Range; basin of the Zeya (River Namuga); Northern Mongolia; North America (Hudson Bay, Labrador).

This plant should not be called *C. Halleri* Gunn., as was done by V.I. Krechetovich in the "Flora of the USSR," but *C. norvegica* Retz.

Carex Halleri Gunn. was published (Gunner, l. c.) in the following form: "*Carex (Halleri) spica androgyna: terminali tripartita.* Hall. 1356. Oed. fl. dan. VII, 8, t. 403. Scheuchz. hist. t. II, f. 8, citante Oedero. Habitat in alpibus. Oeder." The diagnosis given is almost completely taken from the work of Haller (Historia plantarum, 1769), and that author is cited first. Thus, Gunner (1772) based his species primarily on the plant which Haller had in mind. In the second source cited (Fl. Danica, 1770, but volume III not VII), the plant here under treatment (i.e., *C. norvegica*) is illustrated. From Haller's diagnosis it is impossible to determine to which species the plant described by him belonged, but Retzius (l. c.), who in all probability saw it, stated in his description of *C. norvegica* that the latter was "a Car. Hall. ft. H. 1356 distinctissima." Thus Retzius made the equivalent to a lectotype designation for *C. Halleri*, a species which was based on at least two different types, Haller's plant and a plant different from it (= *C. norvegica*). That lectotype is Haller's plant, which is unknown to us.

Carex norvegica varies in the degree of scabrousness of the distal margins of the perigynia. Individuals posessing perigynia with scabrous-aciculate margins and beaks occur according to our observations in Scandinavia and on the Kola Peninsula (not uncommonly, but not everywhere). In plants from other parts of the range the perigynia bear isolated acicules or are completely smooth. Similar specimens were referred by Kalela (l. c.) to a separate subspecies, ssp. *inserrulata* Kalela distributed in NE North America and Greenland. In fact the range of this subspecies (if recognized as such) is considerably wider, since nonscabrous or only slightly scabrous perigynia are also found in plants from Siberia and the European part of the USSR (both from inside and outside the Arctic). In Chukotka Kalela (l. c.

30) recognized another subspecies, *C. norvegica* ssp. *conicorostrata* Kalela. In specimens of this subspecies the perigynia taper gradually (without incurvation) to the beak. The perigynia of the other subspecies of *C. norvegica* (i.e., *C. norvegica* s. str. and *C. norvegica* ssp. *inserrulata*) taper rather abruptly to a short beak. *Carex norvegica* ssp. *conicorostrata* was based only on the type material (Bering Strait, St. Lawrence Bay, Luetke Harbour, 1881, Aurel and Arthur Krause, No. 213a. An isotype of this subspecies is present in the herbarium of the Botanical Institute of the USSR Academy of Sciences in Leningrad). Consequently it is hard to say at the present time whether the given type specimens represent a distinct race or a chance deviation. In general the question of intraspecific differentiation in *C. norvegica* needs further investigation.

Carex norvegica is rather well delimited from the closely related *C. angarae* Steud. In certain arctic districts the ranges of the two species partly overlap one another. Specimens intermediate between the two species sometimes occur.

69. ***Carex sabulosa*** Turcz. ex Kunth, Enum. pl. II (1837), 432; Krechetovich in Fl. SSSR III, 275, pro parte (excluding Tibetan plants); Karavayev, Konsp. fl. Yak. 68; Polunin, Circump. arct. fl. 87, 104.

    C. alpina var. β Trev. in Ledebour, Fl. ross. IV (1852), 286.

    C. melanantha var. *sabulosa* Kük., Cyper. Caricoid. (1909), 392; id., Cyper. Sibir. 104.

    Ill.: Fl. SSSR III, pl. XVI, fig. 4.

    Growing on riparian sands.

    Soviet Arctic. Lower reaches of Lena (Chekurovka, mouth of the River Beder, near the mouths of the rivers Beris, Balagannakh and Kissilyakh).

    Foreign Arctic. Not occurring.

    Outside the Arctic. Basin of the upper course of the Yenisey; Prebaikalia; Transbaikalia; Yakutia (basin of the River Vilyuy and Siktyakh settlement on the lower reaches of the Lena); Central Asia (Balkhash region); Northern Mongolia.

70. ***Carex Gmelinii*** Hook. et Arn., Bot. Beech. Voy. III (1832), 118, tab. 27; Treviranus in Ledebour, Fl. ross. IV, 288; Ostenfeld, Fl. arct. 1, 66; Kükenthal, Cyper. Caricoid. 396; id., Cyper. Sibir. 106; Krechetovich in Fl. SSSR III, 277; Hultén, Fl. Al. II, 352, map 283; Polunin, Circump. arct. fl. 87, 102.

    Ill.: Kükenthal, Cyper. Caricoid., fig. 62; Ostenfeld, l. c., fig. 36.

    Growing mainly in maritime meadows and on sea cliffs and sandbars, but sometimes at sites remote from the sea.

    Soviet Arctic. East coast of Chukotka Peninsula (Lawrence Bay and vicinity of Senyavin Hot Springs); district of the mouth of the Anadyr; lower reaches of Penzhina; Bay of Korf.

    Foreign Arctic. Alaska (Kotzebue Sound).

    Outside the Arctic. Coasts of Sea of Okhotsk; Primorskiy Kray; Sakhalin; Kamchatka; Commander and Kurile Islands; Hokkaido; coasts of the Sea of Japan and the Yellow Sea; coasts of Alaska with adjacent islands and of British Columbia.

71. ***Carex holostoma*** Drej., Rev. Car. bor. (1841) 29 (excl. syn.); Ostenfeld, Fl. arct. 1, 62; Kükenthal, Cyper. Caricoid. 387; Krechetovich in Fl. SSSR III, 264; id. in Areal, map 35; Kalela in Ann. Bot. Soc. Vanamo 19, 3, 9; Hultén, Atlas, map 392; Kuzeneva in Fl. Murm. II, 102; A.E. Porsild, Ill. Fl. Arct. Arch. 54, map 89; Böcher & al., Grønl. Fl. 261; Hultén, Amphi-atl. pl. 224, map 206; Polunin, Circump. arct. fl. 94, 102.

    Ill.: Ostenfeld, l. c., fig. 32; Kükenthal, l. c., fig. 60, F, G; Polunin, l. c. 95.

    Arctic species with disjunct range. Growing in moist tundra.

    Soviet Arctic. Murman [recorded by Hultén (Atlas) for the Pechenga district]; mouth of the Yenisey (Krechetovich in Areal); Chaun Plain; Chukotka Mountains (River Kuekvun); SE part of Chukotka Peninsula (Lorino); Anadyr Basin.

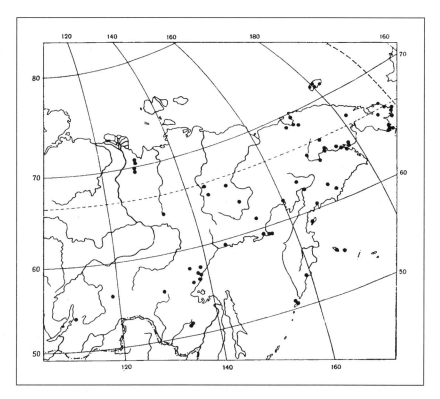

MAP III–51 Distribution of *Carex podocarpa* R. Br.

    Foreign Arctic. Arctic Alaska (Kotzebue Sound); western part of arctic coast of Canada; Southern Baffin Island; Labrador; SW Greenland; Northern Iceland (single locality); Arctic Scandinavia.
    Outside the Arctic. Extreme north of Norway and Sweden; central part of Kola Peninsula [according to Hultén (Atlas) and Krechetovich (Areal)]; Verkhoyansk Range (basins of the rivers Bryungada, Nerganchan and Sartang); basins of upper right-bank tributaries of the Kolyma (valleys of the Rivers Tala and Buyunda); Alaska; Great Bear Lake and the western side of Hudson Bay.

72. ***Carex podocarpa*** R. Br. in Richardson, Bot. App. Franklin Journ. (1823), 751; Ostenfeld, Fl. arct. 1, 63; Kükenthal, Cyper. Caricoid. 411; Krechetovich in Bot. mat. Gerb. Bot. inst. AN SSSR IX, 19; Hultén, Fl. Al. II, 362, map 295; Polunin, Circump. arct. fl. 94, 104; Derviz-Sokolova in Bot. mat. Gerb. Bot. inst. AN SSSR XXI, 73.
    C. koraginensis Meinsh. in Acta Horti Petropol. XVIII, 3 (1901), 351.
    C. behringensis C.B. Clarke in Bull. Misc. Inform. Add., ser. VIII (1908), 81; Krechetovich in Fl. SSSR III, 259.
    C. Tolmiei var. *nigella* (Boott) Kük., l. c. 411, pro parte; id., Cyper. Sibir. 109, pro parte.
    C. Tolmiei var. *invisa* (Bailey) Kük., Cyper. Caricoid. (1909), 109 (respecting Dybovskiy's plants).
    C. pauxilla V. Krecz. in Fl. SSSR III (1935), 260 and 601; Karavayev, Konsp. fl. Yak. 67.
    Ill.: Polunin, l. c. 95.
        Growing in shrub-sedge tundra on mountain slopes, on shores of rivers and streams, and in moist meadows and marshes.
    Soviet Arctic. Middle course of the Kharaulakh River; basin of Chaun Bay; Wrangel Island; Chukotka Mountains; Chukotka Peninsula from Uelen to the Bay of Krest (inclusively); Anadyr Basin; the Penzhina and the Bay of Korf. (Map III–51).

Foreign Arctic. Not occurring.

Outside the Arctic. Barguzinskiy, Yablonovyy (River Udyum Basin) and Kalarskiy Ranges; basin of the middle course of the River Maya; Verkhoyansk-Kolymsk mountain country; coasts of the Sea of Okhotsk; Kamchatka; Commander and Kurile Islands (Shumshu Island); Alaska; western part of Canada and the United States.

Carex podocarpa was described from NW Canada (Point Lake). The perigynia of the type specimen of this species, received by us from Kew, possess smooth margins. In plants from Soviet territory usually referred to *C. podocarpa*, the margins of the perigynia vary from being completely smooth to more or less finely scabrous-aciculate on their distal half. In plants from the Chukotka Peninsula the perigynia are predominantly smooth or with only isolated acicules, but in specimens from the Anadyr Basin the perigynia are almost always distinctly aciculate. In districts outside the Arctic (Transbaikalia, the Verkhoyansk Range, the Kolyma Basin, and the coasts of the Sea of Okhotsk), specimens occur with both smooth and scabrous perigynia. In all probability *C. podocarpa* consists of two subspecies, a Chukotkan-American subspecies with smooth perigynia and an Anadyr-Okhotsk subspecies with more or less scabrous perigynia, but this question can only be decided after study of more or less extensive American material of this species.

Extremely close to *C. podocarpa* is *C. koraginensis* Meinsh. described from Kamchatka (Koraginskiy Island). This is characterized by its describer as having aciculate perigynia and acuminate scales. Study of the type specimens of *C. koraginensis* and comparison of them with the type of *C. podocarpa* has not revealed any substantial differences. The scales in *C. podocarpa* are lanceolate or elongate-ovate, acute or more rarely slightly acuminate or obtusish. In authentic *C. koraginensis* the scales are lanceolate, acute or with very short-awned tips (awns not more than 0.3–0.5 mm long), with scales of both types present at the same level on the same spikelet. Plants from other districts of Kamchatka, for example the basin of the River Avacha, possess scales with longer awnlike tips (0.7–1.5 mm long). Specimens with acute or obtusish scales are also not uncommon in Kamchatka. Specimens with slightly awn-tipped scales also occur in the Penzhina and Anadyr Basins and in Transbaikalia. Thus, the differences between *C. koraginensis* and *C. podocarpa* are practically nonexistent, and the two species are identical to one another. Possibly the Kamchatka plants with awned tips of the scales more strongly expressed than in the type of *C. koraginensis* represent a very weakly differentiated race of *C. podocarpa* s. l. or, more probably, a variety or form of the latter. Specimens from the Anadyr Basin and Chukotka determined as *C. koraginensis* by V.I. Krechetovich are in no way different from specimens recognized by us as *C. podocarpa*, nor from the type of the latter.

73. **Carex Tolmiei** Boott in Hook., Fl. bor.-amer. II (1840), 224; id., Illustr. Carex II, 100; Kükenthal, Cyper. Caricoid. 411, pro parte; Krechetovich in Bot. mat. Gerb. Bot. inst. AN SSSR IX, 20.

C. nesophila H.T. Holm in Amer. Journ. Sci., ser. IV, XVII (1904), 315; Hultén, Fl. Al. II, 361, map 194; Polunin, Circump. arct. fl. 94, 103.

C. arakamensis C.B. Clarke in Bull. Misc. Inform. Add., ser. VIII (1908), 82; Krechetovich in Fl. SSSR III, 265.

C. melanostoma Fisch. ex V. Krecz. in Fl. SSSR III (1935), 261 and 602 (respecting typical and Anadyr-Chukotkan plants).

Ill.: Polunin, l. c. 95.

Growing on rocky slopes of hills, in sedge-moss-dwarf shrub meadows, and on streamsides.

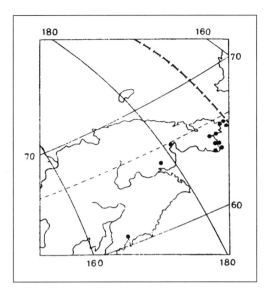

MAP III–52 Distribution of *Carex Tolmiei* Boott.

Soviet Arctic. Chukotka Peninsula (eastern and southern parts from Cape Dezhnev to the Bay of Krest inclusively); Ratmanov Island; basin of left-bank tributaries of the Anadyr; Bay of Korf (River Kultushnaya). (Map III–52).

Foreign Arctic. Arctic coast of Alaska (Lake Peters at 69°N) and the Seward Peninsula.

Outside the Arctic. Alaska and adjacent islands; Oregon and Montana in the USA.

74. ***Carex macrochaeta*** C.A. Mey. in Mém. Acad. Sci. St.-Pétersb. Sav. Étr. I (1831), 224, tab. XIII; Treviranus in Ledebour, Fl. ross. IV, 305; Kükenthal, Cyper. Caricoid. 412; id., Cyper. Sibir. 110; Hultén, Fl. Kamtch. I, 195; Krechetovich in Fl. SSSR III, 259; Mackenzie in N Amer. Fl. XVIII, 6, 352; Hultén, Fl. Al. II, 356, map 290; Polunin, Circump. arct. fl. 94.

C. excurrens Cham. ex Steud., Syn. Cyper. (1855) 228.

Ill.: Kükenthal in Engler, Pflanzenreich IV, 20, fig. 65; Fl. SSSR III, pl. XVII, fig. 8.

Growing in moist places in tundra.

Soviet Arctic. Chukotka Peninsula (Lawrence Bay, Bay of Krest); Anadyr Basin (the crossing to the River Utesiki).

Foreign Arctic. Not occurring.

Outside the Arctic. ? Kamchatka; Kurile and Commander Islands; South and SW Alaska with adjacent islands; Yukon; Vancouver Island.

75. ***Carex stylosa*** C.A. Mey. in Mém. Acad. Sci. St.-Pétersb. Sav. Étr. I (1831), 222, pl. XII; Treviranus in Ledebour, Fl. ross. IV, 305; Ostenfeld, Fl. arct. 1, 62; Kükenthal, Cyper. Caricoid. 395; id., Cyper. Sibir. 105, pro parte (excluding Kolyma plants); Krechetovich in Fl. SSSR III, 262; Hultén, Fl. Al. II, 352, map 282; Polunin, Circump. arct. fl. 93, 105.

C. nigritella Drej., Rev. Car. bor. (1841), 32.

C. beringiana Cham. ex Steud., Syn. Cyper. (1855), 229.

C. stylosa var. *nigritella* (Drej.) Fern. — Böcher & al., Grønl. Fl. 260.

Ill.: Polunin, l. c. 93.

Growing in marshy places.

Soviet Arctic. Chukotka Peninsula (Lawrence Bay).

Foreign Arctic. Arctic Alaska; extreme SW and SE Greenland.

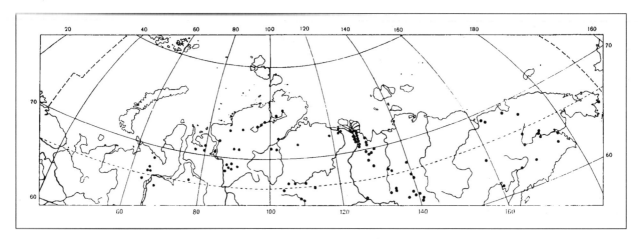

Map III–53 Distribution of *Carex melanocarpa* Cham. ex Trautv.

Outside the Arctic. Kamchatka; western part of North America from Alaska and adjacent islands to Washington State; Labrador and NW Newfoundland.

76. ***Carex melanocarpa*** Cham. ex Trautv., Fl. taim. (1847) 21, tab. 4; Treviranus in Ledebour, Fl. ross. IV, 302; Ostenfeld, Fl. arct. 1, 84; Scheutz, Pl. jeniss. 179; Krylov, Fl. Zap. Sib. III, 460; Krechetovich in Fl. SSSR III, 312; Karavayev, Konsp. fl. Yak. 68; Polunin, Circump. arct. fl. 94, 103.
    C. brachyphylla Turcz. in Bull. Soc. Nat. Mosc. XI, 1 (1838), 104 (nomen nudum).
    C. inornata Turcz., Fl. baic.-dahur. II, 1 (1856), 282.
    C. ericetorum ssp. *melanocarpa* (Cham.) Kük., Cyper. Caricoid. (1909), 440; id., Cyper. Sibir. 117.
    Ill.: Trautvetter, l. c., tab. 4; Ostenfeld, l. c., fig. 62; Polunin, l. c. 95.

    Growing on stony or rocky sites in dry moss-lichen-dwarf birch tundras and in open larch woods with mosses and lichens, also in boulder fields among thickets of cedar-pine stlanik.

    Soviet Arctic. Polar Ural; Karskaya Tundra; the south of the Obsko-Tazovskiy Peninsula; Gydanskaya Tundra; lower reaches of Yenisey; coast and islands of Yenisey Bay; Taymyr; lower reaches of the Khatanga and Olenek; lower reaches and delta of the Lena; Buorkhaya Bay (Tiksi Bay); Northern Anyuyskiy Range; district of Chaun Bay (Baranikha); Chukotka Peninsula (Lawrence Bay); basins of the Anadyr and Penzhina (Palmatkina Range). (Map III–53).

    Foreign Arctic. Not occurring.

    Outside the Arctic. Prepolar Ural (basin of the River Voykar); Central Siberian Plateau (north of 62°N); Central Yakutia; basins of the rivers Kalar, Chara and Aldan; Eastern Sayan (Tunkinskiye Barrens); Verkhoyansk Range; upper right-bank tributaries of the Kolyma; coast of the Sea of Okhotsk (Shantar Islands).

77. ***Carex ericetorum*** Pall., Hist. pl. Palat. II (1777), 580; Treviranus in Ledebour, Fl. ross. IV, 303; Kükenthal, Cyper. Caricoid. 440, pro parte; id., Cyper. Sibir. 117; Krylov, Fl. Zap. Sib. III, 485; Perfilev, Fl. Sev. I, 119; Krechetovich in Fl. SSSR III, 312; Hultén, Atlas, map 376; Kuzeneva in Fl. Murm. II, 116, map 37.
    C. ericetorum var. *strictifolia* Kryl., l. c. 485.
    C. ericetorum ssp. *baicalensis* Gorodk. ex V. Krecz. in Fl. Zabayk. II (1931), 126.
    C. approximata auct. non All. — Krechetovich in Fl. SSSR III, 313.
    Ill.: Fl. SSSR III, pl. XIX, fig. 7; Fl. Murm. II, pl. XXXVI.

    Plant of the forest zone, just penetrating the Arctic.

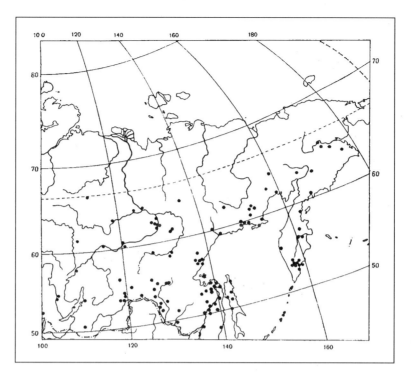

MAP III–54 Distribution of *Carex Vanheurckii* Muell. Arg.

Growing on sands, on dry slopes, in pine woods and heather stands, and in meadows in dry valleys.

Soviet Arctic. Murman (where the Atchernok River enters the Ponoy); lower reaches of Pechora.

Foreign Arctic. Not occurring.

Outside the Arctic. Western Europe; European part of the USSR (except extreme south and Crimea); Precaucasus (Stavropol); Middle and Southern Urals; West Siberia (south of 65°N); Altay; basin of the upper course of the Yenisey (to the latitude of Krasnoyarsk); Lower Tunguska near Yerbogachen; Prebaikalia; Transbaikalia.

78. Carex Vanheurckii Muell. Arg. in Van Heurck, Observ. bot. pl. nov. I (1870), 30; Krechetovich in Fl. SSSR III, 316; Karavayev, Konsp. fl. Yak. 69.

C. amblyolepis Trautv. et Mey., Fl. ochot. phaen. (1856) 99; Komarov, Fl. Kamch. I, 253; Hultén, Fl. Kamtch. I, 198; non Peterm. (1844).

C. pennsylvanica var. *amblyolepis* (Trautv. et Mey.) Kük. in Öfvers. Finska Vet. Soc. Förhandl. XLV, 8 (1902–1909), 8; id., Cyper. Caricoid. 446; id., Cyper. Sibir. 120.

C. pilulifera auct. non L. — Treviranus in Ledebour, Fl. ross. IV, 302.

Ill.: Russk. bot. zhurn. 3/6 (1911), fig. 101 (under *C. pennsylvanica* var. *amblyolepis*).

Growing in dry moss-shrub or moss-lichen tundra, on sandy bluffs along rivers, and on vegetated gravelbars.

Soviet Arctic. Anadyr and Penzhina Basins; Bay of Korf (Kultushnoye settlement). (Map III–54).

Foreign Arctic. Not occurring.

Outside the Arctic. Eastern Sayan (Okinskiy Range); Central Siberian Plateau; Prebaikalia; Central Yakutia; southern part of Verkhoyansk Range; basin of the upper course of the Kolyma; Preamuria; Primorskiy Kray; coasts of Sea of Okhotsk;

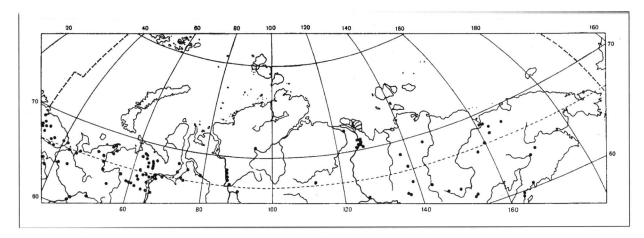

MAP III–55 Distribution of *Carex globularis* L.

Kurile Islands; Sakhalin; Japan (islands of Hokkaido and Honshu); Northern Mongolia; NE China.

79. **Carex globularis** L., Sp. pl. (1753), 976; Treviranus in Ledebour, Fl. ross. IV, 304; Scheutz, Pl. jeniss. 179; Kükenthal, Cyper. Caricoid. 437; id., Cyper. Sibir. 116; Krylov, Fl. Zap. Sib. III, 483; Perfilev, Fl. Sev. I, 120; Krechetovich in Fl. SSSR III, 317; Leskov, Fl. Malozem. tundry 37; Hultén, Atlas, map 373; Kuzeneva in Fl. Murm. II, 118, map 38; Karavayev, Konsp. fl. Yak. 69.

C. mitsuriokensis Lévl. et Vaniot in Bull. Acad. Intern. Géogr. Bot. XIX (1909), 33.

Ill.: Kükenthal, Cyper. Caricoid., fig. 69; Fl. Murm. II, pl. XXXVII.

Species widely distributed in the taiga zone of Eurasia. Growing in the Arctic in moist peaty shrub-moss tundras, in boggy mossy open forests (larch or cedar-pine), on ashes, and on the margins of sphagnum bogs.

Soviet Arctic. Murman [extreme western part (Hultén, Atlas) and Ponoy]; Kanin (southern part); Northern Timan (Perfilev, l. c.); Malozemelskaya Tundra; lower reaches of Pechora; Bolshezemelskaya Tundra; Polar Ural; lower reaches of the Ob (Salekhard); Ob Sound, Obsko-Tazovskiy Peninsula; lower reaches of Yenisey; Taymyr forest-tundra (Volochanka); lower reaches of the Olenek and Lena; lower reaches of Kolyma; Anadyr and Penzhina Basins; Bay of Korf. (Map III–55).

Foreign Arctic. Arctic Scandinavia (isolated occurrences).

Outside the Arctic. Scandinavia; Finland; Poland; northern and central districts of the European part of the USSR (south to approximately 52°N); all Siberia and the whole of the Far East; Northern Mongolia; NE China; the north of the Korean Peninsula.

80. **Carex sabynensis** Less. ex Kunth, Enum. pl. II (1837), 440; Krylov, Fl. Zap. Sib. III, 492; Perfilev, Fl. Sev. I, 120; Krechetovich in Fl. SSSR III, 329; Karavayev, Konsp. fl. Yak. 69; Ohwi, Cyper. Japon. 352.

C. obliqua Turcz. ex Bess. in Flora XVII, 1 (1834), 27 (nomen nudum).

C. Brenneri Christ in Scheutz, Pl. jeniss. (1888), 178; Ostenfeld, Fl. arct. 1, 84.

C. umbrosa ssp. *sabynensis* (Less.) Kük., Cyper. Caricoid. (1909), 468, pro parte; id., Cyper. Sibir. 128.

C. Sadae Lévl. et Vaniot in Bull. Acad. Intern. Géogr. Bot. XIX (1909), 33.

C. Cordouei Lévl. in Bull. Acad. Intern. Géogr. Bot. XIX (1909), 34.

C. eriandrolepis Lévl., l. c. 34.

C. pisiformis var. *subebracteata* Kük., Cyper. Caricoid. (1909), 447; id., Cyper. Sibir. 132 (respecting plants from Maximowicz and Fr. Schmidt).

Ill.: Ostenfeld, l. c., fig. 61 (under *C. Brenneri* Christ); Kükenthal, Cyper. Sibir., fig. 107.
   Growing in moist meadows among riparian shrubbery, in dwarf birch tundras, on vegetated mountain slopes, in open forests, and on limestone rocks.
Soviet Arctic. Eastern part of Bolshezemelskaya Tundra; Karskaya Tundra; Polar Ural; lower reaches of Yenisey.
Foreign Arctic. Not occurring.
Outside the Arctic. Urals (mainly the Northern and Middle); Altay; Sayans; Norilsk Mountains; basin of the Lower Tunguska; Yenisey Ridge; Yakutia north of the Vilyuy; basin of the upper course of the Aldan; Prebaikalia; Preamuria; Primorskiy Kray; coast of the Sea of Okhotsk (Ayan); Sakhalin; Kurile Islands; Japan; Korean Peninsula; NE China; Northern Mongolia.

81. *Carex Trautvetteriana* Kom., Fl. Manchzh. I (1901), 393 (in nota); Krechetovich in Fl. SSSR III, 331; Karavayev, Konsp. fl. Yak. 69.
C. ebracteata Trautv. in Acta Horti Petropol. V (1877), 125; non Philip. (1864).
C. pisiformis var. *ebracteata* (Trautv.) Kük., Cyper. Caricoid. (1909), 447; id., Cyper. Sibir. 133.
   Growing in dry *Dryas-Kobresia*-small sedge tundras, on carbonate stony slopes, and in open larch forests with dwarf shrubs, mosses and lichens.
Soviet Arctic. Lower reaches of the Anabar, Olenek and Lena.
Foreign Arctic. Not occurring.
Outside the Arctic. Northern part of Central Siberian Plateau [River Velingna (a tributary of the Olenek) and the Markha Basin]; upper reaches of the Aldan (Evota Barren); basin of the River Moma.

82. *Carex livida* (Wahlb.) Willd., Sp. pl. IV (1850), 285; Treviranus in Ledebour, Fl. ross. IV, 292; Kükenthal, Cyper. Caricoid. 510; Komarov, Fl. Kamch. I, 260; Hultén, Fl. Kamtch. I, 202; Krylov, Fl. Zap. Sib. III, 503; Perfilev, Fl. Sev. I, 123; Krechetovich in Fl. SSSR III, 343; Hultén, Fl. Al. II, 371, map 308; id., Atlas, map 381; id., Amphi-atl. pl. 214, map 196; Kuzeneva in Fl. Murm. II, 120.
C. limosa δ *livida* Wahlb. in Kongl. Vet. Akad. Nya Handl. XXIV (1803), 162.
Ill.: Fl. Murm. II, pl. XXXVIII.
   Growing in moss-sedge fens.
Soviet Arctic. Murman [reported by Hultén (Atlas) for the Rybachiy Peninsula, the Pechenga district, and the lower reaches of the Kola, Iokanga and Ponoy].
Foreign Arctic. Arctic coast of Canada (single locality between the Mackenzie and Anderson Rivers); Northern Iceland; Arctic Scandinavia.
Outside the Arctic. Scandinavia; Finland; Kola Peninsula; Karelia; vicinity of Leningrad; Arkhangelsk Oblast (Solza settlement); West Siberia (single locality in the basin of the Upper Pur); Southern Kamchatka; Kurile Islands; the north of the Korean Peninsula; Hokkaido; west coast of North America from Alaska to 38°N; Southern Canada and the northeastern states of the USA.

83. *Carex vaginata* Tausch in Flora IV (1821), 557; Treviranus in Ledebour, Fl. ross. IV, 291; Krylov, Fl. Zap. Sib. III, 503; Perfilev, Fl. Sev. I, 123; Krechetovich in Fl. SSSR III, 344; Hultén, Fl. Al. II, 372; id., Atlas, map 384; Kuzeneva in Fl. Murm. II, 122; A.E. Porsild, Ill. Fl. Arct. Arch. 56, map 93; Böcher & al., Grønl. Fl. 265; Polunin, Circump. arct. fl. 98, 105; Hultén, Circump. pl. I, 88 (excl. *C. falcata* Turcz.), map 79, pro parte.
C. panicea β *sparsiflora* Wahlb., Fl. Lapp. (1812), 236.
C. sparsiflora (Wahlb.) Steud., Nomencl. bot., ed. 2, 1 (1841), 296; Ostenfeld, Fl. arct. 1, 81; Kükenthal, Cyper. Caricoid. 511, pro parte; id., Cyper. Sibir. 146, pro parte; Andreyev, Mat. fl. Kanina 162; Leskov, Fl. Malozem. tundry 40.
C. quasivaginata C.B. Clarke in Bull. Misc. Inform. Add., ser. VIII (1908), 79.

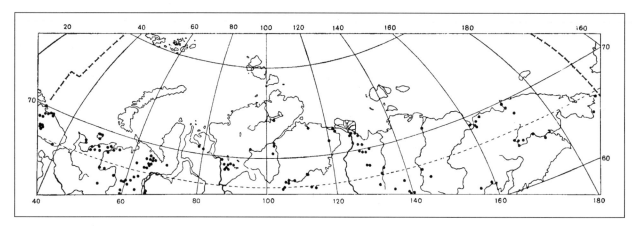

MAP III–56 Distribution of *Carex vaginata* Tausch.

C. melanocarpa auct. non Cham. ex Trautv. — Tolmachev, Fl. Taym. I, 103.
C. algida Turcz. ex V. Krecz. in Fl. SSSR III (1935), 345, 612; Karavayev, Konsp. fl. Yak. 69.

Ill.: Fl. Murm. II, pl. XL.

Plant of the forest zone, also more or less widespread in temperate districts of the Arctic. Growing in moss-shrub tundras, lowland and alpine sedge-cottongrass tundras, moist mixed herb meadows, *Hypnum* fens, river valleys, and open birch or larch forests.

Soviet Arctic. Murman; Kanin; Timanskaya, Malozemelskaya and Bolshezemelskaya Tundras; Kolguyev Island; Polar Ural; Yamal; Gydanskaya Tundra; lower reaches of Yenisey; Taymyr (on SE coast of Lake Taymyr); right bank of the Khatanga; Anabar and Olenek Basins; lower reaches of Lena; Indigirka Valley; lower reaches of the Kolyma (Pokhodsk, Nizhniye Kresty); Northern Anyuyskiy Range (Urney Mountain); basin of Chaun Bay; Chukotka Peninsula; Anadyr Basin. (Map III–56).

Foreign Arctic. Arctic Alaska; arctic coast of Canada; Labrador; southern islands of the Canadian Arctic Archipelago; East Greenland (between 72° and 78°N); Iceland; Arctic Scandinavia.

Outside the Arctic. Great Britain (northern part); Scandinavia; Finland; the north of Central Europe east of 10°E; mountains of Central Europe; European part of the USSR north of 50°N; Precaucasus and Southern Transcaucasus; Urals; West Siberia; Sayans; Central Siberian Plateau; middle course of the Yenisey (vicinities of Yeniseysk and Krasnoyarsk); Verkhoyansk Range; North America north of 42°N.

Carex algida, described by V.I. Krechetovich from the Sayans and reported by him in the "Flora of the USSR" for arctic districts of Europe and Siberia, is identical with *C. quasivaginata* C.B. Clarke described from Arctic Norway. Plants referred to these species are in our opinion no more than an arctic form of *C. vaginata*. Individuals of *C. vaginata* from arctic districts and from the alpine zone of mountains (Sayans, Urals, Verkhoyansk Range) differ from plants from the forest zone only in their shorter stature and the more intense colour of their scales and perigynial beaks (sometimes the scales are completely dark).

Clarke (l. c.) characterized *C. quasivaginata* by the smaller size of the plants and additionally by their few-flowered spikelets. The latter character is indeed found in arctic specimens, but in far from all.

V.I. Krechetovich (l. c.) characterized *C. algida* by its having smaller perigynia than in *C. vaginata* (3–3.5 mm long, not 4 mm) with slightly emarginate beaks; but these differences, as well as several others suggested (spikelet size, colour of perigynia), have not been confirmed in studies of herbarium material. Ripe perigynia with developed achenes are 4–4.5(5) mm long in *C. algida*, that is the same as in *C. vaginata*. Only unripe perigynia with undeveloped or abortive achenes are usually smaller, but not less than 3.5 mm long (perigynia only 3 mm long occur extremely rarely). Among specimens of *C. algida*, plants have been found in which ripe perigynia of normal size and smaller unripe perigynia occur in the same spikelet. In the Arctic, apparently stunted individuals whose perigynia are all abortive with undeveloped achenes turn up rather frequently. Probably such specimens were taken for a distinct species. With respect to the degree of emargination of the beak, arctic plants are also completely indistinguishable from nonarctic. The beak varies in both (it may be emarginate only anteriorly or both anteriorly and posteriorly, in the latter case becoming shallowly bidentate).

In the Anadyr Basin and outside the Arctic in the Sayans, the Verkhoyansk Range and the upper reaches of the Kolyma, *C. vaginata* meets the species which replaces it in Southern Siberia and the Far East, *C. falcata* Turcz. The characters of these species (or perhaps they are really subspecies) overlap (see the key). And in some cases it is impossible to determine whether a given plant belongs to *C. falcata* or to *C. vaginata*.

84. *Carex falcata* Turcz., Fl. baic.-dahur. II, 2 (1856), 276; id. in Bull. Soc. Nat. Mosc. XXVIII, 1 (1855), 341 (nomen nudum); Krechetovich in Fl. SSSR III, 346; Karavayev, Konsp. fl. Yak. 69.

C. sparsiflora var. *Petersii* Kük., Cyper. Caricoid. (1909), 513.
C. sparsiflora var. *falcata* (Turcz.) Kük., Cyper. Sibir. (1911), 146.
C. vaginata auct. non Tausch — Hultén, Circump. pl. I, 88, map 79, pro parte.
Ill.: Fl. SSSR III, pl. XX, fig. 7.

Species mainly of the forest zone, in the northeast of its range penetrating the Arctic where it occurs in open forests of the forest-tundra, fescue-*Dryas* tundras with mosses and lichens, bog margins, and cedar-pine thickets with lichens and *Sphagnum*.

Soviet Arctic. Anadyr and Penzhina Basins.
Foreign Arctic. Not occurring.
Outside the Arctic. Central Siberian Plateau (basin of the upper course of the Vilyuy); Prebaikalia; Transbaikalia; Central Yakutia; Verkhoyansk Range; basin of the upper tributaries of the Kolyma; coasts of the Sea of Okhotsk; Kamchatka; Sakhalin; Preamuria; Primorskiy Kray; Japan; Korean Peninsula; Northern Mongolia; NE China.

85. *Carex pediformis* C.A. Mey. in Mém. Acad. Sci. St.-Pétersb. Sav. Étr. I (1831), 219, tab. X; Treviranus in Ledebour, Fl. ross. IV, 290, pro parte; Kükenthal, Cyper. Caricoid. 490, pro parte; id., Cyper. Sibir. 135, pro parte; Krylov, Fl. Zap. Sib. III, 493, pro parte; Krechetovich in Fl. SSSR III, 368; Karavayev, Konsp. fl. Yak. 69; Polunin, Circump. arct. fl. 97, 104, pro parte.

Ill.: Russk. bot. zhurn. 3/6 (1911), fig. 111; Fl. SSSR III, pl. XIX, fig. 3.

Steppe, meadow-steppe and forest-steppe plant, distributed mainly in central and southern districts of Siberia. Replaced in forested districts of the European part of the USSR, Southern Fennoscandia and Central Europe by the very closely related *C. rhizina* Blytt, a species that perhaps should be considered a subspecies or variety of *C. pediformis*. Just penetrating arctic limits, where it occurs in river valleys on steppelike south-facing rocky slopes and limestone cliffs in the zone of open larch forests.

Soviet Arctic. Lower reaches of the Lena (Chekurovka village, valleys of the Rivers Atyrkan and Neleger); Chaun Plain; Anadyr Basin (River Medvezhka).
Foreign Arctic. Not occurring.
Outside the Arctic. The northeast of the Komi ASSR (Adak on the River Usa); Central

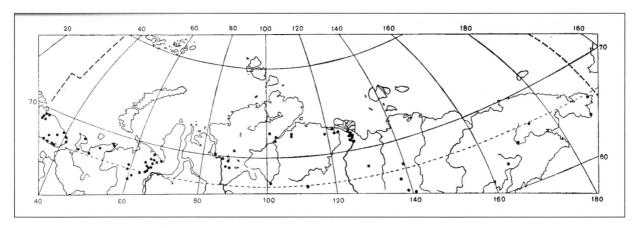

MAP III–57 Distribution of *Carex glacialis* Mackenz.

Russian Hills and eastern districts of the European part of the USSR; Middle and Southern Urals; Altay; Southern West Siberia; Northern Kazakhstan; Yenisey Basin south of 62°N; Central Sayan; Prebaikalia; Transbaikalia; upper course of the Vilyuy; Central Yakutia; Verkhoyansk-Kolymsk mountain country; Amur Basin; Primorskiy Kray; Mongolia; Northern and NE China.

86. **Carex supina** Wahlb. ssp. **spaniocarpa** (Steud.) Hult., Fl. Al. II (1942), 365; id., Circump. pl. I, 188, map 178; A.E. Porsild, Ill. Fl. Arct. Arch. 55, map 90; Böcher & al., Grønl. Fl. 256.
   C. spaniocarpa Steud., Syn. Cyper. (1855) 225; Krechetovich in Fl. SSSR III, 375; Yurtsev in Bot. zhurn. XLVII, 3, 318, map 1.
   C. supina L. s. l. — Ostenfeld, Fl. arct. 1, 86; Kükenthal, Cyper. Caricoid. 455, pro parte; Polunin, Circump. arct. fl. 95, 105.
   Ill.: Porsild, l. c., fig. 20, d.

   Subarctic cryophilic steppe race of the steppe species *C. supina* s. l. Distributed extremely sporadically in the Arctic and only on its southern fringe; here it grows on dry sandy or rocky south-facing slopes, sometimes together with other steppe plants.

   Differing from *C. supina* L. s. str. in having narrower [3–3.5(4) mm long, about 1.5 mm wide] ovoid perigynia (consequently less abruptly transformed into the beak) that are reddish brown distally, and scales usually equalling or almost equalling the perigynia; in *C. supina* s. str. the perigynia are globose-ovoid, 3–3.5(4) mm long and 2–2.3 mm wide, abruptly tapering to the beak, stramineous yellow with brown beak, and the scales usually shorter than the perigynia. Subspecies *spaniocarpa* was considered by V.I. Krechetovich in the "Flora of the USSR" to be a separate species, *C. spaniocarpa* Steud., but since the differences between the latter and *C. supina* are too minor, it is more correct to rank them as subspecies.

   Soviet Arctic. Lower reaches of the Yenisey (vicinity of Dudinka); lower reaches of the Anabar (Lake Ulakhan-Kyuyel); district of Chaun Bay (Baranikha); basin of the River Belaya, a tributary of the Anadyr; Bay of Korf. Reported by Hultén (Circump. pl. I, map 178) for the lower reaches of the Lena, Alazeya and Kolyma.
   Foreign Arctic. Arctic coast of Canada; Baffin Island; Southampton Island; SW and East Greenland.
   Outside the Arctic. Southern part of the Verkhoyansk-Kolymsk mountain country (basins of the Yana and Indigirka); Alaska and the northern part of Canada.

87. ***Carex glacialis*** Mackenz. in Bull. Torr. Bot. Club XXXVII (1910), 244; id. in N Amer. Fl. XVIII, 4, 221; Krechetovich in Fl. SSSR III, 374; Hultén, Fl. Al. I, 266, map 301; id., Atlas, map 407; Kuzeneva in Fl. Murm. II, 126, map 41; A.E. Porsild, Ill. Fl. Arct. Arch. 56, map 91; Böcher & al., Grønl. Fl. 257; Karavayev, Konsp. fl. Yak. 70; Polunin, Circump. arct. fl. 97, 102; Hultén, Circump. pl. I, 30, map 23.

C. pedata Wahlb., Fl. Lapp. (1812) 239; Treviranus in Ledebour, Fl. ross. IV, 236; Scheutz, Pl. jeniss. 177; Ostenfeld, Fl. arct. 1, 87; Kükenthal, Cyper. Caricoid. 495; id., Cyper. Sibir. 138; Krylov, Fl. Zap. Sib. III, 496; Perfilev, Fl. Sev. I, 122; Leskov, Fl. Malozem. tundry 39; non L. (1763).

Ill.: Fl. Murm. II, pl. XLI, fig. 2; Polunin, l. c. 96.

Circumpolar arctic species, avoiding High Arctic districts with severe climatic conditions but rather common in high subarctic mountains.

A characteristic plant of poorly vegetated dry stony slopes with little snow cover derived from limestones or sandstones. Thus, in the lower reaches of the Lena this species is very common on the mountainous right side of the river within the Tuora-Sis Range, where thick layers of Lower Paleozoic limestones are exposed; here it grows in dry or patchy *Dryas-Kobresia*-sedge glades in larch forest (with *Caragana jubata* and *Rhododendron Adamsii*) and in similar patchy *Dryas-Kobresia*-sedge tundras (with *Carex macrogyna, C. Trautvetteriana, C. rupestris, Kobresia Bellardii, K. filifolia* and *K. simpliciuscula*); higher up it is very common on semivegetated limestone screes in beds of flowering plants (together with *Dryas punctata, Salix berberifolia* ssp. *fimbriata* and *Carex rupestris*). The species also occurs under similar conditions further north on the mountainous right side of the Lena (opposite Tit-Ary Island), as well as on the mountainous right side of the Olenek. On the unforested mountain plateaux on the left side of the Lena (Rivers Ayakit, Tigiya, etc.), the species occurs in dry patchy *Dryas* tundras on the loose alluvium of sandbars.

Soviet Arctic. Murman [mouth of the River Voroney; also according to Hultén (Atlas) in the Pechenga district and the lower reaches of the Kola and Ponoy Rivers]; Kanin; Malozemelskaya Tundra; eastern part of the Bolshezemelskaya Tundra (basin of the River Usa, Vorkuta); Karskaya Tundra; Polar Ural; lower reaches of Yenisey; southern part of Taymyr (lower reaches of the River Dudypta, basin of the River Novaya), Popigay Basin; lower reaches of the Olenek and Lena; Chukotka (Teakachin Mound, River Kuvet, River Chegitun, Cape Chaplin); Anadyr Basin (the Medvezhi Hills and the Nelti Range in the basin of the Belaya River); Penzhina Basin (River Oklan). (Map III–57).

Foreign Arctic. Arctic coast of Canada (western part); Canadian Arctic Archipelago; SW and East Greenland; Iceland; Arctic Scandinavia.

Outside the Arctic. Scandinavia; Finland (northern part); Kola Peninsula; Prepolar, Northern and Middle Urals; lower reaches of the Bolshoy Pur; northern part of the Central Siberian Plateau; Eastern Sayan (Kitoyskiye Barrens); Barguzinskiy Range; Verkhoyansk Range; Cherskiy Range; Kolyma Mountains; basins of the Olekma and Aldan; Alaska and Northern Canada.

88. ***Carex flava*** L., Sp. pl. (1753) 975; Treviranus in Ledebour, Fl. ross. IV, 299; Kükenthal, Cyper. Caricoid. 671; Perfilev, Fl. Sev. I, 126; Krechetovich in Fl. SSSR III, 387; id. in Mat. ist. fl. rastit. SSSR I, 66, map 19; Hultén, Atlas, 101, map 402; id., Fl. Al. II, 377, map 315; id., Amphi-atl. pl. 62, map 43; Kuzeneva in Fl. Murm. II, 127, map 43.

Ill.: Fl. SSSR III, pl. XXI, fig. 1; Fl. Murm. II, pl. XLII, fig. 1,

Plant of the forest zone, penetrating the Arctic only in Northern Europe.

Growing on shores of streams, in moist or marshy meadows, and in grass-sedge fens.

Soviet Arctic. Murman (Pechenga district, Rybachiy Peninsula, Kildin Island and Yarnyshnaya Bay in the Teriberka district).

Foreign Arctic. Iceland; Arctic Scandinavia.

Outside the Arctic. Western Europe; Scandinavia; Finland; European part of the USSR (mainly western half); Northern Caucasus; East Siberia (basin of the River Kudara); Pacific coast of Alaska; Western and Eastern Canada; NW and NE states of the USA; North Africa; ? India.

89. ***Carex Oederi*** Retz., Fl. Scand. Prodr. (1779) 179; Kükenthal, Cyper. Caricoid. 673, pro parte; id., Cyper. Sibir. 173; Krylov, Fl. Zap. Sib. III, 517, pro parte; Perfilev, Fl. Sev. I, 125; Krechetovich in Fl. SSSR III, 390; Hultén, Atlas, map 400; id., Amphi-atl. pl. 11; id., Circump. pl. I, 170, map 162; Kuzeneva in Fl. Murm. II, 128.

    C. flava δ Trev. in Ledebour, Fl. ross. IV (1852), 300.

    Ill.: Fl. SSSR III, pl. XXI, fig. 4; Fl. Murm. II, pl. XLII, fig. 2.

    Growing in marshy or moist sites, in grass-sedge or more rarely mossy fens.

    Soviet Arctic. Murman (Pechenga district and between Teriberka and Kharlovka).

    Foreign Arctic. Iceland; Arctic Scandinavia.

    Outside the Arctic. Western Europe; Scandinavia; Finland; European part of the USSR (mainly western half); Southern Ural; West Siberia (Tomsk Oblast in the vicinity of the Salairskiy mine); Altay; basin of the upper course of the Yenisey (to the mouth of the Angara); Central Sayan; Prebaikalia; North Africa; Azores; northern Atlantic districts of North America.

    In Central Asia (mountains of East Kazakhstan, the Pamiro-Alay and the Tien Shan), *C. Oederi* is replaced by the very closely related *C. philocrena* V. Krecz., and in Southern Kamchatka, Sakhalin, Japan and North America by another closely related species, *C. viridula* Michx. Additionally, in Western Europe and Eastern Labrador another species closely approaching these species occurs, *C. tumidocarpa* Anderss. (= *C. demissa* Hornem.). The entire complex of species related to *C. Oederi* requires special investigation.

90. ***Carex Williamsii*** Britt. in Bull. N.Y. Bot. gard. II, 6 (1901), 159; Hultén, Fl. Al. II, 376, map 314; id., Circump. pl. I, map 48; Raymond in Natur. Canad. 77, 222–227, map 1; A. Löve, D. Löve & Raymond in Canad. Journ. Bot. 35, 745; Polunin, Circump. arct. fl. 100, 105; Yegorova in Nov. sist. vyssh. rast. (1964), 31.

    C. Novograblenovii Kom. in Byull. Bot. sada SSSR XXX, 1/2 (1932), 199; Krechetovich in Fl. SSSR III, 432.

    C. Peshemskyi Malysch. in Bot. mat. Gerb. Bot. inst. AN SSSR XXI (1961), 462.

    Ill.: Polunin, l. c. 101.

    Growing in sedge-moss marshy tundra, sedge-sphagnum bogs, *Hypnum* fens, and on vegetated maritime gravelbars. Occurring very sporadically with large gaps.

    Soviet Arctic. Polar Ural (sources of the River Khadata and the basin of the Sob); Obsko-Tazovskiy Peninsula (Nyda); Buorkhaya Bay (vicinity of Tiksi Bay and Kharaulakh Bay); district of Chaun Bay (Pevek, Mounts Ionay and Elveney); Arakamchechen Island; Anadyr Basin (River Belaya near Ust-Dvukh settlement, mouth of the River Yablonovaya) and Penzhina Basin (upper reaches of the Rivers Oklan and Slovutnaya); Bay of Korf.

    Foreign Arctic. Arctic Alaska; western part of the arctic coast of Canada; Labrador.

    Outside the Arctic. Central Sayan (Idarskoye Belogore); Southern Transbaikalia (basin of the River Chikoy); Verkhoyansk Range (basins of the Rivers Tompo and Bryungada); upper reaches of the Kolyma (basins of the Rivers Tala and Buyunda); sources of the Bureya; Kamchatka (Pinachevo-Nalachevskiy Pass); Alaska; Yukon; the west of Hudson Bay; Northern Labrador.

91. ***Carex capillaris*** L., Sp. pl. (1753) 977; Treviranus in Ledebour, Fl. ross. IV, 295 (excl. var. β); Ostenfeld, Fl. arct. 1, 90, pro parte; Kükenthal, Cyper. Caricoid. 590, pro parte; id., Cyper. Sibir. 157, pro parte; Krylov, Fl. Zap. Sib. III, 511, pro parte;

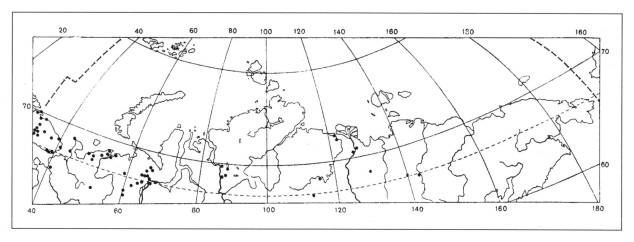

MAP III–58 Distribution of *Carex capillaris* L.

Perfilev, Fl. Sev. I, 124; Krechetovich in Fl. SSSR III, 428, pro parte; Leskov, Fl. Malozem. tundry 36; Hultén, Atlas, map 408; Kuzeneva in Fl. Murm. II, 132, map 45; A.E. Porsild, Ill. Fl. Arct. Arch. 58, map 97, pro parte; A. Löve, D. Löve & Raymond in Canad. Journ. Bot. 35, 748 (excl. *C. fuscidula* V. Krecz.); Böcher & al., Grønl. Fl. 263; Polunin, Circump. arct. fl. 100, 102, pro parte; Hultén, Circump. pl. I, map 47, pro parte.

C. chlorostachys Stev. in Mém. Soc. Nat. Mosc. IV (1813), 68; Krechetovich, l. c. 428.

C. capillaris ssp. *chlorostachys* A. Löve, D. Löve et Raymond in Canad. Journ. Bot. 35 (1957), 749.

Ill.: Fl. Murm. II, pl. XLIV.

Species distributed mainly in the boreal forest and the forest zone of mountains, but also occurring in the forest-tundra, tundra and alpine zones.

Growing in moist or marshy meadows, on the shores of rivers and streams, on the margins of marshes and in marshy shrubbery.

Soviet Arctic. Murman (scattered along the entire coast); Kanin (Cape Konyushin); Timanskaya and Malozemelskaya Tundras; lower reaches of Pechora; Bolshezemelskaya Tundra; Polar Ural; lower reaches of the Yenisey (Dudinka, Khantayka); lower reaches of the Olenek and Lena. (Map III–58).

Foreign Arctic. East Greenland; Iceland; Arctic Scandinavia. It is not yet possible to give more or less accurate information on the distribution of *C. capillaris* L. in the arctic part of North America, where the closely related species *C. fuscidula* grows as well as *C. capillaris*. The former has not been distinguished from *C. capillaris* by investigators of the American flora, and consequently distributional records of the two species have not been separated. According to the literature, *C. capillaris* s. l. occurs throughout the arctic part of North America.

Outside the Arctic. Western Europe; northern half of the European part of the USSR; Caucasus; Urals; the south of West Siberia; ? Altay; Yenisey Basin south of Yeniseysk; Prebaikalia (rare); Central Yakutia; Verkhoyansk Range; Preamuria (Urkan Basin); Canada; northern Atlantic states of the USA (in the western states of Wyoming, Colorado, Idaho, etc. a species close to *C. Karoi* Freyn. apparently occurs; we have not found the true *C. capillaris* among herbarium material from the Western United States).

92. **Carex fuscidula** V. Krecz. ex Egor. in Nov. sist. vyssh. rast. (1964), 36; *C. fuscidula* V. Krecz. in Areal I (1952), 32 (nomen nudum).

C. capillaris auct. non L. — Kükenthal, Cyper. Caricoid. 590, pro parte; Krechetovich

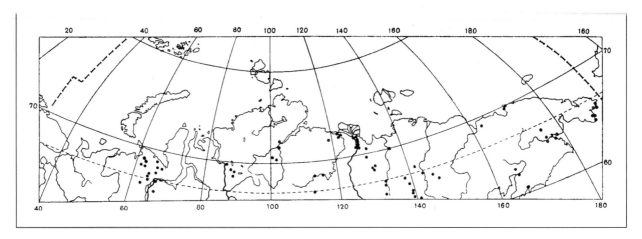

MAP III–59 Distribution of *Carex fuscidula* V. Krecz. ex Egor.

in Fl. SSSR III, 428 (respecting Chukotkan plants); A.E. Porsild, Ill. Fl. Arct. Arch. 59, map 97, pro parte; Polunin, Circump. arct. fl. 100, 102, pro parte; Hultén, Circump. pl. I, map 47, pro parte; Derviz-Sokolova in Bot. mat. Gerb. Bot. inst. AN SSSR XXI, 77.

C. lenaensis auct. non Kük. — Krechetovich in Fl. SSSR III, 429, pro parte (respecting plants from the Siberian Arctic and Chukotka).

Growing in moss-sedge and *Dryas* tundras, more rarely in sparse open larch woods with dwarf shrubs, mosses and lichens.

Carex fuscidula is close to *C. capillaris*, but well differentiated from it morphologically and geographically. Starting with the Bolshezemelskaya Tundra, *C. fuscidula* distinctly predominates over *C. capillaris* and completely replaces it east of the Lena.

Soviet Arctic. Eastern part of Bolshezemelskaya Tundra; Karskaya Tundra; Polar Ural; lower reaches of the Ob (near Salekhard); lower reaches of the Yenisey (basin of the River Dudinka); southern fringe of the Taymyr Peninsula (basins of the Rivers Romanikha and Khatanga); Olenek Basin; lower reaches of Lena; lower course of Yana; Northern Anyuyskiy Range; district of Chaun Bay; east and SE part of Chukotka Peninsula; Arakamchechen Island; basins of the Anadyr and of the lower course of the Penzhina; Bay of Korf. (Map III–59).

Foreign Arctic. Within the arctic part of North America, so far as we can judge on the basis of herbarium specimens we have seen, *C. fuscidula* occurs in Alaska, Ellesmere Island, Southampton Island, and western and eastern parts of Greenland. It is possible that *C. fuscidula* replaces *C. capillaris* throughout the arctic territory of North America.

Outside the Arctic. Prepolar, Northern and Middle Urals; northern and central part of the Central Siberian Plateau (basins of the rivers Kotuy, Chibichete and Markha); east coast of Baykal (Chivyrkuyskiye Barrens); Kalarskiy Range; upper reaches of Aldan; Verkhoyansk Range; Suntar-Khayata Range; basin of upper right-bank tributaries of the Kolyma; North America (Colorado, Manitoba, the east of Hudson Bay, and the White Mountains in New Hampshire).

93. Carex Krausei Boeck. in Engler, Bot. Jahrb. VII (1886), 279; Hultén, Fl. Al. II, 376; id., Circump. pl. I, map 48; A. Löve, D. Löve & Raymond in Canad. Journ. Bot. 35, 747.

C. capillaris var. *nana* f. *Krausei* (Boeck.) Kük., Cyper. Caricoid. (1909), 591.

C. capillaris auct. non L. — Polunin, Circump. arct. fl. 102, pro parte.

C. lenaensis auct. non Kük. — Krechetovich in Fl. SSSR III, 429.

Growing in sandy places on tundra and on river shores.

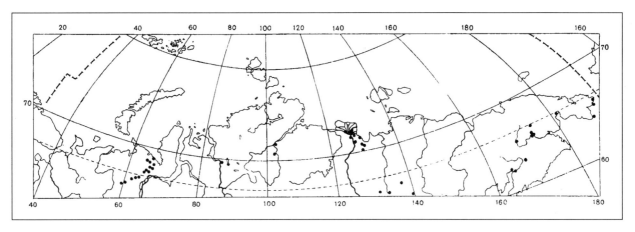

Map III–60 Distribution of *Carex Ledebouriana* C.A. Mey. ex Trev.

Carex Krausei was described from SE Alaska. V.I. Krechetovich in the "Flora of the USSR" mistakenly identified this species with *C. lenaensis* Kük., which is in fact a synonym of *C. Ledebouriana* C.A. Mey. ex Trev. Subsequently specimens belonging to *C. Krausei* were determined by V.I. Krechetovich as *C. Chamissonis*. However, these specimens along with a series of others (including some from the lower Lena) have proved to be identical to an isotype of *C. Krausei* discovered by us in the herbarium of the Botanical Institute of the USSR Academy of Sciences. We note that this type specimen of *C. Krausei* originally determined as *C. capillaris* L. was also erroneously identified with *C. Chamissonis* by V.I. Krechetovich. *Carex Krausei* has been divided by A. and D. Löve (A. Löve, D. Löve & Raymond, l. c.) into two subspecies, ssp. *Krausei* and ssp. *Porsildiana* (Polunin) A. et D. Löve. The latter differs from the typical subspecies in having smaller perigynia (1.3–1.8 mm long). According to the data of these authors, *C. Krausei* ssp. *Krausei* is distributed only in Alaska, while ssp. *Porsildiana* is found on the coast of Hudson Bay, the Ungava Peninsula, Greenland and Iceland.

Within the Soviet Arctic, *C. Krausei* is represented by the typical subspecies. *Carex Krausei* is very close to *C. Boecheriana* A. et D. Löve et Raymond, which in the opinion of these authors is endemic to Greenland. Morphological differences between these species are found solely in the shape of the perigynia: in *C. Krausei* the perigynia are lanceolate, in *C. Boecheriana* ovoid. More significant are the karyological differences: in *C. Krausei* the chromosome number is 2n=38, in *C. Boecheriana* 2n=56.

Soviet Arctic. Lower reaches of the Yenisey (Tolstyy Nos); lower reaches of the Lena (mouth of the River Beder and Tit-Ary Island).

Foreign Arctic. Labrador, Greenland, Iceland.

Outside the Arctic. Northern edge of the Central Siberian Plateau (basin of the River Medvezhya); Khatanga; Altay; Sayans; Transbaikalia; North America (Alaska, Hudson Bay).

94. *Carex Ledebouriana* C.A. Mey. ex Trev. in Bull. Soc. Nat. Mosc. XXXVI, 1 (1863), 540; Krechetovich in Fl. SSSR III, 431; Karavayev, Konsp. fl. Yak. 71; A. Löve, D. Löve & Raymond in Canad. Journ. Bot. 35, 746; Hultén, Circump. pl. I, map 48, pro parte.

C. capillaris β Trev. in Ledebour, Fl. ross. IV (1852), 295.

C. lenaensis Kük. in Öfvers. Finska Vet. Soc. Förhandl. XLV, 8 (1903), 10.

C. capillaris var. *Ledebouriana* (C.A. Mey. ex Trev.) Kük., Cyper. Caricoid. (1909), 591; id., Cyper. Sibir. 157; Krylov, Fl. Zap. Sib. III, 511.

C. capillaris auct. non L. — Polunin, Circump. arct. fl. 102, pro parte.

Arctic-alpine (predominantly alpine) Siberian species. Occurring in the Arctic on rocky or stony slopes in lichen, *Dryas*-lichen and mossy tundras.

Carex lenaensis Kük. described from the lower reaches of the Lena (Tit-Ary Island) is identical with *C. Ledebouriana*. *Carex lenaensis* in the sense of V.I. Krechetovich in the "Flora of the USSR" consisted of two different species, *C. fuscidula* V. Krecz. ex Egor. and *C. Krausei* Boeck.

Soviet Arctic. Eastern part of the Bolshezemelskaya Tundra (River Silova and the upper course of the River Usa); upper reaches of the River Kara; Polar Ural; lower reaches of the Yenisey (Dudinka) and of the Khatanga; lower reaches and delta of the Lena; Buorkhaya Bay (Tiksi Bay) and the Kharaulakh River; Chukotka Peninsula; Anadyr and Penzhina Basins. (Map III–60).

Foreign Arctic. Not occurring.

Outside the Arctic. Prepolar and Northern Urals; Altay; Kuznetskiy Alatau; basin of the upper course of the Yenisey; Sayans; Transbaikalia; northern part of the Central Siberian Plateau (basin of the River Kotuy); Verkhoyansk Range; Preamuria (east of the River Bolshaya Bira); Northern Sikhote-Alin; coast of Sea of Okhotsk (Dzhugdzhur Range and River Lantar); Northern Mongolia.

95. **Carex lasiocarpa** Ehrh. in Hannover. Magaz. IX (1784), 132; id., Beitr. III, 73; Kükenthal, Cyper. Caricoid. 747, pro parte; id., Cyper. Sibir. 193; Komarov, Fl. Kamch. I, 269; Hultén, Fl. Kamtch. I, 212, map 230; Krylov, Fl. Zap. Sib. III, 532; Krechetovich in Fl. SSSR III, 416; Hultén, Atlas, map 420; id., Circump. pl. I, 76, map 67; Kuzeneva in Fl. Murm. II, 130, map 44; Karavayev, Konsp. fl. Yak. 70.

C. filiformis Good. in Trans. Linn. Soc. II (1794), 172; Treviranus in Ledebour, Fl. ross. IV, 319; Perfilev, Fl. Sev. I, 129.

Ill.: Fl. Murm. II, pl. XLIII.

Characteristic plant of sphagnum bogs and grass-sedge fens in the forest zone of Eurasia and North America. Entering the Arctic only in the European part of its range. Growing in sphagnum bogs, grass-sedge fens and flarks, and on quaking marshy shores of waterbodies; often forming beds.

Soviet Arctic. Murman; SE part of Bolshezemelskaya Tundra.

Foreign Arctic. Arctic Scandinavia.

Outside the Arctic. The whole of Northern and Central Europe; European part of the USSR (except southern districts); Caucasus and the Southern Ural; Kazakhstan (in the west, the extreme north, and the northern Balkhash region); West Siberia south of 64°N, and further north in the basin of the Sukhoy Poluy; East Siberia south of 62°N; Amur Basin; Primorskiy Kray; southern half of Kamchatka; Sakhalin; islands of Hokkaido and Honshu; the north of the Korean Peninsula; found in North America in Southern Alaska and approximately between 40° and 60°N, mainly in the east.

In Hultén's opinion (Circump. pl.), *C. lasiocarpa* is divisible across its range into three very weakly differentiated races: *C. lasiocarpa* s. str. of Eurasia, ssp. *occultans* (Franch.) Hult. from Sakhalin and Japan, and ssp. *americana* (Fern.) Hult. from North America. Since only *C. lasiocarpa* s. str. grows in the Arctic, critical study of other forms of *C. lasiocarpa* was outside our task.

96. **Carex rhynchophysa** C.A. Mey. in Suppl. Index Semin. Hort. Bot. Petropol. IX (January–February 1844), 9; Treviranus in Ledebour, Fl. ross. IV, 318; Perfilev, Fl. Sev. I, 128; Krechetovich in Fl. SSSR III, 440; Hultén, Fl. Al. II, 380, map 319; id., Atlas, map 413; Kuzeneva in Fl. Murm. II, 134; Karavayev, Konsp. fl. Yak. 71.

C. laevirostris (Blytt ex Fries) Fries in Bot. Notis. 1/2 (March 1844), 24; Kükenthal, Cyper. Caricoid. 724; id., Cyper. Sibir. 182; Krylov, Fl. Zap. Sib. III, 523.

C. bullata β *laevirostris* Blytt ex Fries, Novit. Fl. Suec. Mant. II (1839), 59.

Ill.: Russk. bot. zhurn. 3/6 (1911), fig. 150 (under *C. laevirostris* Blytt).

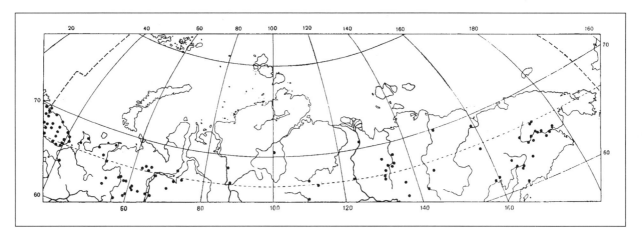

MAP III–61 Distribution of *Carex rostrata* Stokes.

Species widespread in the forest zone of Eurasia and just penetrating arctic districts. Growing on river shores, in flarks, sedge fens and marshy open forests.

Soviet Arctic. Lower reaches of the Pechora (Perfilev, l. c.); Bolshezemelskaya Tundra (River Adzva); River Nyda; Anadyr and Penzhina Basins; Bay of Korf.

Foreign Arctic. Reported from near Lake Inari in Finnish Lappland.

Outside the Arctic. Fennoscandia (mainly the eastern part) and the European part of the USSR (except southern districts); Urals; the whole of West Siberia; basin of the upper course of the Yenisey; Central Siberian Plateau south of the Arctic Circle; Prebaikalia; Central Yakutia; basins of the Indigirka and Kolyma; coasts of Sea of Okhotsk; Preamuria; Primorskiy Kray; Kamchatka; Kurile Islands; Sakhalin; islands of Hokkaido and Honshu; Northern Mongolia; NE China; the north of the Korean Peninsula; Southern Alaska; Canada (the Yukon and the basin of the Mackenzie River).

97. *Carex rostrata* Stokes in Wither., Arrang. Brit. Pl., ed. 2, II (1787), 1059; Ostenfeld, Fl. arct. 1, 93; Kükenthal, Cyper. Caricoid. 720, pro parte; id., Cyper. Sibir. 181; Krylov, Fl. Zap. Sib. III, 521; Mackenzie in N Amer. Fl. XVIII, 7, 456; Leskov, Fl. Malozem. tundry 39; Hultén, Fl. Al. II, 378, map 317; id., Atlas, map 414; id., Circump. pl. I, 104, map 95; Böcher & al., Grønl. Fl. 266; Polunin, Circump. arct. fl. 99, 104.

C. ampullacea Good. in Trans. Linn. Soc. II (1794), 207; Treviranus in Ledebour, Fl. ross. IV, 318; Perfilev, Fl. Sev. I, 128.

C. utriculata auct. non Boott — Perfilev, l. c. 128; Krechetovich in Fl. SSSR III, 444; Karavayev, Konsp. fl. Yak. 71.

C. rostrata var. *utriculata* (Boott) L.H. Bailey in Proc. Amer. Acad. XXII (1886), 67; Kükenthal, Cyper. Caricoid. 722, pro parte; Krylov, l. c. 521.

C. inflata auct. non Huds. — Krechetovich in Fl. SSSR III, 442; Kuzeneva in Fl. Murm. II, 134, map 46; Karavayev, l. c. 71.

Ill.: Russk. bot. zhurn. 3/6 (1911), fig. 149; Fl. SSSR III, pl. XXIII, fig. 3; Fl. Murm. II, pl. XLIV, fig. 2.

Circumpolar species, very widespread in the forest zone of Eurasia and North America, also occurring not uncommonly in temperate arctic districts. Growing on vegetated shores of waterbodies, in sedge fens, on the margins of sphagnum bogs, in flarks, and in marshy river floodplains; forming extensive beds.

Soviet Arctic. Murman (everywhere, but rather sporadic); Timanskaya Tundra (Perfilev, l. c.); Malozemelskaya Tundra; lower reaches of Pechora; SE part of Bolshezemelskaya Tundra; Polar Ural; lower reaches of the Ob (Salekhard); Ob Sound (Nakhodka Bay); middle course of the River Nyda; lower reaches of the

Yenisey (Khantayka), lower reaches of the Lena (Bulun), Indigirka Basin, lower reaches of the Kolyma (Kolymskaya settlement); Anadyr and Penzhina Basins; Bay of Korf. (Map III–61).

Foreign Arctic. Arctic Alaska (isolated occurrences); extreme south of Greenland; Iceland; Arctic Scandinavia.

Outside the Arctic. Europe; Caucasus; the whole of West Siberia; Altay; Kazakhstan; almost the whole of East Siberia; Soviet Far East; the north of the Korean Peninsula; Northern Mongolia; North America (except southeastern states of the USA).

Carex rostrata is a rather variable species. In particular, it varies in the shape of the leaf blade in cross-section and in the degree of divergence of the ripe perigynia from the spikelet rachis. Throughout the arctic range of the species individuals occur with the leaves either flat or somewhat channelled or more or less folded lengthwise, and with perigynia that diverge from the spikelet rachis either obliquely or almost horizontally (as in *C. rhynchophysa*). Plants with flat leaves of varying width and obliquely divergent perigynia occur most frequently. Correlations between the stated characters, as well as between them and other variable characters (colour and shape of the scales, shape and size of the perigynia and the degree of abruptness of their transition to the beak, width of the leaf blade) have not been observed. In the literature plants of *C. rostrata* with flat leaves are sometimes called *C. utriculata* Boott or *C. rostrata* var. *utriculata* (P.N. Krylov, I.P. Perfilev, V.I. Krechetovich, etc.).

Carex utriculata Boott was described from Western Canada. Study of the type of *C. utriculata* Boott received from Kew has shown that this plant differs from *C. utriculata* auct. Authentic specimens of *C. utriculata* possess scales with scabrous awns, with the length of the awns on the lower scales reaching 3 mm. In all Eurasian and American plants erroneously determined as *C. utriculata* studied by us, the scales are acute. Furthermore, the type of *C. utriculata* in distinction from *C. utriculata* auct. possesses broader pistillate spikelets (1.3 cm wide). Plants identical with *C. utriculata* Boott do not occur in Eurasia. The Far Eastern *C. rostrata* referred to *C. utriculata* by V.I. Krechetovich also do not possess the characters of that species.

98. ***Carex rotundata*** Wahlb. in Kongl. Vet. Akad. Nya Handl. XXIV (1803), 153; Treviranus in Ledebour, Fl. ross. IV, 300; Ostenfeld, Fl. arct. 1, 94; Tolmatchev, Contr. Fl. Vaig. 128; Andreyev, Mat. fl. Kanina 162; Krylov, Fl. Zap. Sib. III, 522; Perfilev, Fl. Sev. I, 127; Krechetovich in Fl. SSSR III, 443; Tolmachev, Obz. fl. N.Z. 150; Leskov, Fl. Malozem. tundry 40; Hultén, Fl. Al. II, 379, map 318; id., Atlas, map 415; Kuzeneva in Fl. Murm. II, 136, map 48; Karavayev, Konsp. fl. Yak. 71; Polunin, Circump. arct. fl. 100, 104; Hultén, Circump. pl. I, map 11.

C. rostrata ssp. *rotundata* (Wahlb.) Kük., Cyper. Caricoid. (1909), 723; id., Cyper. Sibir. 182.

C. rotundata f. *Sommieri* Christ in Sommier, Fl. Ob infer. (1896), 203; Kükenthal, Cyper. Caricoid. 723; id., Cyper. Sibir. 182; Krylov, l. c. 523 (pro var. *Sommieri*).

C. melozitnensis Porsild in Rhodora 41 (1939), 209.

Ill.: Fl. Murm. II, pl. XLV; Polunin, l. c. 101.

Species widespread in the Arctic, possessing an almost circumpolar range. Also occurring in adjacent districts outside the Arctic; in East Siberia and the Far East penetrating far to the south in mountains.

Growing on tundra in sphagnous peatlands, sedge-moss fens, and flarks. Very characteristic of level areas of marshy tundra with soft moss carpet, where it often grows in profusion.

Soviet Arctic. Murman (everywhere); Kanin; Timanskaya and Malozemelskaya Tundras; lower reaches of Pechora; Bolshezemelskaya and Karskaya Tundras;

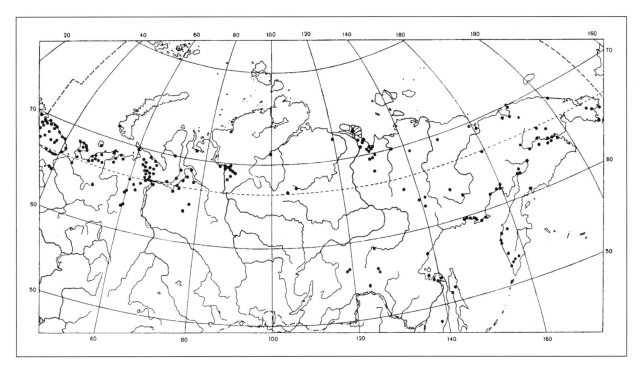

Map III–62 Distribution of *Carex rotundata* Wahlb.

Polar Ural; Vaygach Island; Novaya Zemlya (Gusinaya Zemlya on the South Island); Yamal; lower reaches of Ob; Obsko-Tazovskiy Peninsula; Gydanskaya Tundra; lower reaches of the Yenisey and Olenek; lower reaches and delta of the Lena; Buorkhaya Bay; lower reaches of Kolyma; Chukotka (from Chaun Bay to Chaplino); Anadyr and Penzhina Basins; Bay of Korf. (Map III–62).

Foreign Arctic. Arctic Alaska; arctic coast of Canada; Arctic Scandinavia.

Outside the Arctic. Fennoscandia (mainly in the north); coasts of White Sea; Prepolar and Northern Urals; northern part of West Siberia (north of the middle course of the Ob); northern part of the Central Siberian Plateau; southern part of the Verkhoyansk Range; basins of the Yana, Kolyma and Aldan; upper reaches of Indigirka; Kalarskiy Range; northern edge of Amur Basin; Sikhote-Alin Range; coasts of Sea of Okhotsk; Sakhalin; Kamchatka; Kurile Islands; the north of the Korean Peninsula; Alaska; subarctic districts of Canada.

In the "Flora of the USSR" *C. rotundata* is reported for the Altay, but this record is not confirmed by herbarium material. Nor is there any information on the occurrence of *C. rotundata* there in special literature.

Carex rotundata hybridizes with *C. rostrata*. Hybrids between them are always sterile and usually deviate towards *C. rotundata*, differing from it in having broader and sometimes flattish leaves, somewhat larger pistillate spikelets, narrower scales (elongate-elliptical or elongate-ovoid), and paler perigynia with longer beaks. In these respects the hybrid specimens have some resemblance to *C. rostrata*.

V.I. Krechetovich (l. c.) and O.I. Kuzeneva (l. c.) referred similar plants with intermediate characters to *C. stenolepis* Less., and accordingly put forward the hypothesis that the latter species was the hybrid between *C. rotundata* and *C. rostrata*. *Carex stenolepis* Less. [Reise Loffod. (1831) 301] was described from Norway (Tronheim). The hybrid progeny of *C. rostrata* and *C. rotundata* encountered by us, especially on the Kola Peninsula, and accepted as *C. stenolepis* agree completely in their characters with the original description of that species.

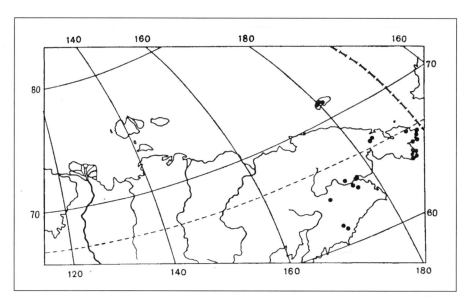

Map III-63 Distribution of *Carex membranacea* Hook.

Therefore, we can assert their identity with *C. stenolepis* with a certain degree of confidence. But these infertile hybrid forms cannot be accepted as a separate species.

Carex stenolepis is also not accepted as a species by contemporary foreign authors (Hultén, etc.). In Kükenthal's monograph (Cyper. Caricoid.) *C. stenolepis* stands in synonymy with *C. rostrata* var. *borealis* Hartm. [Handb. Scand. Fl., ed. I (1820), 39], characterized by the same characters as *C. stenolepis*.

99. ***Carex membranacea*** Hook. in Parry, Journ. Second Voy. App. (1825), 406; Krechetovich in Fl. SSSR III, 449; Hultén, Fl. Al. II, 382, map 321; A.E. Porsild, Ill. Fl. Arct. Arch. 59, map 101; Polunin, Circump. arct. fl. 100, 103.

    C. compacta R. Br., Bot. app. in Ross, Voyage of discovery (1819), 143 (nomen nudum).
    C. saxatilis var. *compacta* Dew. in Amer. Journ. Sci XI (1826), 310.
    C. membranopacta L.H. Bailey in Bull. Torr. Bot. Club XX (1893), 428.
    C. physochlaena H.T. Holm in Amer. Journ. Sci., ser. IV, XVII (1904), 317.
    C. vesicaria ssp. *saxatilis* var. *compacta* Dew. — Kükenthal, Cyper. Caricoid. 728.
    Ill.: A.E. Porsild, l. c., fig. 22, b.

    Beringian-American, predominantly arctic species. Growing in moss-sedge, sedge-cottongrass and willow tundras, on river shores, and in moist meadows.
    Soviet Arctic. Wrangel Island; Chukotka (the Kuvet, Vankarem and Chegitun Rivers, and the east and SE coast); Anadyr Basin; Koryakia. (Map III–63).
    Foreign Arctic. Arctic Alaska; arctic coast of Canada; Labrador; Canadian Arctic Archipelago; Hudson Bay.
    Outside the Arctic. Canada north of 60°N; in Labrador approximately north of 54°N.

100. ***Carex vesicaria*** L., Sp. pl. (1753) 979 (excl. var. β); Treviranus in Ledebour, Fl. ross. IV, 319; Kükenthal, Cyper. Caricoid. 725, pro parte; id., Cyper. Sibir. 183; Krylov, Fl. Zap. Sib. III, 525; Perfilev, Fl. Sev. I, 127; Mackenzie in N Amer. Fl. XVIII, 7, 451; Krechetovich in Fl. SSSR III, 445; Leskov, Fl. Malozem. tundry 41; Hultén, Atlas, map 417; id., Circump. pl. I, 68, 244, map 60, pro parte; Kuzeneva in Fl. Murm. II, 138, map 49; Karavayev, Konsp. fl. Yak. 71.

    Ill.: Fl. SSSR III, pl. XXIII, fig. 1; Fl. Murm. II, pl. XLV, fig. 3.

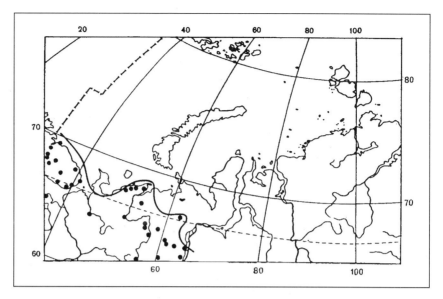

MAP III–64 Distribution of *Carex vesicaria* L.

Species mainly characteristic of the forest zone of Eurasia and North America; just penetrating the Arctic. Growing on the shores of rivers, lakes and streams, and in marshy willow carr, grass-sedge fens, and oxbows.

Soviet Arctic. Murman (very rare); Malozemelskaya Tundra; Southern Bolshezemelskaya Tundra (River Shapkina and Sivaya Maska). (Map III–64).

Foreign Arctic. Arctic Scandinavia.

Outside the Arctic. Europe; the whole of West Siberia; Kazakhstan; basins of the Lower Tunguska and of the upper course of the Yenisey; Prebaikalia (mainly the northern part); Southern Yakutia (very rare, in the lower reaches of the Olekma and the vicinity of Tommot). Replaced in North America by very closely related taxa.

101. Carex vesicata Meinsh. in Acta Horti Petropol. XVIII, 3 (1901), 367; Komarov, Fl. Kamch. I, 266; Hultén, Fl. Kamtch. I, 210; Krechetovich in Fl. SSSR III, 446; Karavayev, Konsp. fl. Yak. 71; Hultén, Circump. pl. I, 244.

C. vesicaria var. *tenuistachya* Kük. in Bot. Centralbl. 77 (1899), 58; id., Cyper. Sibir. 185.

C. vesicaria f. *tenuistachya* (Kük.) T. Koyama in Journ. Fac. Sci. Univ. Tokyo III, 8 (1962), 242.

Ill.: Fl. SSSR III, pl. XXIII, fig. 2.

Growing on the shores of rivers, lakes and overgrown waterbodies.

Soviet Arctic. Anadyr and Penzhina Basins; Bay of Korf.

Foreign Arctic. Not occurring.

Outside the Arctic. Central Siberian Plateau (basin of the Kemkem, a tributary of the Vilyuy); Prebaikalia; Aldan Basin; southern part of Verkhoyansk Range; upper reaches of Indigirka; Amur Basin; Primorskiy Kray; basin of Sea of Okhotsk; Kamchatka; Kurile Islands; Sakhalin; NE China; Northern Mongolia; the north of the Korean Peninsula; Japan. Apparently also occurring in Western North America.

Carex vesicata is not recognized as a species by foreign authors, such as Hultén (Circump. pl.) and Koyama (l. c.). However, this is a rather distinctive taxon replacing *C. vesicaria* east of Baykal. The characters distinguishing *C. vesicata* from *C. vesicaria* (the smaller overall size of the plant, the smaller pistillate spikelets and smaller perigynia which are usually reddish brown) are maintained in almost all herbarium specimens of *C. vesicata* from Siberia and the Far East. Very rarely in Primorskiy Kray, Kamchatka and Sakhalin specimens are found with large broad perigynia; these possibly belong to another closely related race. In support of recognizing *C. vesicata* as a separate species, it may be added that individuals with short spikelets (as in *C. vesicata*) sometimes occur within the range of *C. vesicaria*, but they still have large perigynia. Forms transitional between these species have not been observed.

102. *Carex sordida* Heurck et Muell. Arg. in Heurck, Observ. bot. pl. nov. I (1870), 33; Krechetovich in Fl. SSSR III, 462.

C. hirta var. γ Trev. in Ledebour, Fl. ross. IV (1852), 319.
C. hirta var. *glabrata* Turcz. in Bull. Soc. Nat. Mosc. XI, 1 (1838), 104 (nomen nudum).
C. orthostachys var. *hirtaeformis* Maxim., Prim. fl. Amur. (1859), 316.
C. akanensis Franch. in Bull. Soc. Philom. Paris, sér. 8, 7 (1895), 51.
C. amurensis Kük. in Bot. Centralbl. 77 (1899), 94; Krechetovich in Fl. SSSR III, 457; Karavayev, Konsp. fl. Yak. 73.
C. burejana Meinsh. in Acta Horti Petropol. XVIII, 3 (1901), 368, pro maxima parte.
C. drymophila var. *akanensis* (Franch.) Kük., Cyper. Caricoid. (1909), 756; id., Cyper. Sibir. 119.
C. drymophila var. *glabrata* (Turcz.) Ohwi, Cyper. Japan (1936), 507.
C. drymophila var. *abbreviata* (Kük.) Ohwi, Fl. Japan (1953), 216; Koyama in Journ. Fac. Sci. Univ. Tokyo III, 8, 246.

Ill.: Fl. SSSR III, pl. XXIV, fig. 5.

Growing in floodplain shrubbery and meadows.

Soviet Arctic. Anadyr Basin (Markovo village and the River Algan); Penzhina Basin; Bay of Korf (Kultushnoye village).

Foreign Arctic. Not occurring.

Outside the Arctic. Transbaikalia; Aldan Mountains; Amur Basin; Primorskiy Kray; coasts of Sea of Okhotsk; Kamchatka; Sakhalin; Northern Mongolia; NE China; Korean Peninsula; Japan.

In the majority of specimens known from the Arctic the leaf blades are bare and the leaf sheaths normally pubescent only at their mouths (more rarely over their whole surface); the perigynia are mostly without hairs. Individuals with similar characters also predominate in Kamchatka. Plants from the more southern districts of the range of *C. sordida* usually possess more or less strongly pubescent leaf blades, sheaths and perigynia.

103. *Carex saxatilis* L. s. l.

This species consists of two subspecies, ssp. *saxatilis* (= *C. saxatilis* L. s. str.) and ssp. *laxa* (Trautv.) Kalela. These subspecies are characterized by a highly variable complex of distinguishing characters, of which the most fundamental is the length of the pedicel of the pistillate spikelets. In Eastern North America, Greenland, Scandinavia, Finland and the Kola Peninsula plants possessing the characters of *C. saxatilis* s. str. (erect pistillate spikelets, almost sessile or on pedicels up to 1 cm long, more rarely up to 1.5 cm) are encountered most frequently. East of Novaya Zemlya and the Urals (as far as in Western North America) plants with drooping pistillate spikelets on elongate capillary pedicels mostly 2–3.5 cm long distinctly predominate; the latter belong to ssp. *laxa*. At the same time, plants indistinguishable from the other subspecies occur within the range of each subspecies, as well as specimens which could with equal justification be referred to either of the subspecies.

1. Lower pistillate spikelets (1)1.5–2.5(3) cm long, on capillary pedicels (1)1.5–3.5(5) cm long, mostly drooping or divergent, more rarely erect. Perigynia normally ovoid or elongate-ovoid, more rarely elongate-elliptical,

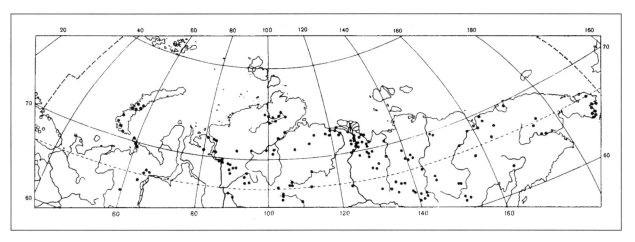

MAP III–65 Distribution of *Carex saxatilis* L. ssp. *saxatilis* (open circles) and ssp. *laxa* (Trautv.) Kalela (black circles).

(3)3.5–4.5(5) mm long. Usually 2 staminate spikelets. ....
................................C. SAXATILIS ssp. LAXA (TRAUTV.) KALELA.
- Lower pistillate spikelets 0.8–2 cm long, on somewhat thicker pedicels 0.3–1.5(3) cm long (usually 0.3–1 cm), mostly erect, rarely drooping. Perigynia ovoid, more rarely elongate-ovoid, (2.5)3–4 mm long. Usually 1 staminate spikelet. ....C. SAXATILIS ssp. SAXATILIS.

a. *Carex saxatilis* ssp. *saxatilis*

C. saxatilis L., Sp. pl. (1753) 976; Perfilev, Fl. Sev. I, 127, pro parte; Krechetovich in Fl. SSSR III, 448, pro parte (respecting European plants); Hultén, Atlas, map 416; Kuzeneva in Fl. Murm. II, 140, map 51; A.E. Porsild, Ill. Fl. Arct. Arch. 58 (excl. var. *miliaris*), map 100 (*C. saxatilis* var. *rhomalea*); Böcher & al., Grønl. Fl. 266; Hultén, Circump. pl. I, 22, pro parte, map 16 (*C. saxatilis*).

C. pulla Good. in Trans. Linn. Soc. III (1797), 78; Treviranus in Ledebour, Fl. ross. IV, 308; Ostenfeld, Fl. arct. 1, 95, pro parte.

C. saxatilis var. *rhomalea* Fernald in Rhodora 3 (1901), 50.

C. vesicaria ssp. *saxatilis* (L.) Kük., Cyper. Caricoid. (1909) 727, pro parte; id., Cyper. Sibir. 186, pro parte.

C. rhomalea Mackenz. in Bull. Torr. Bot. Club XXXVII (1910), 246.

Ill.: Fl. Murm. II, pl. XLVI.

Growing in sedge-moss and sedge-cottongrass tundras, in sedge fens, and on the shores of streams, rivers and lakes.

Soviet Arctic. Murman (from the Rybachiy Peninsula to Iokanga, rare); Kanin (northern part). (Map III–65).

Foreign Arctic. Eastern part of Canadian Arctic Archipelago; Arctic Labrador; NW, SW and East Greenland; Iceland; Spitsbergen; Arctic Scandinavia.

Outside the Arctic. Fennoscandia; the north of England; Faroe Islands; Middle Ural (Konzhakovskiy Rock); Eastern Canada (Manitoba, Hudson Bay).

Carex Grahami Boott described from Scotland is very close to *C. saxatilis* s. str. According to O.I. Kuzeneva (l. c. 139, map 49), one locality for *C. Grahami* is situated within the arctic part of the Kola Peninsula between Kharlovka and Varzino. But in Hultén's opinion (Amphi-atl. pl. 11) *C. Grahami* occurs only in Scotland where it is very rare; plants from other districts of Europe identified as *C. Grahami* are hybrids between *C. saxatilis* and *C. vesicaria*.

Indeed, plants agreeing with the description of *C. Grahami* seen by us from nonarctic districts of the Kola Peninsula appear somewhat intermediate between the above two species. They differ from *C. saxatilis* only in having thicker pistillate spikelets about 1 cm wide (these are 0.6–0.8 cm wide in *C. saxatilis*), longer spikelet pedicels, and paler scales; in these characters they approach *C. vesicaria*.

Some of the specimens identified as *C. Grahami* are sterile. This fact, along with the rare occurrence of specimens of the *C. Grahami* type, testifies in favour of a hybrid origin of these plants. Contemporary investigators of the British flora list *C. Grahami* as one of the synonyms of *C. stenolepis* (see the comments above under *C. rotundata*), which in turn is the hybrid between *C. rotundata* and *C. rostrata*.

b. **Carex saxatilis** ssp. **laxa** (Trautv.) Kalela in Ann. Bot. Soc. Vanamo 14, 2 (1941), 15; Hultén, Circump. pl. I, 22, map 16.

? *C. physocarpa* Presl, Reliq. Haenk. I (1830), 205; Hultén, Fl. Al. II, 380, map 320; A.E. Porsild, Ill. Fl. Arct. Arch. 58, map 98; Polunin, Circump. arct. fl. 88, 104.
C. saxatilis var. *major* Olney in S. Wats., Bot. King's Explor. (1871), 370.
C. pulla var. *laxa* Trautv. in Acta Horti Petropol. V (1877), 130.
C. pulla f. *pedunculata* Kjellman in Vega-Exped. Vet. Jaktag. Stockholm 1 (1882), 560.
C. pulla var. *sibirica* Christ in Vet. Acad. Handl. Stockholm XXII (1887), 181.
C. vesicaria ssp. *saxatilis* (L.) Kük. f. *laxa* (Trautv.) Kük., Cyper. Caricoid. (1909), 728; id., Cyper. Sibir. 186.
C. vesicaria ssp. *saxatilis* var. *physocarpa* (Presl) Kük., Cyper. Caricoid. (1909), 728.
C. pulla auct. non Good. — Perfilev, Fl. Sev. I, 127, pro parte.
C. procerula V. Krecz. in Fl. SSSR III, 449, 622; Karavayev, Konsp. fl. Yak. 72.
C. saxatilis var. *laxa* (Trautv.) Ohwi in Journ. Jap. Bot. XI (1935), 408.

Ill.: Fl. SSSR III, pl. XIII, fig. 6 (under *C. procerula* V. Krecz.); Porsild, l. c., fig. 22, c (under *C. physocarpa*).

Growing in sedge-moss and sedge-cottongrass tundras, in fens and on the shores of waterbodies. Often occurring along streams, at incompletely vegetated sites, and on more or less rocky slopes. More common in districts with mountains or cliffs.

Soviet Arctic. Polar Ural; Yugorskiy Peninsula (Khabarovo and Cape Sokoliy); Vaygach Island; Novaya Zemlya (South Island and the southern half of the North Island); northern part of Gydanskiy Peninsula; lower reaches of the Yenisey and the district of Yenisey Bay; Taymyr; River Khatanga; lower reaches of the Anabar and Olenek; lower reaches and delta of the Lena; lower reaches of the Yana and Kolyma; Chukotka; Anadyr and Oklan Basins. (Map III–65).

Foreign Arctic. Arctic Alaska; western part of the arctic coast of Canada and of the Canadian Arctic Archipelago.

Outside the Arctic. Prepolar and Northern Urals; Central Siberian Plateau (mainly the northern part); Eastern Sayan (Tunkinskiye and Kitoyskiye Barrens); Transbaikalia; Aldan Basin; Verkhoyansk Range; Indigirka and Kolyma Basins; coast of the Sea of Okhotsk (Okhotsk, Ayan); Dusse-Alin Range; Kamchatka (Anaun); Kurile Islands; Western North America from Alaska to Colorado.

Kalela (l. c.) accorded var. *laxa* Trautv. the rank of subspecies, but did not designate a type of this new taxon. We have selected as the lectotype of ssp. *laxa* one of the specimens cited by Trautvetter (l. c.) in the description of var. *laxa*. [Lectotypus *C. saxatilis* ssp. *laxa* (Trautv.) Kalela: Sibiria orientalis, ad fl. Tomba superiorem, 25 VI 1874, Czekanowski et Müller. In Herb. Botan. Inst. Acad. Sci. URSS (Leningrad) conservatur].

We do not have complete confidence that ssp. *laxa* is identical with *C. physocarpa* Presl (described from Western Canada), since we have not seen the type of the latter. However, *C. physocarpa* in the sense of American authors, so far as can be judged from relevant literature and herbarium material, does not differ from ssp. *laxa*.

FAMILY XV

Lemnaceae Dumort.

DUCKWEED FAMILY

GENUS 1   **Lemna** L. — DUCKWEED

AQUATIC PLANTS, freely floating or submersed on the bottom of shallow waterbodies, with small leaflike thallus ("frond") from whose lower surface arises a slender rootlet. The much reduced flowers rarely develop. Growing in standing waterbodies in a submersed state or on the surface, in the latter case with their dense aggregations forming a veritable film covering the surface of the water. Occurring frequently and abundantly in the northern temperate zone. Not very characteristic of the tundra zone and occurring only in its more temperate parts, where as a rule they grow less abundantly. The available information on the distribution of duckweeds in the Arctic is extremely sparse. It is possible that these plants have sometimes been overlooked by investigators because they catch the eye only when they grow in masses. But it is undisputed that duckweeds are absent from districts where severe arctic conditions prevail.

1. Fronds *floating on water surface*, elliptical, firm, *thickish, light green*, with entire margin, 3–4.5 mm long, without "stalk." . . . . 1. **L. MINOR** L.
– Fronds *submersed in water*, sometimes lying free on bottom, *thin, slightly translucent, bright green*, oblanceolate, with margin dentate towards tip, 5–9 mm long, with slender "stalk." . . . . 2. **L. TRISULCA** L.

    *1. **Lemna minor** L., Sp. pl. (1753), 970; Ledebour, Fl. ross. IV, 493; Krylov, Fl. Zap. Sib. III, 540; Andreyev, Mat. fl. Kanina 163; Perfilev, Fl. Sev. I, 134; Kuzeneva in Fl. SSSR III, 493; Hultén, Atlas 108 (428); Selivanova-Gorodkova in Fl. Murm. II, 148; Karavayev, Konsp. fl. Yak. 72; Polunin, Circump. arct. fl. 141.
    Ill.: Fl. Murm. II, pl. XLVIII, 2.
        In small standing waterbodies. Rare in the Arctic, only in NE Europe.
    Soviet Arctic. Northern Kanin in lakes in the basin of the Severnyy Nettey River (Andreyev, 10 VIII 1928; Matveyev, 6 VII 1945); lower reaches of the Pechora in the Naryan-Mar district, in pond among sedge stems (Gornovskiy and Dorogostayskaya, 13 VIII 1958).
    Foreign Arctic. Not occurring.
    Outside the Arctic. Possessing an almost cosmopolitan distribution, but not penetrating desert and high alpine regions. In Fennoscandia distributed in general south of the Arctic Circle, but reported in Finland at one point almost at 68°N and on the Kola Peninsula on the southern portion of the Terskiy Shore (lower reaches of the River Strelna). In Arkhangelsk Oblast and the Komi ASSR occurring regularly only south of the 65th parallel, but locally distributed further north; in the Pechora valley apparently more or less continuously distributed as far as the forest-tundra. In West and Central Siberia ranging north to 61–62°, in Yakutia reaching the watershed between the Vilyuy and the Olenek and penetrating the basins of the Yana and Indigirka.

2. **Lemna trisulca** L., Sp. pl. (1753), 970; Ledebour, Fl. ross. IV, 17; Scheutz, Pl. Jeniss. 163; Krylov, Fl. Zap. Sib. III, 540; Andreyev, Mat. fl. Kanina 163; Perfilev, Fl. Sev. I, 134; Kuzeneva in Fl. SSSR III, 493; Leskov, Fl. Malozem. tundry 11; Hultén, Atlas 108 (429); Selivanova-Gorodkova in Fl. Murm II, 148; Karavayev, Konsp. fl. Yak. 72; Polunin, Circump. arct. fl. 111.

Ill.: Fl. SSSR III, pl. XXVI, 7; Fl. Murm. II, pl. XLVIII, 3.

Growing in small ponds, often on the bottom; sometimes abundant in oxbows with rather murky water. Not uncommon in tundra waterbodies among sedge beds. Penetrating arctic limits to a greater degree than does *L. minor*, especially in NE Europe.

Soviet Arctic. Northern Kanin in valley of Bolshaya Kambalnitsa River (Andreyev, 9 VIII 1928), and in pond in district of Mar-Sed Hill (Matveyev, 5 VII 1945); River Pesha (according to Perfilev); Malozemelskaya Tundra in pond on Korovinskiy Range west of the Martyshikha location (Leskov, 18 IX 1931); lower reaches of Pechora River in vicinity of town of Naryan-Mar (Gornovskiy and Dorogostayskaya, 15 VIII 1958; reported as a common plant sometimes growing in great abundance in oxbows of the Pechora); western part of Bolshezemelskaya Tundra (Vangurey) in tundra lakes in basins of rivers Payyakha and Khylchuyu (Andreyev and Savkina, 27 VII, 22 VIII, 9 IX 1930); lower reaches of Ob within the forest-tundra (vicinity of Salekhard); lower reaches of Yenisey (furthest north site: Lukovaya Channel, V. Tugarinova, 13 VIII 1907).

Foreign Arctic. Not occurring.

Outside the Arctic. In Fennoscandia mainly south of the 65th parallel, but reaching 68°N in the valley of the River Torne-Älv and 67°N in Karelia; not found on the Kola Peninsula. Apparently more or less ubiquitous in the forest zone of Arkhangelsk Oblast and the Komi ASSR. In the Siberian North known only from isolated localities which attest to the distribution of the species north to the lower reaches of the Ob and Yenisey, to the watershed between the Vilyuy and the Olenek, and to the basins of the Yana and Indigirka. Further south found throughout Europe including the Caucasus, throughout Siberia and temperate districts of the Far East, in Southern Asia, in northern and tropical Africa, and in North America; disjunctly in Australia.

FAMILY XVI

Juncaceae Vent.

RUSH FAMILY

SMALL FAMILY CONTAINING two extensive and six oligotypic genera. Native mainly to extratropical regions of the Northern and Southern Hemispheres and to high mountains of the tropical zone. Both the large genera, *Juncus* and *Luzula*, are distributed in both hemispheres and include representatives characteristic of the arctic flora. Additionally, the Arctic is penetrated by a series of species mainly distributed in the temperate zone.

1. Leaf sheaths open, with more or less hyaline margins. Leaves bare, sometimes reduced or one of them (involucral bract) appearing to be a straight continuation of the stem. Capsule trilocular, many-seeded. . . . .
. . . . .1. **JUNCUS** L. — RUSH
– Leaf sheaths closed. Leaves usually hairy on margins, always normally developed. Capsule unilocular, three-seeded. . . . .
. . . . .2. **LUZULA** DC. — WOODRUSH

GENUS 1

Juncus L. — RUSH

EXTENSIVE GENUS (including up to 300 species), distributed predominantly in temperate and partly in subtropical latitudes of both (Northern and Southern) hemispheres. Represented in tropical countries by a certain number of predominantly high alpine species. In truly arid regions only isolated species confined to moist habitats occur. More than 60 species (some of doubtful taxonomic status) have been recorded within the USSR, predominantly in forested regions of the country.

In the arctic flora the genus *Juncus* is represented by a considerable number of species, but the majority of them are not very typical of its composition. These are the widespread boreal species, native to the temperate zone of the Northern Hemisphere and often widespread there but only marginally penetrating the Arctic (*J. filiformis, J. nodulosus, J. bufonius*, etc.). Approaching these are certain species associated with sea coasts (*J. balticus*, etc.). A second series of rushes occurring in the Arctic consists of arctic-alpine species. Some of these (especially *J. biglumis, J. triglumis, J. castaneus*) also penetrate arctic districts at high latitude and belong among the plants characteristic of the Arctic. The genus does not include any plants endemic to the Arctic.

With respect to habitat conditions, species of *Juncus* almost invariably gravitate towards areas with adequate moisture and little or no turf formation. Some of them grow on sandy or clayey stretches along rivers (*J. arcticus, J. filiformis*), and certain species are associated with sea shores (for example, *J. Haenkei*).

Such species as *J. biglumis*, *J. triglumis* and *J. castaneus* are true tundra plants, growing mainly on patches of exposed loamy soil in so-called patchy tundras. A special position with respect to its association with dry stony habitats is occupied by *J. trifidus*, characteristic of the Atlantic Arctic.

The distribution of species of *Juncus* in the Soviet Arctic is such that a certain portion of them (mainly the true tundra species) are more or less uniformly distributed throughout its sectors. Species penetrating the Arctic from adjacent temperate areas are most often found either in the European Arctic or in the Far East near the Pacific coast.

1. *Annual* low-growing plant (no more than 10–15 cm high) with fibrous roots and numerous obliquely ascending or almost prostrate stems. Inflorescence more or less loose, branched, with rather numerous solitary florets that each possess small bracteoles. [SUBGENUS **TENAGEIA** (DUM.) O. KTZE.]. . . . . .1. **J. BUFONIUS** L. (S. L.) — **TOAD RUSH**
   - *Perennial plants with long creeping rhizomes* that bear more or less numerous stems arranged in rows; or *rhizomes abbreviated, branched, forming small tufts*. Stems straight (or sometimes slightly curved at base), unbranched. . . . . .2.

2. Inflorescence *terminal or slightly displaced to the side*; in the latter case involucral bract short and more or less flat (not cylindrical) at base. Basal and stem leaves *possessing normally developed long leaf blades*. . . . . .3.
   - Inflorescence *lateral, displaced to the side due to the cylindrical or subulate involucral bract being directed straight upwards and appearing to be a continuous extension of the stem*. Basal and stem leaves *without blades*; only elongate leaf sheaths present. [SUBGENUS **JUNCOTYPUS** (DUM.) O. KTZE.]. . . . . .13.

3. Stem leaves *displaced to upper part of stem* and (together with the 1–2 involucral bracts) crowded at the base of the small dense few-flowered inflorescence (most often with 2-3 florets, rarely only one) which appears sessile in the angle formed by the bases of the bracts. Leaves long and narrow, setiform. Short grey-green plant (10–15, rarely up to 20 cm high), forming small dense stiffish tufts. [SUBGENUS **PSEUDOTENAGEIA** V. KRECZ. ET GONTSCH.]. . . . . .2. **J. TRIFIDUS** L.
   - Stem leaves *situated on lower part of stem or not developed at all*. Involucral bracts slightly exceeding or shorter than inflorescence. . . . . . .4.

4. Inflorescence *more or less branched, consisting of numerous small heads or solitary florets*. Branches of inflorescence of nonuniform length, often with secondary branching. . . . . .5.
   - Inflorescence *consisting of solitary dense head or 2-3 heads situated on unbranched erect-standing pedicels*. [SUBGENUS **STYGIOPSIS** (GDGR.) O. KTZE.]. . . . . .9.

5. Inflorescence paniculate or corymbose-paniculate, with slender branchlets. Florets numerous, arranged solitarily or sometimes clustered in

groups but not forming dense heads. Involucral bract always well developed. Plants green, greyish green or yellowish green. [SUBGENUS **PSEUDOTENAGEIA** V. KRECZ. ET GONTSCH.]. . . . . .6.

– Inflorescence compound, umbellate-paniculate, usually with numerous, sometimes thickish branchlets. Florets aggregated in clusters (heads) at tips of branchlets. Involucral bract usually weakly developed and not exceeding half height of inflorescence. Stems sometimes with reddish tint on basal part. [SUBGENUS **OZOPHYLLUM** (DUM.) O. KTZE. — **J. ALPINUS** VILL., S. L.]. . . . . .8.

6. Plant green. Inflorescence corymbose-paniculate, more or less crowded, with numerous florets arranged in rather dense clusters at the tips of the branched branchlets; pedicels 0.5–5 mm long. Perianth segments 2–2.5 mm long, ferruginous or greenish brown, broad. Anthers not much longer than stamen filaments. Capsule almost globose, up to 3 mm long, with very short beak. . . . . .3. **J. COMPRESSUS** JACQ.

– Plant yellowish green or greyish green. Inflorescence paniculate, more or less loose, with relatively few uncrowded florets at the tips of the branched branchlets; pedicels 1–10 mm long. Perianth segments 2.2–4 mm long, oblong-elliptical, reddish brown or blackish brown (to almost black). Anthers several times longer than filaments. [**J. GERARDII** LOIS., S. L.]. . . . .7.

7. Perianth segments *brown or reddish brown, not more than 3 mm long*. Capsule broadly elliptical, 3–3.5 mm long, light castaneous or greenish brown, with beak up to 0.4 mm long. . . . . .4. **J. GERARDII** LOIS., S. STR.

– Perianth segments *blackish brown, sometimes almost black*, usually with purplish tint, somewhat lighter on midrib, usually *longer than 3 mm* (*up to 4 mm*). Capsule elliptical or oblong-elliptical, with very short beak, dark purplish brown, sometimes almost black. . . . . .4a. **J. GERARDII** SSP. **ATROFUSCUS** (RUPR.) TOLM.

8. All branches of inflorescence directed obliquely upwards so that the inflorescence as a whole is attenuate. Heads with few florets (2–6, rarely up to 8) on pedicels of different lengths, due to which the obconical heads are rather loose. . . . . .5. **J. NODULOSUS** WAHLB.

– Branches of inflorescence more or less widely divaricate (in fruit often diverging at right angles), so that the inflorescence as a whole is diffuse. Heads with numerous (5–12) tightly sessile florets, so that the heads as a whole are roundish. . . . . .6. **J. ARTICULATUS** L. (**J. LAMPOCARPUS** EHRH.)

9. *Low-growing* (*up to 15 cm high*) *plants* with slender (not more than 1 mm in diameter) stems, *forming small tufts*. Inflorescence dense, capitate, usually solitary, few-flowered (1–4 florets). Florets 2–4 mm long. . . . . . . .10.

– *Larger* (*15–25 cm, sometimes up to 35 cm high*) *plants, not forming tufts*. Stems mostly 1–2.5 mm in diameter. Inflorescence consisting of 1–3 heads, each containing 2–8 florets. Florets 4–5 mm long. . . . . .12.

10. Inflorescence always *solitary*, sessile, somewhat displaced to the side (*leaning to the side of the almost erect involucral bract which is lanceolate, flat basally, subulately compressed at tip* and 1½–2 times as long as the inflorescence). *Perianth segments 2–3 mm long, obtuse*, covering only up to half of capsule. Capsule with small notch distally. ......7. **J. BIGLUMIS** L.

– Inflorescence consisting *of 1 or more rarely 2 heads, erect* or (if 2 heads present) one of them slightly leaning to the side. *Perianth segments 3–4 mm long, acute*, covering more than half of capsule. Capsule tapering distally or truncate (without notch)...................................11.

11. Inflorescence consisting of 1 or 2 heads, in the latter case one of them on slender pedicel. Involucral bract *subulate, 1½–2 times as long as inflorescence or equalling it*. Florets pale. Capsule yellowish, tapering apically. ....8. **J. STYGIUS** L.

– Inflorescence always consisting of one terminal head. Involucral bract *scalelike* (resembling the bracteoles), *not projecting above the erect head*. Florets pinkish or rusty brown. Capsule brown, with broad truncate tip. ... .9. **J. TRIGLUMIS** L.
[In eastern districts of the Soviet Arctic and the American Arctic 9a, **J. TRIGLUMIS** SSP. **ALBESCENS** (LGE.) HULT., occurs; this differs from the typical form in having narrower and lighter perianth segments.]

12. Leaves *involute-channelled*, about 2 mm wide. Lower involucral bract *broadened at base, brownish, membranous, abruptly tapering and narrowly linear distally, just exceeding the inflorescence or not projecting above it*. Heads sessile or on short pedicels. Bracteoles *brownish*, almost equalling the florets in length. Perianth segments *castaneous, the outer acute, the inner obtusish*, 4–4.5 mm long. Capsule dark castaneous, obtuse, 6–7.5 mm long, with short beak. Plant (10–)15–30 cm high............... ....10. **J. CASTANEUS** SM.

– Leaves more or less *flattened-channelled*, up to 5 mm wide. Lower involucral bract *up to 12 cm long, green, with almost parallel margins, gradually tapering apically, considerably exceeding inflorescence*. Second involucral bract similar but shorter. Individual heads of inflorescence sometimes on long (up to 4 cm) pedicels. Bracteoles *albescent or slightly ferruginous*, somewhat shorter than florets. Perianth segments *light castaneous, lanceolate, all acute*, 4–5 mm long. Capsule rusty-rufous, pale basally, 8–10 mm long, tapering to extended beak. Plant 18–35 cm high. ....11. **J. LEUCOCHLAMYS** ZING. VAR. **BOREALIS** TOLM.

13. Involucral bract *of approximately same length as stem* (sometimes even slightly longer), so that the inflorescence is situated *at about the middle of the total height of the plant*. Rhizome rather slender. Stems slender (about 1 mm thick), closely approximated, sometimes immediately adjacent to one another. Inflorescence fasciculate, consisting of 3–12 florets; pedicels of nonuniform length, but generally short. Plant greyish green. ....12. **J. FILIFORMIS** L.

– Involucral bract *considerably shorter than stem* (at most half as long, sometimes shorter). Inflorescence situated *considerably above middle of total height of plant* (at two-thirds of this height or higher). . . . . .14.

14. Rhizome slender, *much branched, with its relatively short branches forming a dense tuft*. Stems slender, short (not more than 25–28 cm high), closely approximated, slightly longitudinally furrowed. Involucral bract short (2–8 cm). Inflorescence few-flowered. . . . . .15.
- Rhizome elongate, *ascending or horizontal, bearing smooth* (not furrowed) *stems arranged in a row*. Involucral bract short (3–5 cm) and soft, or longer (5–20 cm) and stiff with prickly tip. In the latter case the plant is large (30–65 cm high) and greyish green. . . . . .16.

15. Inflorescence very meagre, consisting of 1–3 florets. Florets 4.5 mm long, greenish, rather pale. Stems less than 1 mm thick. Involucral bract 3–5(8) cm long. . . . . .13. **J. BRACHYSPATHUS** MAXIM.
- Inflorescence consisting of 2–6 (usually 3–4) florets. Florets dark castaneous, 5 mm long. Stems 1.5–2 mm in diameter, yellowish. Involucral bract very short (2–4 cm). . . . . .14. **J. BERINGENSIS** BUCH.

16. Plants *15–30 cm high, yellowish green*. Rhizome horizontal, *bearing uniformly spaced, not very closely approximated stems that form a long sparse row*. Inflorescence more or less dense, fasciculate, with 3–8 florets. Plants of tundra and river valleys. . . . . .17.
- Plants *30–65 cm high, greyish green*. Rhizome obliquely ascending or horizontal. *Stems closely approximated, forming a more or less dense row*. Involucral bract stiff with prickly tip, 5–20 cm long. Inflorescence many-flowered, fasciculate or fasciculate-paniculate. Plants of sea coasts. . . . .18.

17. Inflorescence *condensed, almost capitulate*. Pedicels 3–5 mm (rarely up to 8 mm) long. *Capsule length 4–5 mm*. . . . . .15. **J. ARCTICUS** WILLD.
- Inflorescence *more or less loose, forming short raceme*. Pedicels of individual florets very different in length, fluctuating from 3–4 to 8–12 mm. *Capsule length 3.5–4 mm*. . . . . .15a. **J. ARCTICUS** SSP. **ALASKANUS** HULT.

18. Perianth segments of about same length as capsule. Inflorescence fasciculate-paniculate, loose, up to 3 cm long. Stems 1.3–2 mm thick. Total height of plant 30–50 cm. . . . . .16. **J. BALTICUS** WILLD.
- Perianth segments longer than capsule. Inflorescence many-flowered, fasciculate, usually rather condensed, 1–2 cm long (sometimes up to 3 cm in fruit). Stems firm, 1.5–2.5 mm thick. Total height of plant 40–65 cm. . . . .
. . . . .17. **J. HAENKEI** E. MEY.

1. ***Juncus bufonius*** L., Sp. pl. (1753), 328; E. Meyer in Ledebour, Fl. ross. IV, 231; Scheutz, Pl. jeniss. 171; Lange, Consp. fl. groenl. 125; Gelert, Fl. arct. 1, 26; Buchenau in Pflanzenreich IV, 36, 105; Krylov, Fl. Zap. Sib. III, 561; Perfilev, Fl. Sev. I, 138; Gröntved, Pterid. Spermatoph. Icel. 188; Hultén, Fl. Al. III, 422; id., Atlas 111 (442); Böcher & al., Grønl. Fl. 230; Polunin, Circump. arct. fl. 115. — **Toad Rush**.
J. ambiguus Guss. — Shlyakov in Fl. Murm. II, 155.

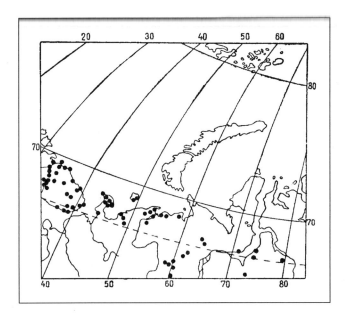

MAP IV-1 Distribution of *Juncus trifidus* L.

J. nastanthus V. Krecz. et Gontsch. in Fl. SSSR III, 517.

Plant widespread in the northern temperate zone, just penetrating arctic limits and scarcely persistent under such conditions. Most frequently reported as an obvious introduction near the fringes of the Arctic. Growing along roads and ditches in settlements and their immediate vicinity, as well as on riverine sands where it is carried by rivers flowing from the south.

The collective species *J. bufonius* L. is usually divided by contemporary systematists into a series of subordinate forms of doubtful status, that are often treated as species and contrasted with *J. bufonius* s. str. Because toad rush generally occurs as an introduction in the Arctic and arctic material cannot provide the basis for critical study of the *J. bufonius* complex, we refrain from distinguishing these forms and treat *J. bufonius* in a wide sense. For particular sites from which material has been studied by specialists, we indicate their identification of the relevant specimens.

Soviet Arctic. In Murman on the southern part of Kola Bay within the forest-tundra (*J. ambiguus* Guss.); vicinity of Vorkuta (introduced); in the lower reaches of the Yenisey at Old Zaostrovskoye (*J. nastanthus* V. Krecz. et Gontsch.) and Nikandrovskiy Island; coast of Bay of Korf at Olyutorki settlement.

Foreign Arctic. Extreme SW Greenland; Iceland; Arctic Scandinavia.

Outside the Arctic. Greater part of Europe, Siberia and Far East; Caucasus; Northern Kazakhstan and districts beneath the mountains of Central Asia; Middle East, Northern India; Mongolia, Northern China, Korean Peninsula, Japan; considerable part of North America; extreme north of Africa. Introduced in Australia.

2. ***Juncus trifidus*** L., Sp. pl. (1753), 326; E. Meyer in Ledebour, Fl. ross. IV, 233; Ruprecht, Fl. samojed. cisur. 59; Lange, Consp. fl. groenl. 123; Gelert, Fl. arct. 1, 26; Buchenau in Pflanzenreich IV, 36, 109; Krylov, Fl. Zap. Sib. III, 563; Perfilev, Fl. Sev. I, 140; Tolmachev, Fl. Kolg. 40; id., Obz. fl. N.Z. 150; Leskov, Fl. Malozem. tundry 44; Devold & Scholander, Fl. pl. SE Greenl. 109; Gröntved, Pterid. Spermatoph. Icel. 190; Hultén, Atlas 113 (449); id., Amphi-atl. pl. 46 (28); Shlyakov in Fl. Murm. II, 168; A.E. Porsild, Ill. Fl. Arct. Arch. 62; Böcher & al., Grønl. Fl. 230; Polunin, Circump. arct. fl. 113.

Ill.: Fl. Murm. II, pl. LIV; Porsild, l. c., fig. 23d, 24d.

Characteristic plant of moderately northern parts of the Atlantic Arctic. Occurring almost exclusively on stony or sandy sites devoid of closed vegetational cover or covered with a low carpet of lichens. Sometimes colonizing rocks, where it roots in crevices or in the accumulations of soil that fill holes. In mountainous districts (in Murman and locally in the Urals) this is a generally typical and abundantly occurring plant of rocky sites. Outside the more significant uplands it is more often found on open, sometimes windblown sandy ridges and hillocks. Occasionally encountered in hummocky peaty tundras, where it grows on surfaces of peat mounds where the vegetation has died.

Soviet Arctic. Murman (ubiquitous and frequent); Kanin (common, especially in the north); Timanskaya and Malozemelskaya Tundras (common); Kolguyev Island (rare); lower reaches of Pechora; Bolshezemelskaya Tundra (recorded for a few sites both in the west and east; absent from many sites); Polar Ural (common); shores of Ob Sound and Tazovskaya Bay north to the 68th parallel; lower reaches of River Pur. (The record for the Anabar River Basin is not confirmed by herbarium material; the old record for Novaya Zemlya is evidently erroneous.) (Map IV–1).

Foreign Arctic. Labrador; Southern Baffin Island; both coasts of Greenland south of 72°N; Iceland; Arctic Scandinavia.

Outside the Arctic. Northern Great Britain and adjacent islands; mountains of Scandinavia; forest zone of Finnish Lappland and the Kola Peninsula; occasional on the shores of the Gulf of Bothnia and on the Karelian sea coast; mountains of Central and Southern Europe; throughout the high Urals; middle course of Ob (Surgut); Altay, Kuznetskiy Alatau, Western and Central Sayans; Yenisey Ridge; Atlantic districts of North America from Labrador to the extreme northeast of the USA.

3. *Juncus compressus* Jacq., Stirp. Vindob. (1762), 60, 235; E. Meyer in Ledebour, Fl. ross. IV, 229; Buchenau in Pflanzenreich IV, 36, 111; Perfilev, Fl. Sev. I, 139; Krechetovich & Goncharov in Fl. SSSR III, 527; Hultén, Atlas 112 (446); id., Amphi-atl. pl. 170 (152); Shlyakov in Fl. Murm. II, 165.

Species widespread in the temperate zone of Eurasia, but in its typical form apparently not penetrating arctic limits. The most northern reliable localities are in the central part of the Kola Peninsula (basin of Lake Imandra). Perfilev's and Leskov's records of this species as growing in the Timanskaya Tundra (Indiga) refer to *J. Gerardii* ssp. *atrofuscus* (Rupr.) Tolm.

Outside the Arctic. Widespread in Central Europe, almost throughout the European territory of the USSR, in the Caucasus, and in southern parts of Siberia (north to Yeniseysk and Yakutsk).

4. *Juncus Gerardii* Lois., Journ. de Bot. II (1809), 284; E. Meyer in Ledebour, Fl. ross. IV, 229; Buchenau in Pflanzenreich IV, 36, 112; Perfilev, Fl. Sev. I, 139 (pro parte); Krechetovich & Goncharov in Fl. SSSR III, 528; Hultén, Atlas 113 (450) (pro parte); id., Amphi-atl. pl. 172 (154) (pro parte).

Species widespread in the temperate zone of Europe, Southern West Siberia and Central Asia. In more northern regions confined to sea coasts, along which it penetrates the European Arctic where it occurs in coastal meadows.

Soviet Arctic. West Murman (Yekaterininskiy Island and Sayda-Guba on Kola Bay). Records of occurrence on the arctic coast east of the White Sea refer to ssp. *atrofuscus*.

Foreign Arctic. Arctic Scandinavia.

Outside the Arctic. Coasts of Northern, Western and Southern Europe; sporadically in the interior of the European continent, the Caucasus, the Aral-Caspian depression, and Southern Siberia; east coast of North America.

4a. Juncus Gerardii ssp. atrofuscus (Rupr.) comb. n.
Juncus atrofuscus Rupr., Fl. samojed. cisur. (1845), 59; Krechetovich & Goncharov in Fl. SSSR III, 529; Shlyakov in Fl. Murm. II, 168.
J. Gerardii var. *atrifuscus* Trautvetter in Bull. Soc. nat. Mosc. XL, 2 (1867), 110; Buchenau in Pflanzenreich IV, 36, 113.
J. Gerardii Lois. — Perfilev, Fl. Sev. I, 139; Leskov, Fl. Malozem. tundry 44; Hultén, Atlas 113 (450) (pro parte); id., Amphi-atl. pl. 172 (154) (pro parte).

Under northern conditions a typical plant of sea coasts, occurring only in the European Arctic.

Soviet Arctic. Murman (ubiquitous); southern part of Kanin; coast of Cheshskaya Bay; coast of Timanskaya Tundra. There are records of its growing at the mouth of the Pechora (Perfilev).
Foreign Arctic. Arctic Scandinavia (?) (relations to *J. Gerardii* s. str. unclear).
Outside the Arctic. Coasts of White Sea. Recorded (doubtfully!) for the south of West and Central Siberia, Eastern Kazakhstan, Northern Mongolia and Central Yakutia.

5. Juncus nodulosus Wahlb., Fl. Upsal. (1820), 114; Gröntved, Pterid. Spermatoph. Icel. 191; Shlyakov in Fl. Murm. II, 169.
J. alpinus Vill. — E. Meyer in Ledebour, Fl. ross. IV, 224 (pro parte); Scheutz, Pl. jeniss. 171; Lange, Consp. fl. groenl. 124; Gelert, Fl. arct. 1, 24 (pro parte); Buchenau in Pflanzenreich IV, 36, 214; Krylov, Fl. Zap. Sib. III, 575; Perfilev, Fl. Sev. I, 139; Krechetovich & Goncharov in Fl. SSSR III, 537; Karavayev, Konsp. fl. Yak. 73; Polunin, Circump. arct. fl. 115.
J. alpinus ssp. *nodulosus* Lindm. — Hultén, Fl. Al. 417; Böcher & al., Grønl. Fl. 230.
J. alpinus var. *alpestris* et var. *rariflorus* — Hultén, Atlas 109 (432), 110 (435).
Ill.: Fl. Murm. II, pl. LV, 1.

Moist river shores and marshes in the temperate north and locally in the southern fringe of the Arctic.

Soviet Arctic. Murman (apparently common); western fringe of Bolshezemelskaya Tundra (River Shapkina); lower reaches of River Pur; lower reaches of Yenisey (southern tundra and forest-tundra at Zaostrovskoye, Khantayka, etc.).
Foreign Arctic. Extreme SW Greenland, Iceland, Arctic Scandinavia.
Outside the Arctic. Considerable part of Europe; Siberia, Northern Kazakhstan, Kamchatka, temperate north of North America.

The statement in the "Flora of the USSR" that *J. alpinus* grows in all arctic districts of the USSR was evidently based on some kind of misunderstanding.

6. Juncus articulatus L., Sp. pl. (1753) (s. str.); Trautvetter, Pl. Sib. bor.; Gröntved, Pterid. Spermatoph. Icel. 186; Hultén, Atlas 110 (438).
J. lampocarpus Ehrh. — Buchenau in Pflanzenreich IV, 36, 217; Krylov, Fl. Zap. Sib. III, 538; Perfilev, Fl. Sev. I, 139–140; Karavayev, Konsp. fl. Yak. 73.

Poorly vegetated river shores, moist silty meadows.

Species widespread in the temperate north, locally entering subarctic districts but only exceptionally penetrating arctic limits.

Soviet Arctic. Coast of the Arctic Ocean at the mouth of the Kolyma; River Alakit in the Olenek River Basin (forest-tundra).
Foreign Arctic. Iceland, Norwegian coast north to 70°.
Outside the Arctic. Europe, including the Mediterranean and the greater part of the European territory of the USSR; middle zone and south of Siberia; Middle East and Central Asia; northern fringe of Africa; NE North America. As an introduction in Eastern Australia and New Zealand.

7. Juncus biglumis L., Sp. pl. (1753); E. Meyer in Ledebour, Fl. ross. IV, 233; Trautvetter, Fl. taim. 23; Ruprecht, Fl. samojed. cisur. 60; Schmidt, Fl. jeniss. 122; Kjellman,

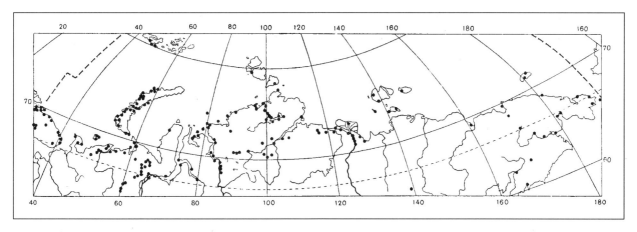

MAP IV-2 Distribution of *Juncus biglumis* L.

Phanerog. sib. Nordk. 120; Scheutz, Pl. jeniss. 171; Lange, Consp. fl. groenl. 122; Gelert, Fl. arct. 1, 25; Buchenau in Pflanzenreich IV, 36, 233; Lynge, Vasc. pl. N.Z. 89; Tolmatchev, Contr. Fl. Vaig. 128; Krylov, Fl. Zap. Sib. III, 577; Tolmachev & Pyatkov, Obz. rast. Diksona 156; Tolmachev, Fl. Kolg. 40; id., Fl. Taym. I, 104; id., Obz. flory N.Z. 150; Perfilev, Fl. Sev. I, 141; Krechetovich & Goncharov in Fl. SSSR III, 521; Hanssen & Lid, Fl. pl. Franz Josef L. 32; Leskov, Fl. Malozem. tundry 44; Devold & Scholander, Fl. pl. SE Greenl. 108; Seidenfaden & Sørensen, Vasc. pl. NE Greenl. 104; Hultén, Fl. Al. III, 421; id., Atlas 111 (441); Tikhomirov, Fl. Zap. Taym. 30; Grøntved, Pterid. Spermatoph. Icel. 188; A.E. Porsild, Vasc. pl. W Can. Arch. 97; id., Ill. Fl. Arct. Arch. 61; Shlyakov in Fl. Murm. II, 160; Karavayev, Konsp. fl. Yak. 71; Korotkevich, Rast. Sev. Zemli; Böcher & al., Grønl. Fl. 232; Polunin, Circump. arct. fl. 115.

Ill.: Porsild, Ill. Fl., fig. 24e.

The species of *Juncus* most characteristic of the Arctic and most widespread within arctic limits. In the Soviet Arctic it occurs more or less ubiquitously (although not uniformly common everywhere), ranging from the southern fringes of the tundra zone as far as the most northern capes and islands. It is more common in districts at relatively high latitudes, where it is often one of the plants regularly encountered. But it is nowhere particularly abundant.

Growing in moderately moist places, on bare clayey patches in polygonal tundras, and on clayey or stony slopes, normally at sites somewhat protected by snow cover in winter. It does not shun deeply snow-covered sites, but is not characteristic of them. Densely vegetated sites and sites subject to strong desiccation are avoided.

Soviet Arctic. Murman (common); Kanin, Timanskaya and Malozemelskaya Tundras (common); Kolguyev and Vaygach Islands; Pay-Khoy, Polar Ural (common); Bolshezemelskaya Tundra (everywhere); throughout Novaya Zemlya (more or less common everywhere); Franz Josef Land; Yamal; shores of Ob Sound and Tazovskaya Bay; Gydanskaya Tundra; lower reaches of Yenisey; islands in Yenisey Bay and the part of the Kara Sea off Taymyr; throughout Taymyr from the forest-tundra to Cape Chelyuskin (common in the central part of the peninsula); Severnaya Zemlya; arctic coast and tundras of Yakutia, and the lower reaches of the rivers which transect it; New Siberian Islands; polar and Beringian coasts of Chukotka; Wrangel Island; Anadyr Basin; Koryakia. (Map IV-2).

Foreign Arctic. Arctic Alaska, arctic coast of Canada, Labrador; throughout the Canadian Arctic Archipelago (including Northern Ellesmere Island); Greenland, including Peary Land (absent from extreme SE Greenland); Iceland; Spitsbergen; Arctic Scandinavia.

Outside the Arctic. Mountains of Northern Great Britain and Scandinavia; Northern Ural, Altay, Sayans, Mount Sokhondo in Transbaikalia, mountains of NE Yakutia; Kamchatka; subarctic districts of Alaska and Canada; Rocky Mountains south to 50°N.

8. ***Juncus stygius*** L. in Syst. nat. ed. 10, 2 (1759), 987; E. Meyer in Ledebour, Fl. ross. IV, 232; Buchenau in Pflanzenreich IV, 36, 225; Krylov, Fl. Zap. Sib. III, 578; Krechetovich & Goncharov in Fl. SSSR III, 523; Perfilev, Fl. Sev. I, 140; Hultén, Atlas 115 (457); id., Amphi-atl. pl. 248 (230); Shlyakov in Fl. Murm. II, 164; Karavayev, Konsp. fl. Yak. 73.

Boreal species just penetrating arctic limits. The majority of records of its occurrence in the Arctic contained in the literature are based on inaccurate identifications and in fact refer to other species, usually *J. biglumis* or *J. triglumis*. This applies, for instance, to the plants from the River Boganida in the Taymyr forest-tundra referred to *J. stygius* by Trautvetter (Fl. boganid. 150). The herbarium of the Botanical Institute in Leningrad contains a certain number of plants identified as *J. stygius* from the portion of the Bolshezemelskaya Tundra near the Urals and from the Anadyr River Basin. All these belong either to *J. biglumis* or (in lesser part) to *J. triglumis*.

Soviet Arctic. West Murman (near the Norwegian frontier and at the entrance of Kola Bay).

Foreign Arctic. Arctic Scandinavia (rare).

Outside the Arctic. Distributed mainly in marshes of the northern forest zone of Eurasia. Within the USSR the northern boundary of the species range (except for the stated localities in Murman) extends from the southern edge of the Kola Peninsula to the upper reaches of the Mezen and Vychegda, then to the middle course of the Ob and Southern Yakutia. East of the White Sea *J. stygius* occurs nowhere even close to the forest-tundra. In the Far East it is reported from Kamchatka and the northern part of Primorskiy Kray. Southwards it does not cross the boundaries of the taiga zone. Found outside the USSR in mountain forest districts of Central Europe and widely in Fennoscandia. Replaced in North America by subspecies *americanus* (Buch.) Hult., to which our Far Eastern plants also belong according to Hultén.

9. ***Juncus triglumis*** L., Sp. pl. (1753), 938; E. Meyer in Ledebour, Fl. ross. IV, 233; Schmidt, Fl. jeniss. 122; Scheutz, Pl. jeniss. 171; Gelert, Fl. arct. I, 25; Buchenau in Pflanzenreich IV, 36, 224; Krylov, Fl. Zap. Sib. III, 576; Andreyev, Mat. fl. Kanina 164; Tolmachev, Mat. fl. Mat. Shar 287; id., Nov. dan. fl. Vayg. 90; id., Obz. fl. N.Z. 150; Perfilev, Fl. Sev. I, 141; Krechetovich & Goncharov in Fl. SSSR III, 522; Gröntved, Pterid. Spermatoph. Icel. 192; Hultén, Atlas 116 (460); Shlyakov in Fl. Murm. II, 162; Böcher & al., Grønl. Fl. 232; Karavayev, Konsp. fl. Yak. 71; Polunin, Circump. arct. fl. 115.

Relatively rare plant, distributed in moderately northern parts of the Arctic. Occurring in small quantity and not everywhere. It is possible that this species is less rare than it seems, but has been mistaken for the more common *J. biglumis* L. on superficial inspection. Growing in clayey patchy tundras on patches of unvegetated soil or at the edges of the bands of closed vegetation which border these patches.

Soviet Arctic. Murman; Northern Kanin; northern part of Malozemelskaya Tundra; NW and SE Bolshezemelskaya Tundra; Polar Ural; Vaygach Island; Novaya Zemlya (the west coast of the South Island and the southeast of the North Island); Gydanskaya Tundra; lower reaches of Yenisey; Taymyr-Khatanga forest-tundra;

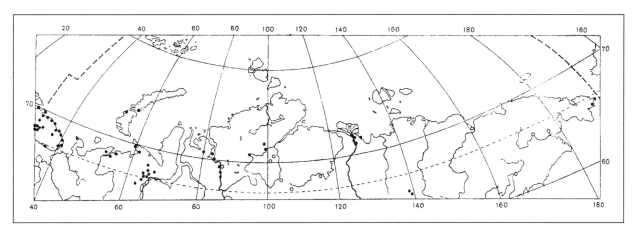

MAP IV–3 Distribution of *Juncus triglumis* L.
Black circles — J. triglumis s. str.; *open circles — J. triglumis* ssp. *albescens* (Lge.) Hult. (Correction: Localities on the coast of the Bering Strait actually refer to *J. triglumis* ssp. *albescens*).

lower reaches of Lena; district of Tiksi Bay. [Replaced further east by *J. triglumis* ssp. *albescens* (Lge.) Hult.]. (Map IV–3).

Foreign Arctic. Iceland; Spitsbergen; Arctic Scandinavia.

Outside the Arctic. Mountains of Scandinavia and Central Europe; Northern Ural; Caucasus; mountains of Southern Siberia.

9a. Juncus triglumis ssp. **albescens** (Lge.) Hult., Circump. pl. I (1962), 241.

J. triglumis var. *albescens* Lge., Consp. fl. Groenl. (1880), 123.

J. albescens Fernald in Rhodora 26 (1924), 202; A.E. Porsild, Vasc. pl. W Can. Arch. 96; id., Ill. Fl. Arct. Arch. 62; Polunin, Circump. arct. fl. 113.

J. triglumis L. — Hultén, Fl. Al. III, 431.

J. Schischkini Kryl. et Sumn. — Krechetovich & Goncharov in Fl. SSSR III, 523-524 (pro parte).

Ill.: Porsild, Ill. Fl., fig. 24b.

In contemporary literature *J. triglumis* var. *albescens*, originally described from Greenland, is either simply not distinguished (with corresponding plants treated indiscriminately as specimens of *J. triglumis*) or else ranked as the separate species *J. albescens* considered to replace the basic Eurasian species in Greenland and North America. Study of herbarium material has convinced us that this contrast is incorrect. Plants of the *J. triglumis* type from the Anadyr and Kolyma Basins, the Okhotsk Coast and a large part of Yakutia undoubtedly tend more towards the American form than towards the West Eurasian type (*J. triglumis* s. str.). But the differences between the type and the American (or more accurately American-East Siberian) form are too minor and obscure to justify the separation of the latter as a species distinct from typical *J. triglumis*. We therefore consider *albescens* to be a geographical race of *J. triglumis*.

Soviet Arctic. Lower reaches of Yenisey (Dudinka); at the edge of the Central Siberian Plateau in the Khatanga Basin; lower reaches of Kolyma; Beringian coast of Chukotka; lower course of Anadyr; Koryakia. (Map IV–3).

Foreign Arctic. West coast of Alaska; Arctic Canada; Labrador; Canadian Arctic Archipelago (north to 80°); Greenland (mainly in the west, on the east coast south of 70°N).

Outside the Arctic. Subarctic East Siberia (in the basins of the Olenek, Vilyuy and Aldan); Verkhoyansk Range; Kolyma-Okhotsk watershed and north coast of Sea of

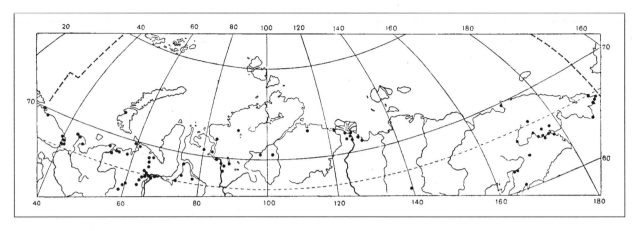

MAP IV-4. Distribution of *Juncus castaneus* Sm.

Okhotsk; subarctic districts of North America. It is possible that plants from the mountains of Southern Siberia identified as *J. triglumis* should be referred to ssp. *albescens*.

10. ***Juncus castaneus*** Sm., Fl. brit. 1 (1800), 383; E. Meyer in Ledebour, Fl. ross. IV, 232; Schmidt, Fl. jeniss. 123; Scheutz, Pl. jeniss. 171; Lange, Consp. fl. groenl. 123; Gelert, Fl. arct. 1, 24; Buchenau in Pflanzenreich IV, 36, 293; Lynge, Vasc. pl. N.Z. 136; Tolmatchev, Contr. Fl. Vaig. 128; Perfilev, Mat. fl. N.Z. Kolg. 17; id., Fl. Sev. I, 140; Andreyev, Mat. fl. Kanina 164; Krylov, Fl. Zap. Sib. III, 580; Leskov, Fl. Malozem. tundry 44; Krechetovich & Goncharov in Fl. SSSR III, 525; Tolmachev, Obz. fl. N.Z. 150; Gröntved, Pterid. Spermatoph. Icel. 187; Hultén, Fl. Al. III, 423; id., Atlas 112 (445); Sørensen, Vasc. pl. E Greenl. 159; Seidenfaden & Sørensen, Vasc. pl. NE Greenl. 104; A.E. Porsild, Vasc. pl. W Can. Arch. 97; id., Ill. Fl. Arct. Arch. 61; Shlyakov in Fl. Murm. II, 164; Karavayev, Konsp. fl. Yakut. 71; Böcher & al., Grønl. Fl. 232; Polunin, Circump. arct. fl. 115.

Ill.: Fl. Murm. II, pl. L, 4; Porsild, Ill. Fl. Arct. Arch., fig. 23a, 24a.

Circumpolar species widespread in the Arctic, native mainly to relatively temperate parts of the region but ranging far to the north in certain sectors (for example, the Thule district in Greenland). Growing on poorly vegetated sites, most frequently on clayey (also on clayey-stony or sandy) soil under conditions of adequate moisture. Occurring locally (for example, on Vaygach Island) in the immediate vicinity of the sea shore, where it is occasionally reached by seawater spray. It usually grows in groups, something undoubtedly associated with its ease of producing rhizomes which then develop numerous aerial shoots and give rise to separate plants. In lowland tundras of temperate arctic districts this species is often found on high riverbanks or on riverside slopes, i.e., where for some reason or other bare soil accumulates. It sometimes follows watercourses along riverbanks and at the foot of shoreline slopes, readily colonizing fresh alluvia.

Soviet Arctic. Murman; Kanin (northern part); Northern Malozemelskaya Tundra (rare); (not recorded for Kolguyev Island); lower reaches of Pechora; greater part of Bolshezemelskaya Tundra (common both in the western part and in the east in the district of the upper reaches of the Usa); Polar Ural; Pay-Khoy; Vaygach Island; west coast of the South Island of Novaya Zemlya (district of Moller Bay); lower reaches of Ob; Obsko-Tazovskiy Peninsula and lower reaches of River Pur; Gydanskaya Tundra; lower reaches of Yenisey (from the northern parts of the forest-tundra to the river mouth); southern part of Taymyr (on the lower course of the Pyasina reaching the mouth of the River Tareya slightly north of 73°; absent from the district of Lake Taymyr); Kheta Valley on the northern edge of the Central Siberian Plateau; lower reaches of Anabar; lower reaches of the Lena and the shores of channels in its delta; coasts of Buorkhaya Bay; in Chukotka near the Bay of Chaun and in the far east; Anadyr Basin (common more or less everywhere); Koryakia (Penzhina Basin and coast of Bering Sea). (Map IV–4).

Foreign Arctic. Arctic Alaska and Canada, Labrador; Canadian Arctic Archipelago; both coasts of Greenland; Iceland; Spitsbergen; Arctic Scandinavia.

Outside the Arctic. Mountains of Scotland and Scandinavia; mountains of Central Europe; Urals south to the 60th parallel; Verkhoyansk Range; coasts (northern and western) of Sea of Okhotsk; Kamchatka; subarctic districts of North America.

Replaced by the very closely related *J. triceps* Rostk. in the Altay, Sayans and the mountains of Prebaikalia. Very closely related species also grow in the Tien Shan, Pamiro-Alay and Himalayas, namely *J. macrantherus* Krecz. et Gontsch., *J. sphacellatus* Buch. and *J. himalensis* Klotzsch.

11. *Juncus leucochlamys* Zing. var. *borealis* Tolm., var. n.

J. leucochlamys — Karavayev, Konsp. fl. Yak. 73.

Differt a *J. leucochlamyde* typico capitulis paucioribus (1–3), compactis, interdum paucifloribus.

Habitat in Sibiria orientale arctica et subarctica.

Typus: In valle fluminis Medveshja, infra ostium fl. Bychy, 6 aug. 1935, Th. Sambuk.

The presence among northern (Yakutian) plants usually identified as *J. castaneus* Sm. of a form closer to the Transbaikalian *J. leucochlamys* was first brought to notice by V.I. Krechetovich and N.F. Goncharov in their work on *Juncus* material for the "Flora of the USSR" (1935). Subsequent authors identified as *J. leucochlamys* their collections from the northern edge of the Central Siberian Highlands (F.V. Sambuk), from the lower course of the Lena (B.N. Gorodkov and B.A. Tikhomirov), and from the Verkhoyansk Range (I.D. Kildyushevskiy). The presence of *J. leucochlamys* in Northern Yakutia was accepted, for instance, in the review by M.N. Karavayev (1958).

Inspection of numerous specimens of *J. castaneus* s. l. from the Siberian Arctic has revealed plants, belonging to the form here described, in collections from the lower reaches of the Yenisey, from the Khatanga Basin and especially from the lower course of the Lena and adjacent tundra districts. It has also been collected in northern districts of Kamchatka (Komarov, 1909; Novograblenov, 1930) and more recently by I.D. Kildyushevskiy (1960) on the Penzhina River.

The form here described differs from the widespread *J. castaneus* (s. str.) in having better developed flat basal leaves which reach or exceed the middle of the stem. One of the stem leaves almost reaches the inflorescence. The lower involucral bract is long (up to 12 cm), straight, gradually tapering apically and usually considerably exceeding the inflorescence. The heads are partly on long (up to 4 cm) pedicels (except for poorly developed plants possessing only one head). The bracteoles are albescent or ferruginous, longer than the perianth. The perianth segments are relatively light, narrowly lanceolate and acute (however, those of *J. castaneus* are also partly acute). The capsule gradually tapers apically and is light ferruginous. The total height of the plant varies from 18 to 38 cm.

In all characters distinguishing them from *J. castaneus*, our plants approach *J. leucochlamys*. But it should be emphasized that they also differ rather sharply from the typical Transbaikalian *J. leucochlamys*, occupying a position in some ways intermediate between it and *J. castaneus*. Thus, the Transbaikalian type possesses an inflorescence consisting of 5–12 heads, while in our form there are no more than 3. The individual florets and fruits in the heads of typical *J. leucochlamys* are often somewhat divaricate, something not observed in our form. It deserves atten-

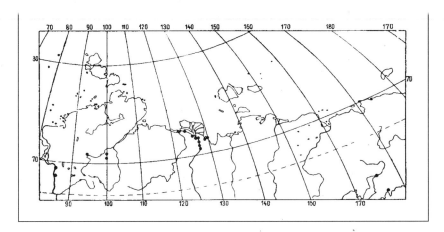

MAP IV–5 Distribution of *Juncus leucochlamys* Zing. var. *borealis* Tolm.

tion that *J. leucochlamys* var. *borealis* normally occurs where *J. castaneus* also occurs, which is suggestive of some kind of genetic connection between them. Could this form be the product of hybridization between *J. leucochlamys* and *J. castaneus*?

It should be emphasized that our variety cannot be associated with the Southern Siberian alpine *J. triceps* Rostk., which apparently represents one of the geographical races of *J. castaneus* s. l.

Juncus leucochlamys grows on shoreline sands and gravelbars, on poorly vegetated slopes, at the outflows of springs, and on patches of clayey soil in open larch forests. It occurs on the southern fringe of the tundra zone and in the forest-tundra, as well as in the mountains of subarctic East Siberia, from the lower reaches of the Yenisey in the west to the Penzhina and Kamchatka in the east.

Soviet Arctic. Lower reaches of Yenisey (rather numerous collections from Dudinka, also Vershininskoye and Luzino); open forest of the Khatanga Basin (vicinity of Volochanka on the Kheta River); northern edge of Central Siberian Plateau in the valley of the River Medvezhya; lower reaches of Lena (from Kyusyur to Tit-Ary on the Lena and along the Olenekskaya Channel all the way to its mouth; occurring apparently more frequently than *J. castaneus*); coast of Buorkhaya Bay; lower reaches of Penzhina River; Bay of Korf. (Map IV–5).

Foreign Arctic. Not occurring.

Outside the Arctic. Our form occurs in the Verkhoyansk Range (Tompo district) and on the northern part of the Kamchatka Peninsula. Typical *J. leucochlamys* Zing. is distributed in Transbaikalia and Northern Mongolia.

12. ***Juncus filiformis*** L., Sp. pl. (1753), 326; E. Meyer in Ledebour, Fl. ross. IV, 223; Ruprecht, Fl. samojed. cisur. 59; Scheutz, Pl. jeniss. 171; Lange, Consp. fl. groenl. 124; Gelert, Fl. arct. 1, 22; Buchenau in Pflanzenreich IV, 36, 127; Krylov, Fl. Zap. Sib. III, 568; Perfilev, Fl. Sev. I, 141; Krechetovich & Goncharov in Fl. SSSR III, 552; Leskov, Fl. Malozem. tundry 44; Gröntved, Pterid. Spermatoph. Icel. 190; Hultén, Fl. Al. III, 427; id., Atlas 113 (449); Shlyakov in Fl. Murm. II, 175; Böcher & al., Grønl. Fl. 231; Polunin, Circump. arct. fl. 112.

Ill.: Fl. Murm. II, pl. LVI, 1.

On silty river shores, in lowland marshes and in marshy meadows on the southern fringes of the Arctic.

Soviet Arctic. Murman (more or less ubiquitous); Kanin, Timanskaya and Malozemelskaya Tundras (more or less common); southern districts of the Bolshezemelskaya Tundra; Polar Ural (not everywhere); lower reaches of Ob and

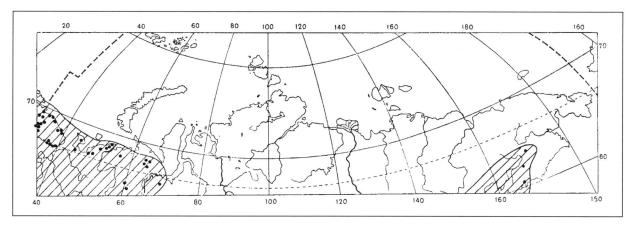

Map IV–6 Distribution of *Juncus filiformis* L.
(Correction: The occurrence in forest-tundra on the Yenisey has been omitted).

SE coast of Yamal; forest-tundra on the lower reaches of the Yenisey (Khantayka); Anadyr Basin (rare, isolated discoveries); Koryakia (Penzhina Basin, Bay of Korf). (Map IV–6).

Foreign Arctic. Extreme south of Greenland; Iceland; Arctic Scandinavia.

Outside the Arctic. Europe, except the Mediterranean proper; forest zone of the European territory of the USSR; Caucasus; greater part of Siberia and the temperate Far East; temperate north of North America. There are records of the discovery of this species in Patagonia.

13. *Juncus brachyspathus* Maxim., Prim. fl. amur. (1859), 293; Buchenau in Pflanzenreich IV, 36, 128; Krylov, Fl. Zap. Sib. III, 572; Perfilev, Fl. Sev. I, 141; Krechetovich & Goncharov in Fl. SSSR III, 553; Karavayev, Konsp. fl. Yak. 72.

Occurring on the southern fringe of the Arctic Region on sandbars and siltbars in river valleys, sometimes beneath a canopy of floodplain willows. Apparently forming rather dense aggregations.

Soviet Arctic. Lower reaches of the Ob and Poluy (mainly near Salekhard), and around the mouth of the Taz; lower reaches of Kolyma; Anadyr Basin (widespread and apparently common); Koryakia.

Foreign Arctic. Not occurring.

Outside the Arctic. Widespread in the forest zone of Siberia, but over much of its extent not reaching the northern forest boundary. In the west penetrating the forested part (right bank) of the Pechora Basin, in the east reaching the shores of the Sea of Okhotsk and Sakhalin, in the south penetrating Primorskiy Kray, NE China and Northern Mongolia.

14. *Juncus beringensis* Buch. in Engl. Bot. Jahrb. XII (1890), 226; id. in Pflanzenreich IV, 36, 129; Komarov, Fl. Kamch. I, 278; Krechetovich & Goncharov in Fl. SSSR III, 556.

Ill.: Buchenau in Pflanzenreich IV, 36, 129, fig. 69.

Characteristic mountain plant of Kamchatka and islands of the Bering Sea, penetrating Koryakia where it occurs (near the coast of the Bay of Korf) on slopes with late snowmelt in the valleys of rivers and streams flowing from the mountains. Locally common.

Soviet Arctic. Koryakia in the district of the Bay of Korf (near Tilichiki settlement, Tolmachev, Stepanova and Fedorova, 1960; near Kultushnoye settlement, Katenin, 1960).

Foreign Arctic. Absent.

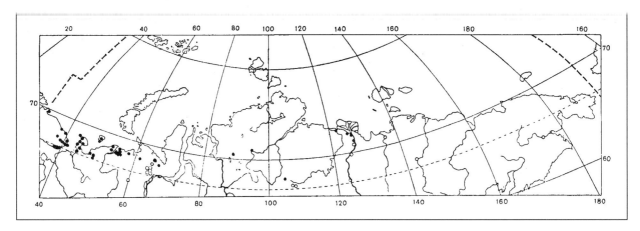

MAP IV-7 Distribution of *Juncus arcticus* Willd.
Black circles — *J. arcticus* s. str.; *open circles* — *J. arcticus* ssp. *alaskanus* Hult.
(Correction: Localities in the lower reaches of the Yenisey and the Gydanskaya Tundra have been omitted).

Outside the Arctic. Known from Koraginskiy Island (whence it was described), the northern coast and alpine districts of Kamchatka, the north coast of the Sea of Okhotsk, the Commander Islands, and the mountains of Northern Japan.

15. ***Juncus arcticus*** Willd., Sp. pl. 2 (1799), 206; E. Meyer in Ledebour, Fl. ross. IV, 223; Ruprecht, Fl. samojed. cisur. 59; Lange, Consp. fl. groenl. 124; Gelert, Fl. arct. 1, 24; Buchenau in Pflanzenreich IV, 36, 147; Tolmachev, Fl. Kolg. 16, 40; Andreyev, Mat. fl. Kanina 164; Krylov, Fl. Zap. Sib. III, 572 (pro parte); Perfilev, Fl. Sev. I, 142; Krechetovich & Goncharov in Fl. SSSR III, 555; Leskov, Fl. Malozem. tundry 43; Gröntved, Pterid. Spermatoph. Icel. 185; Hultén, Atlas 110 (437); Shlyakov in Fl. Murm. II, 176; A.E. Porsild, Ill. Fl. Arct. Arch. 61; Böcher & al., Grønl. Fl. 231; Polunin, Circump. arct. fl. 112.

J. balticus — Schmidt, Fl. jeniss. 123 (pro parte); Scheutz, Pl. jeniss. 171 (pro parte).
Ill.: Fl. Murm. II, pl. LVI, 1; Porsild, l. c., fig. 22f, g, 24c.

Common plant of poorly vegetated sandy riverside stretches which are flooded only at maximal water levels or not flooded at all. It spreads by forming regular rows of stems extending along the buried rhizomes. Occurring frequently in moderately northern parts of the tundra zone, but only on the very northern edge of the forest-tundra. Sometimes also growing on silty river shores, where it may be submerged both during the spring flood and after summer rains. But under these conditions it is usually represented by less typical plants.

Soviet Arctic. Murman; Kanin (throughout the Peninsula, common); Timanskaya and Malozemelskaya Tundras; Kolguyev Island; NW and central parts of Bolshezemelskaya Tundra; Karskaya Tundra and the northern extremity of the Polar Ural; Gydanskaya Tundra; lower reaches of Yenisey (common, but not reaching the river mouth); forest-tundra of Southern Taymyr; lower reaches of Olenek and Lena (frequent).

Along the southern edge of its range starting from the Eastern Bolshezemelskaya Tundra and in extreme NE Asia replaced by the race *J. arcticus* ssp. *alaskanus* Hult. (see below). (Map IV-7).

Foreign Arctic. Arctic Canada (near Hudson Bay); Labrador; Baffin Island; Greenland north to 73–74°; Iceland; Arctic Scandinavia.

Outside the Arctic. Mountains of Scandinavia and Central Europe; coasts of Hudson Bay.

15a. *Juncus arcticus* ssp. *alaskanus* Hult., Fl. Al. III (1943), 418.
   J. balticus — Schmidt, Fl. jeniss. 123 (pro parte); Scheutz, Pl. jeniss. 171 (pro parte).
   J. arcticus — Krylov, Fl. Zap. Sib. III, 572 (pro parte); Karavayev, Konsp. fl. Yak. 74 (pro parte).
   J. balticus var. *alaskanus* Porsild, Ill. Fl. Arct. Arch. 60.
   J. Haenkei — Karavayev, Konsp. fl. Yak. 72.

   Subarctic Siberian and Beringian race of *J. arcticus*, often confused with *J. balticus* (on account of the loose inflorescence). In contrast with the typical form (with which it is connected by intermediates), growing partly on silty river shores flooded by rises in water level during the summer.

   Soviet Arctic. SE Bolshezemelskaya Tundra (upper course of Usa, Vorkuta); Polar Ural; lower reaches of Yenisey (Dudinka and further south in the forest-tundra); lower reaches of Lena (forest-tundra); lower reaches of Yana; lower course of Anadyr. (Map IV–7).

   Plants with characters intermediate between ssp. *alaskanus* and typical *J. arcticus* occur in Kolguyev Island, the extreme northeast of the Malozemelskaya Tundra, a considerable part of the Bolshezemelskaya Tundra, the Gydanskaya Tundra, the lower reaches of the Yenisey, the Khatanga Basin, and the lower reaches of the Lena.

   Foreign Arctic. Arctic Alaska and the western part of Arctic Canada.

   Outside the Arctic. Subarctic districts of East Siberia (in the upper reaches of the Olenek and Vilyuy, the Lena Valley near the Arctic Circle, and the middle course of the Indigirka); middle part of Aldan Basin; high mountains of the Eastern Sayan and SE Altay. Subarctic districts of Western North America.

16. *Juncus Haenkei* E. Mey., Syn. Juncor. (1822) 10; Komarov, Fl. Kamch. I, 278; Krechetovich & Goncharov in Fl. SSSR III, 556.
   J. balticus var. *Haenkei* Buchenau, Monogr. Junc. (1890) 215; id. in Pflanzenreich IV, 36, 145; Gelert, Fl. arct. 1, 23.
   J. balticus var. *sitchensis* (Engelm.) — Hultén, Fl. Al. III, 420.

   Growing on sea shores, on substrates which are to some degree salinized. On coastal sandbars forming almost closed beds which are conspicuous from afar due to the darkish grey-green colour of the stems of this plant. In SE Chukotka confined to sites of the outflows of warm brackish springs.

   Soviet Arctic. Extreme SE Chukotka Peninsula (at the Senyavin, Chaplino and Chaymenskiye Springs); Ugolnaya Bay at the mouth of the Anadyr; shores of the Bay of Korf and Gizhiga Bay.

   Foreign Arctic. Arctic Alaska.

   Outside the Arctic. Coasts of Kamchatka, the Sea of Okhotsk, Primorskiy Kray, Sakhalin, the Kurile and Commander Islands, and Northern Japan; Aleutian Islands and Pacific coast of North America.

17. *Juncus balticus* Willd. in Magaz. Naturf. Fr. Berlin 2 (1809), 298; E. Meyer in Ledebour, Fl. ross. IV, 222; Krechetovich & Goncharov in Fl. SSSR III, 555; Perfilev, Fl. Sev. I, 142 (pro parte); Gröntved, Pterid. Spermatoph. Icel. 187; Hultén, Atlas 110 (437); Shlyakov in Fl. Murm II, 178; Polunin, Circump. arct. fl. 113.
   J. balticus var. *europaeus* Englem. — Gelert, Fl. arct. 1, 23; Buchenau in Pflanzenreich IV, 36, 144.

   Plant of the sea coasts of NW Europe.

   Soviet Arctic. There are only doubtful records for West Murman and the coast of the Timanskaya Tundra.

   Foreign Arctic. Extreme SW Greenland; Iceland; arctic districts of Norway.

   Outside the Arctic. Coasts of the Baltic, North, Norwegian and White Seas.

   Perfilev's record of the occurence of *J. balticus* on the Timan coast was cast in

doubt by Leskov (Fl. Malozem. tundry 44), who did not find a single plant of this species among material from that part of the Arctic preserved in herbaria. Rather numerous specimens, especially from the Bolshezemelskaya Tundra, determined by various botanists at various times as *J. balticus* actually belong to one of the forms of *J. arcticus*. The same applies to plants from the lower reaches of the Yenisey and the Gydanskaya Tundra recorded in the works of Schmidt and Scheutz and mentioned with citation of these authors by Gelert in the "Flora arctica."

GENUS 2 — **Luzula** DC. — WOODRUSH

Genus containing up to 100 species, some polymorphic and differentiated into rather sharply differing varieties and races. The majority of species are distributed in extratropical regions of both hemispheres, as well as in the mountains of tropical South America. Arid regions completely lack *Luzula* species. Numerous species occur in the northern temperate zone, mainly in mountainous districts. In the arctic flora the genus is represented by a whole series of characteristic species. Some of these occur only in the Arctic or only just range beyond its limits (*L. nivalis, L. Wahlenbergii, L. tundricola, L. beringensis*). No less characteristic of the Arctic are certain arctic-alpine species (*L. confusa*), which enjoy a very wide distribution within its limits. Approaching these in some respects are typical subarctic species characteristic of the more temperate districts of our region (*L. parviflora, L. arcuata, L. unalaschkensis*).

It is significant that certain series of closely related species, especially the *Arcticae* V. Krecz. (*L. nivalis, L. tundricola*) and *Arcuatae* V. Krecz. (*L. arcuata* and relatives), are associated predominantly with the Arctic, something which indicates that the association of the relevant groups with the far north is of great antiquity. Similar to the above groups in this respect are species of the series *Parviflorae* V. Krecz. (*L. parviflora, L. Wahlenbergii*). In contrast, the morphologically sharply differentiated *L. spicata*, relatively local in the Arctic, is characterized by a wide distribution in high mountains of the temperate zone.

A certain portion of the *Luzula* species growing in the Arctic must be considered boreal elements in its flora. Of these, only the polymorphic *L. multiflora* is relatively characteristic of more temperate arctic districts and shows a tendency to form specifically northern varieties on the northern edge of its range. The other boreal species just penetrate arctic limits.

Ecologically, the species of *Luzula* occurring in the Arctic are rather diverse. The boreal species penetrating the Arctic are almost entirely confined to meadow associations and grow in river valleys or on euthermal slopes sufficiently protected by snow in wintertime. They and *L. parviflora* often occur among shrubs. Certain species of the genus, for example *L. Wahlenbergii* and *L. arcuata*, are characteristic of drier areas in boggy tundras, where they frequently grow on peat mounds. Finally, those species which penetrate the furthest north, *L. nivalis* and particularly *L. confusa*, are most characteristic of tundra formations which are dry in summer and retain little snow in winter.

Certain species of *Luzula* belong among the characteristic components of particular vegetational associations in the Arctic, playing a substantial role in determining their appearance. This especially applies to alpine tundra species and to species which grow on peat mounds.

The greater part of arctic *Luzula* samples are readily identifiable if we are dealing with populations typical of their species from outside districts where the ranges of closely related species make contact or overlap. However, where the ranges of such species overlap, the species boundaries often become difficult to recognize and the referral of particular specimens to one or other of closely related species (for example, in the case of *L. arcuata* and *L. confusa* in several districts of the European Arctic) proves dependent on the preference given by a particular author to one or other of the basic diagnostic characters, whose "coupling" is broken in many individuals. The situation is complicated by the fact that variation in *Luzula* is evidently affected by ecological conditions (both general climatic and special habitat conditions). Individuals of different species growing under extreme conditions may sometimes acquire much resemblance to one another as a result of parallel ecological variation. And, if this is associated with appreciable deviation from the typical form of the relevant species, recognition of the species to which particular individuals belong is sometimes fraught with difficulties. In the case of species not specifically (or not predominantly) arctic, it is important also to bear in mind that the diagnostic "norms" serving to characterize the species in its typical phenotype may not be applicable without qualification and refinement to arctic material.

The phenomenon of "transgressive variation" is undoubtedly further complicated by the phenomenon of hybridization. The consequences of the latter are recognized as something obvious only when the plants under consideration combine the characters of sharply differing species. This is the case, for example, with the hybrid *L. parviflora* × *nivalis*. Considerably more difficult to judge is the role played by past or present hybridization in "blurring the boundaries" between such species as *L. confusa* and *L. arcuata*, in which case the existence of primary intermediate forms as well as considerable parallel variation may be assumed.

Finally it should be stated that certain forms here treated as distinct species may have originated through hybridization. As examples known to us of such possible interpretations we suggest *L. tundricola* (*L. nivalis* × *Wahlenbergii*?) and *L. beringensis* (*L. confusa* × *unalaschkensis*?).

1. Florets *solitary or arranged in pairs* (rarely groups of 3 florets on some branchlets), at tips of slender branchlets in umbellate or paniculate inflorescence. . . . . .2.
– Florets *aggregated in capitate or spikelike clusters or glomerules*, which are in turn assembled in a more complex (umbellate, capitate or spikelike) inflorescence. [SUBGENUS **GYMNODES** GRISEB.]. . . . . .6.

2. Inflorescence *umbellate*, with more or less *straight unbranched* or weakly branched branchlets that are erect or directed obliquely upwards (sometimes declinate later). *Florets always solitary*. [SUBGENUS **PTERODES** GRISEB.]. . . . . .3.
– Inflorescence *paniculate*, usually somewhat leaning to one side, bearing numerous florets, with slender *curved* branchlets that are *considerably*

branched. Florets solitary or in groups of 2–3. [SUBGENUS ANTHELAEA GRISEB.]. . . . . .4.

3. Leaves *linear*, 2–3 (sometimes up to 4) mm wide. Perianth segments *stramineous* with ferruginous midrib. Anthers *slightly longer* than stamen filaments. Capsule equalling or slightly longer than perianth, triangular-ovoid. . . . . .1. **L. RUFESCENS** FISCH.
– Leaves *linear-lanceolate*, 5–10 mm wide. Perianth segments *castaneous*. Anthers *twice as long as* stamen filaments. Capsule distinctly exceeding perianth, globose-conical. . . . . .2. **L. PILOSA** (L.) WILLD.

4. Forming small, *rather dense tufts*. Basal leaves *prostrately spreading*, 2–4 mm wide, *shining dark green*, often with reddish tint, involute at tip. Stem leaves *2–3, small, narrow* (about 2 mm wide). Stems *numerous*, reddish, 10–20(25) cm high. Inflorescence *very diffuse*; florets on short pedicels. Perianth segments about 2 mm long, brown or reddish brown. Capsule ovoid, equalling or slightly longer than perianth. . . . . .4. **L. WAHLENBERGII** RUPR.
– Stems *solitary* or growing in groups of 2–4, *not forming tufts*, 25–45 cm high. Basal leaves *standing upright*, 6–10 mm wide; stem leaves *3–5*, shorter and narrower than basal; all leaves *light green, opaque*. Inflorescence *strongly asymmetrical, drooping to one side*. . . . . .5.

5. Plant light green. Usually 4–5 stem leaves. Perianth segments broadly lanceolate, bright castaneous, about 2 mm long. Capsule *brown, longer than perianth*. . . . . .3. **L. PARVIFLORA** (EHRH.) DESV.
– Plant light green, often with reddish tint. Usually 3 stem leaves. Perianth segments broadly lanceolate, rusty reddish. Capsule *blackish purple, equalling or just shorter than perianth*. . . . . .3a. **L. PARVIFLORA** SSP. **MELANOCARPA** (MICHX.) TOLM.

6. Lower involucral bract *poorly developed*. Bracteoles *fimbriate-laciniate*. Leaves with acuminate (mucronate) or obtuse tip. . . . . .7.
– Lower involucral bract *well developed, flat*, sometimes exceeding inflorescence. Bracteoles *ciliate, scarcely fimbriate*. Leaves flat, constricted apically to swollen obtuse tip. . . . . .14.

7. Inflorescence *oblong, spikelike*, sometimes somewhat interrupted but not branched, *drooping or leaning*. Silvery bracteoles longer than florets. Stems slender, *in their upper part* (below inflorescence) *distinctly attenuated*. Leaves channelled, narrow. Perianth 2–3 mm long, with finely acuminate segments. Capsule ovoid, equalling or slightly shorter than perianth. Entire plant brownish-reddish. . . . . .5. **L. SPICATA** (L.) DC.
– Inflorescence *erect, not drooping*, capitate or branched. In latter case heads more or less globose, on straight or curved branchlets of nonuniform length. . . . . .8.

8. Leaves *finely acuminate at tip*, channelled or flat. ....9.
- Leaves *with calluslike swelling at tip, obtuse*, always flat. ....13.

9. Inflorescence *dense and unbranched* (forming oblong or irregular head) *or divided into a few* (*not many*) *heads*. In the latter case the branchlets of the inflorescence are rather dense and straight and the head situated at the base of the inflorescence is distinctly larger than the others. Perianth segments broadly lanceolate (ratio of length to width 3:1), mostly about 1.5 mm long. Capsule rounded distally, with very short beak, of about same length as perianth or exserted beyond it. Forming *dense tufts*. Leaves *channelled*, standing upright or sometimes slightly curved with their tips close to the stem. Sheaths at bases of shoots *shining, intensely reddish*. ....6. **L. CONFUSA** LINDB.
- Inflorescence *considerably branched, many-headed*. Its branches of nonuniform length, sometimes branched in their turn, slender, sometimes curved. Heads not large (up to 8 mm in length and width), of approximately the same size as one another. ....10.

10. Forming *very dense* tufts. Leaves *narrow, channelled, with intensely reddish shining sheaths*. Heads of inflorescence numerous, small (sometimes containing only 2–4 florets). Perianth segments narrowly lanceolate (ratio of length to width not less than 4:1), more than 2 mm long. Capsule shorter than perianth, somewhat tapering apically. ....
...................................7. **L. BERINGENSIS** TOLM.
- Tufts *more or less loose*. Leaves *flat, obliquely divergent from stem base*, sometimes curved outwards distally. *Sheaths dull greyish*. ....11.

11. Forming very small loose tufts or sometimes with solitary stems. Basal leaves more or less straight, directed obliquely upwards, long (reaching half stem height or more). Perianth segments *less than 2 mm long, broadly lanceolate* (ratio of length to width 3:1). Capsule *not shorter than perianth*, rounded apically but slightly tapering, with indistinct beak. ....
...........8a. **L. UNALASCHKENSIS** SSP. **KAMTSCHADALORUM** (SAM.) TOLM.
- Forming more or less loose but sometimes rather large tufts. Perianth segments *2–2.5 mm long, narrowly lanceolate* (ratio of length to width 4:1 or 5:1). Capsule slightly *shorter than perianth*, somewhat tapering apically with conspicuous beak. ....12.

12. Basal leaves *curved falcately outwards*, sometimes folded lengthwise. Stems *not rising very far* above leaves (less than twice as high). Leaves light green, shining. ....8. **L. UNALASCHKENSIS** (BUCH.) SATAKE
- Basal leaves *directed obliquely upwards*, straight, flat. Stems *considerably exceeding leaves* (*two or more times as high*), sometimes with well expressed lilac tint. Leaves more or less dull, not light green. ....
..................................9. **L. ARCUATA** (WAHLB.) SW.

13. Inflorescence an unbranched *globose or oblong head*, or with 1–2 lateral heads on short straight branchlets. Entire plant often dark with lilac tint. ..
....10. **L. NIVALIS** LAEST.

– Inflorescence *considerably branched*, with small heads partly on long and slender, sometimes curved branchlets. . . . . .11. **L. TUNDRICOLA** GORODK.

14. Stem and leaves *pale* (slightly dull) *green, without reddish or lilac tint.* Inflorescence *umbellate-paniculate, with 5–15 clearly differentiated, small* (3–4 mm wide), often slightly oblong *pale brown heads.* Florets small, about 2 mm long; perianth segments nonuniform (inner shorter than outer), pale rusty, sometimes whitish with rusty median stripe. Capsule light, of almost same length as perianth. . . . . .16. **L. PALLESCENS** (WAHLB.) BESS.
– Stem and leaves usually *bright green, often with conspicuous reddish or lilac tint.* Inflorescence more or less compact, *densely umbellate or almost capitate, with from 1–2 to 5–7 heads* whose general colour is *dark* (*brown or blackish*). Perianth segments intensely castaneous-rusty or darker (to blackish purple). . . . . .15.

15. Inflorescence *compact, aggregated in a single relatively large head* (up to 1 cm wide) usually of irregular shape; sometimes there is an additional lateral head considerably smaller than the main head; the latter may be imperfectly divided into not fully separate parts. Perianth segments lanceolate or broadly lanceolate, castaneous or blackish brown, with colourless hyaline border on margin and with slender subulate colourless tip. Capsule just shorter than perianth, brown. *Lower involucral bract more or less horizontally divergent* or at first widely divergent then obliquely ascending, often rather broad. Second involucral bract, if well developed, sometimes almost upright on the opposite side to the lower. Leaves most often light green, mostly without reddish tint. . . . . .12. **L. CAPITATA** (MIQ.) NAKAI
– Inflorescence *densely umbellate or almost capitate.* Lower involucral bract *upright or directed obliquely upwards.* . . . . .16.

16. Inflorescence densely umbellate with the heads few and arranged in a more or less dense group. *Florets small,* 2–2.5 mm long, with blackish brown perianth segments of *nonuniform* length (inner shorter than outer). Stems mostly solitary. . . . . .14. **L. SUDETICA** (WILLD.) DC.
– Inflorescence densely umbellate or almost capitate. *Florets 2–2.5–3 mm long,* with perianth segments of *uniform* length. Stems mostly forming small tufts. . . . . .17.

17. Plant *green, completely lacking reddish tint.* Leaves *broad,* 2.5–5 mm wide. Perianth segments lanceolate, castaneous, distally with whitish border on margin. Bracteoles *relatively large,* light castaneous basally, pale distally. Capsule of approximately same length as perianth. . . . . .15. **L. OLIGANTHA** SAM.
– Plant almost always *with distinct reddish or lilac tint.* Bracteoles *less conspicuous,* darker coloured. [13. **L. MULTIFLORA** LEJ., S. L.]. . . . . .18.

18. Perianth segments *broadly lanceolate, castaneous*, with more or less distinct pale hyaline border, *rather abruptly tapering apically to subulate mucro.* Capsule *shorter and often lighter than perianth.* ....
................................13b. **L. MULTIFLORA** SSP. **SIBIRICA** V. KRECZ.
– Perianth segments *lanceolate, gradually transformed apically into subulate mucro, blackish purple or blackish brown*, without or with very narrow hyaline border on margin. Capsule *of approximately same length as perianth*, sometimes slightly exserted above it, shining blackish purple or dark castaneous. ....19.

19. Inflorescence usually densely umbellate, with 2–5–7 small approximately uniform heads on branchlets 0.5–1.5 cm long; more rarely inflorescence compact, almost capitate. Lower involucral bract *almost upright.* Stem leaves usually not more than 2.5–3 mm wide. Plants normally not more than 20 cm high, occasionally up to 25 cm after flowering. ....
........................13a. **L. MULTIFLORA** SSP. **FRIGIDA** (BUCH.) V. KRECZ.
– Inflorescence densely umbellate with small heads on obliquely ascending nonuniform branchlets, sometimes compact and almost capitate. Lower involucral bract *directed obliquely upwards*, sometimes rather widely divergent. Stem leaves in large specimens up to 5 mm wide, shorter and wider than basal leaves. Plants up to 20–25 (sometimes up to 30) cm high.
..............13c. **L. MULTIFLORA** SSP. **KJELLMANIANA** (MIY. ET KUDO) TOLM.

1. ***Luzula rufescens*** Fisch. in Linnaea XXII (1849), 385; E. Meyer in Ledebour, Fl. ross. IV, 215; Trautvetter, Pl. Sib. bor. 116; id., Fl. rip. Kolym. 563; Gelert, Fl. arct. 1, 27; Buchenau in Pflanzenreich IV, 36, 46; Krylov, Fl. Zap. Sib. III, 547; Krechetovich in Fl. SSSR III, 563; Hultén, Fl. Al. 441; Karavayev, Konsp. fl. Yak. 74.
   L. pilosa— Scheutz, Pl. jeniss. 169.
   Plant of forest glades and meadows, widespread in the forest zone of Siberia and the Far East. In the north of its range penetrating the fringes of the Arctic.
   Soviet Arctic. Forest-tundra between the Olenek and Lena; lower reaches of Lena; lower reaches of Kolyma; NW spurs of the Southern Anyuyskiy Range; lower part of Anadyr Basin; upper reaches of River Oklan (Penzhina Basin); coast of Koryakia near the Bay of Korf.
   Foreign Arctic. Not occurring.
   Outside the Arctic. Forested region of Siberia east of the Yenisey and of the Soviet Far East; NE China, Northern Mongolia, the north of the Korean Peninsula, Northern Japan.

2. ***Luzula pilosa*** (L.) Willd., Enum. pl. Horti Berol. (1809), 393; Ruprecht, Fl. samojed. cisur. 58; E. Meyer in Ledebour, Fl. ross. IV, 214; Gelert, Fl. arct. 1, 27; Buchenau in Pflanzenreich IV, 36, 48; Krylov, Fl. Zap. Sib. III, 546; Perfilev, Fl. Sev. I, 135; Krechetovich in Fl. SSSR III, 564; Leskov, Fl. Malozem. tundry 43; Hultén, Atlas (470); Shlyakov in Fl. Murm. II, 182.
   Plant of forest glades and meadows, penetrating the Arctic in its European sector.
   Soviet Arctic. Murman; southern part of Kanin; Timanskaya Tundra.
   Foreign Arctic. Arctic Scandinavia.
   Outside the Arctic. Much of Europe (but absent from the Mediterranean), the forested part of the European territory of the USSR, West Siberia, southern districts of Central Siberia, and the western part of the Caucasus; Eastern North America.

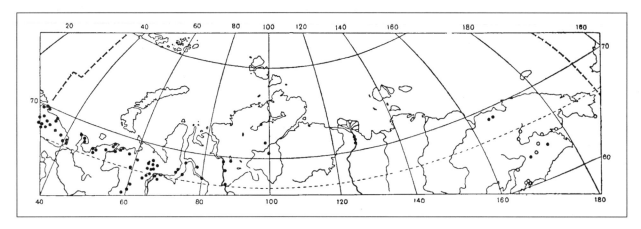

MAP IV–8 Distribution of *Luzula parviflora* (Ehrh.) Desv.
Black circles — *L. parviflora* s. str.; *open circles* — *L. parviflora* ssp. *melanocarpa* (Michx.) Tolm. (Correction: The locality for *L. parviflora* s. str. on Kolyuchin Bay (Pitlekay) has been omitted).

   3. *Luzula parviflora* (Ehrh.) Desv. in Journ. de Bot. 1 (1808), 144; Trautvetter, Fl. boganid. 146; Kjellman, Phanerog. sib. Nordk. 120; Scheutz, Pl. jeniss. 169; Gelert, Fl. arct. 1, 28; Buchenau in Pflanzenreich IV, 36, 61; Andreyev, Mat. fl. Kanina 165; Krylov, Fl. Zap. Sib. III, 548; Perfilev, Fl. Sev. I, 135; Leskov, Fl. Malozem. tundry 42; Hultén, Atlas 118 (469); Shlyakov in Fl. Murm. II, 183; Karavayev, Konsp. fl. Yak. 74; Böcher & al., Grønl. Fl. 234; Polunin, Circump. arct. fl. 117.

   L. spadicea ε *parviflora* — E. Meyer in Ledebour, Fl. ross. IV, 217; Schmidt, Fl. jeniss. 122.

Subarctic-alpine species, distributed predominantly in the southern fringe of the tundra zone and the northern fringe of the forest zone. Growing most frequently among tundra willow thickets, in clearings with mosses and low herbs between denser groups of shrubs at sites with good winter snow cover and more or less protected from wind in summer. Usually encountered in small quantity, but by no means a rarity in a large part of its range. Distinctly more rarely encountered than under the above conditions in dwarf birch tundra or even hummocky tundra. Also encountered on meadowy slopes, but avoiding sites with a taller, better developed herb layer.

Similar in structure to the typical tundra plant *L. Wahlenbergii* (with which it is sometimes confused), but readily distinguished from that species by its opaque lighter green foliage, taller stature, and the greater development and larger size of the stem leaves. The stems are solitary, never forming tufts.

The differences between typical *L. parviflora* and *L. melanocarpa* (Michx.) Desv. occurring in the Far East and North America are not sufficiently constant for a boundary to be drawn between these forms like one between separate species. But, since *L. melanocarpa* in its typical expression differs rather markedly from *L. parviflora* s. str., it would have been incorrect to ignore its existence as a real form of this species. We consider it to be an eastern (Far Eastern and North American) race of this species.

Soviet Arctic. Murman; Kanin; Timanskaya and Malozemelskaya Tundras; Bolshezemelskaya Tundra; Polar Ural; lower reaches of Ob; shores of Ob Sound and Tazovskaya Bay; lower reaches of Yenisey; forest-tundra of Southern Taymyr; Central Taymyr (one locality on the Upper Taymyra River; Middendorff, 1843); lower reaches of Lena; Chukotka Peninsula (Pitlekay); Anadyr Basin (right bank of Velikaya River; apparently occurring rarely). (Map IV–8).

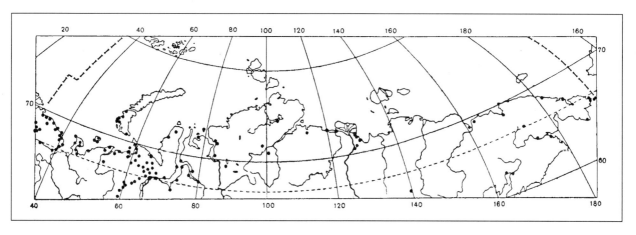

MAP IV–9 Distribution of *Luzula Wahlenbergii* Rupr.

Foreign Arctic. Arctic Scandinavia.
Outside the Arctic. Mountains of Scandinavia; the south of the Kola Peninsula; Northern Ural; the north of the forest zone of West and East Siberia; mountains of Southern Siberia and Northern Mongolia.

3a. *Luzula parviflora* ssp. *melanocarpa* (Michx.) comb. n.
L. melanocarpa (Michx.) Desv. in Journ. de Bot. 1 (1808), 142; Krechetovich in Fl. SSSR III, 567.
L. parviflora var. *melanocarpa* — Buchenau in Pflanzenreich IV, 36, 62.
L. spadicea melanocarpa E. Meyer in Ledebour, Fl. ross. IV, 217, pro parte.
L. parviflora — Lange, Consp. fl. groenl. 125; Hultén, Fl. Al. III, 440.

Growing under habitat conditions generally similar to those of typical *L. parviflora*, mainly among shrub thickets. Sometimes also occurring in areas carpeted with sphagnum. Locally common (for example, in Southern Koryakia).

Soviet Arctic. SE part of Chukotka Peninsula; Anadyr Basin; Koryakia (Penzhina Basin and district of Bay of Korf). (Map IV–8).
Foreign Arctic. Alaska; Labrador; West Greenland north to 72–73°; extreme SE Greenland.
Outside the Arctic. Kamchatka, Commander Islands, Okhotsk Coast, North America in the western mountains and in subarctic (also partly in moderately northern) districts.

4. *Luzula Wahlenbergii* Ruprecht, Fl. samojed. cisur. (1846), 58; Trautvetter, Fl. boganid. 151; Kjellman, Phanerog. sib. Nordk. 120; Scheutz, Pl. jeniss. 169; Gelert, Fl. arct. 1, 28; Lynge, Vasc. pl. N.Z. 92; Tolmatchev, Contr. Fl. Vaig. 128; Tolmachev, Fl. Kolg. 17; id., Fl. Taym. I, 105; id., Obz. fl. N.Z. 151; Andreyev, Mat. fl. Kanina 165; Perfilev, Fl. Sev. I, 135; Krylov, Fl. Zap. Sib. III, 549; Krechetovich in Fl. SSSR III, 567; Leskov, Fl. Malozem. tundry 43; Hultén, Fl. Al. III, 443; id., Atlas 119 (474); Shlyakov in Fl. Murm II, 184; Holmen & Mathiessen in Bot. tidsskr. (1953), 233; Porsild, Ill. Fl. Arct. Arch. 62; Böcher & al., Grønl. Fl. 235.
L. spadicea var. *Kunthii* E. Mey. in Ledebour, Fl. ross. IV, 217; Schmidt, Fl. jeniss. 122.
L. spadicea var. *Wahlenbergii* — Buchenau in Engl. Bot. Jahrb. VII (1885), 171; id. in Pflanzenreich IV, 36, 63.
L. spadicea — Polunin, Circump. arct. fl. 117 (pro maxima parte).
L. parviflora — Perfilev, Mat. fl. N.Z. Kolg. 17.

Almost exclusively arctic plant widespread in the tundras of Eurasia. The typical habitat of *L. Wahlenbergii* is peaty tundra with flat hummocks, in which it grows

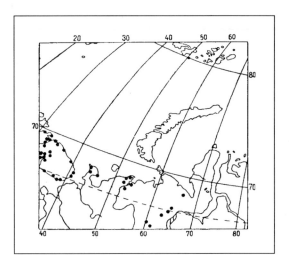

Map IV-10 Distribution of *Luzula spicata* (L.) DC.

on the exposed surfaces of peat mounds, often in considerable quantity. The degree of abundance and distribution of *L. Wahlenbergii* in the Arctic is to a considerable degree determined by its association with a particular type of habitat; wherever there is extensive development of tundra formations with flat-topped hummocks, it is usually common. On moister sites with mossy tundra it occurs more rarely. It avoids the shrub thickets with which the closely related *L. parviflora* is associated.

Luzula Wahlenbergii is readily distinguished from *L. parviflora* by its overall growth form (low stature, formation of tufts), the weak development of stem leaves, the dark colour of the whole plant (brownish lilac stems, shining dark green leaves often with a lilac tint), and partly by the appearance of the inflorescence. None the less cases of confusion of these two species are rather frequent.

Soviet Arctic. Murman (common); Kanin; Northern Malozemelskaya Tundra; Kolguyev Island (common); Bolshezemelskaya Tundra; Polar Ural (common); Pay-Khoy; Vaygach Island (common); South Island of Novaya Zemlya (not everywhere, in restricted quantity); Yamal; lower reaches of Ob; coasts of Ob Sound and Tazovskaya Bay; Gydanskaya Tundra; lower reaches of Yenisey (common); Sibiryakov Island; Southern Taymyr (reaching the SE coast of Lake Taymyr); lower reaches of the Anabar, Olenek and Lena; coast of Buorkhaya Bay; lower reaches of Kolyma; north coast of Chukotka; SE Chukotka Peninsula; Anadyr Basin; coast of Penzhina Bay. (Map IV–9).

Foreign Arctic. West coast of Alaska and neighbouring islands; arctic coast of Canada; Labrador; Southern Baffin Island; East Greenland below 74°38´N; Spitsbergen; Arctic Scandinavia.

Outside the Arctic. Mountains of Fennoscandia; Northern Ural; mountains of NW North America. Records for the mountains of Kamchatka need checking, since some of them definitely refer to *L. parviflora* ssp. *melanocarpa*; plants from Mount Lopatin on Sakhalin Island must be reidentified as *L. unalaschkensis* ssp. *kamtschadalorum* (see below), in which connection doubt also arises about the reliability of records of the occurrence of *L. Wahlenbergii* in the mountains of Northern Japan.

5. ***Luzula spicata*** (L.) DC., Fl. franç. 3 (1805), 161; E. Meyer in Ledebour, Fl. ross. IV, 220; Lange, Consp. fl. groenl. 128; Buchenau in Pflanzenreich IV, 36, 73; M. Porsild, Fl. Disko 62; Krylov, Fl. Zap. Sib. III, 553; Andreyev, Mat. fl. Kanina 165; Leskov, Fl.

Malozem. tundry 43; Devold & Scholander, Fl. pl. SE Greenl. 113; Sørensen, Vasc. pl. E Greenl. 163; Gröntved, Pterid. Spermatoph. Icel. 194; Hultén, Atlas 119 (472); id., Amphi-atl. pl. 236 (218); Shlyakov in Fl. Murm. II, 190; A.E. Porsild, Ill. Fl. Arct. Arch. 63; Böcher & al., Grønl. Fl. 235; Polunin, Circump. arct. fl. 117.

Juncus spicatus L., Sp. pl. (1753), 330.

Ill.: Fl. Murm. II, pl. LXI; A.E. Porsild, Ill. Fl., fig. 23h.

Characteristic plant of mountain tundras of Atlantic districts of the Arctic. Growing in open rocky places, in low-herb meadows, and locally in open sandy areas; common in many districts.

Soviet Arctic. Murman (common); Kanin; Timanskaya and Malozemelskaya Tundras; western edge of the Bolshezemelskaya Tundra (right bank of the lower Pechora, Kuya River); Polar Ural. All records of the occurrence of this species in the Far Eastern Arctic, as well as records for Novaya Zemlya and Vaygach, have proved to be erroneous. (Map IV–10).

Foreign Arctic. Labrador; southern part of Baffin Island; West Greenland north to 72°30´, East Greenland north to 74°; Iceland; Arctic Scandinavia.

Outside the Arctic. Mountains of Central and Southern Europe; Northern Great Britain; mountains of Scandinavia; subarctic north of Fennoscandia; Northern Ural; Caucasus; mountains of Southern Siberia and Central Asia, south to the Himalayas; western mountains of North America from subarctic Alaska to California; temperate NE North America.

6. ***Luzula confusa*** Lindb. in Nya Bot. Notis. (1855), 9; Buchenau in Pflanzenreich IV, 36, 70; M. Porsild, Fl. Disko 62; Lynge, Vasc. pl. N.Z. 90; Andreyev, Mat. fl. Kanina 164; Tolmachev & Pyatkov, Obz. rast. Diksona 157; Tolmachev, Fl. Kolg. 17; id., Mat. fl. Mat. Shar 288; id., Fl. Taym. I, 104; Hanssen & Lid, Fl. pl. Franz Josef L. 32; Krylov, Fl. Zap. Sib. III, 551; Perfilev, Fl. Sev. I, 138; Leskov, Fl. Malozem. tundry 42; Devold & Scholander, Fl. pl. SE Greenl. 111; Hultén, Fl. Al. III, 435; Tikhomirov, Fl. Zap. Taym. 30; A.E. Porsild, Ill. Fl. Arct. Arch. 63; Shlyakov in Fl. Murm. II, 188; Böcher & al., Grønl. Fl. 236; Korotkevich, Rast. Sev. Zemli; Polunin, Circump. arct. fl. 119.

Luzula hyperborea R. Br. — Scheutz, Pl. jeniss. 169.

L. hyperborea var. *major* — Trautvetter, Fl. taim. 24.

L. arcuata var. *confusa* — Sørensen, Vasc. pl. E Greenl. 161; Seidenfaden & Sørensen, Vasc. pl. NE Greenl. 105.

L. arcuata f. *confusa* — Kjellman, Phanerog. sib. Nordk. 120.

L. arcuata var. α et var. η — Schmidt, Fl. jeniss. 122.

L. arcuata — Gelert, Fl. arct. 1, 29 (pro maxima parte); Tolmatchev, Contr. Fl. Vaig. 128.

Ill.: Porsild, Ill. Fl., fig. 23E; Fl. Murm. II, pl. LX, 3.

One of the most characteristic plants of middle and high arctic districts. Growing mainly in dry open mountain tundras, on more or less rocky substrate, forming individual dense tufts or rather loose "patches" which occupy more extensive areas. It does not avoid sites which are highly windswept in wintertime, at which sites the stems and inflorescences of the plant are often evident during the winter when they stick up through the snow cover. At sites where *L. confusa* grows in particular abundance, patches of vegetation alternating with bare areas acquire a brownish hue due to the corresponding colour of the foliage of this species. In more southern districts *L. confusa* also grows on dry areas of tundras with peat mounds, but is here represented by a less typical form.

Luzula confusa appears to be in all respects a "good species" in its typical form, which is universally distributed in high arctic districts and distinguished by its densely tufted growth, the channelled setiform leaves, the distinctly brownish tint of the foliage, and the dense (often undivided) inflorescences. The same may be said of practically all plants of this species from arctic East Siberia (Taymyr and

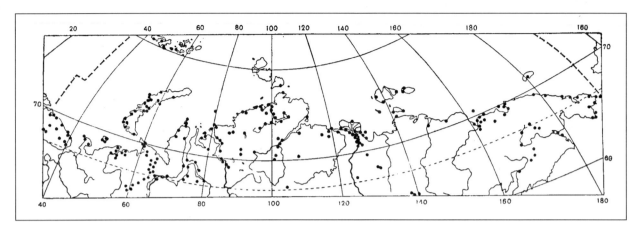

MAP IV–11 Distribution of *Luzula confusa* Lindb.

Yakutia). But in the European Far North the situation is complicated. Here *L. confusa* frequently has a more divided inflorescence and broader leaves, in some respects reminiscent of *L. arcuata* and sometimes confused with it. This may especially be said of plants from peaty tundras of moderately northern districts of the European Arctic. It is entirely possible, since the closely related *L. arcuata* also occurs here locally, that plants whose identity is to some degree doubtful are the product of hybridization between *L. arcuata* and *L. confusa*.

Soviet Arctic. Murman; Northern Kanin; NE Malozemelskaya Tundra; Kolguyev Island (frequent); Western and Eastern Bolshezemelskaya Tundra; Polar Ural (very frequent); Pay-Khoy; Vaygach; Novaya Zemlya all the way to the extreme north (especially abundantly on the coast of the Kara Sea); Franz Josef Land; Yamal; Obsko-Tazovskiy Peninsula; Gydanskaya Tundra; lower reaches of the Yenisey and islands in Yenisey Bay; throughout Taymyr (very common in many districts); Southern Severnaya Zemlya; throughout the whole expanse of Arctic Yakutia from Khatanga Bay to the mouth of the Kolyma (common everywhere); New Siberian Islands; polar and Beringian coasts of Chukotka (common); Wrangel Island; Anadyr Basin; Koryakia. (Map IV–11).

Foreign Arctic. Arctic Alaska and Canada; Labrador; entire Canadian Arctic Archipelago; Greenland from the southern tip to the north coast of Peary Land; Spitsbergen; Arctic Scandinavia.

Outside the Arctic. Mountains of Scandinavia and central part of Kola Peninsula; Northern Ural; Altay; Sayans; certain localities within the system of the Stanovoy Range; the north of the Central Siberian Plateau; Verkhoyansk Range; mountains of NW North America; subarctic part of Canada.

7. ***Luzula beringensis*** Tolm., sp. nova. Perennis densissime caespitosa. Caules stricti erecti, 10–30 cm alti. Vaginae basilares lucidae purpurascentes. Folia angusta, erecta vel subcurvata, canaliculata, apice subulata, flavescente- vel rufescente-viridia. Inflorescentia ramosa, rami gracillimi, recti vel subcurvati. Glomeruli 3–8, compacti, parvi (2–4.5 mm lat.), pauci- (3–6)flori. Bracteae laceroso-ciliatae. Sepala fusca, anguste lanceolata (4plo longiora quam lata), plus quam 2 mm longa. Capsula perianthio brevior, apice vix angustata.

Habitat in tundris lapidosis monticolis terrae Tchuktchorum et Koriakorum, in Asiae parte maxime septentrionali-orientali.

Typus: Terra Koriakorum, in tundra lapidosa in cacumine montis Prodolgovátaia sópka, prope sinum Korfii maris Beringii, 19 viii 1960, A. Tolmatchev, Cl. Stepanova et L. Fedorova.

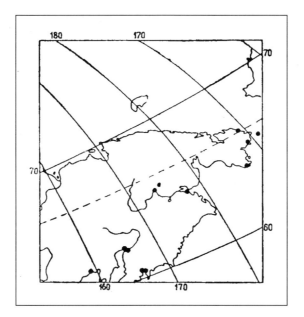

MAP IV–12 Distribution of *Luzula beringensis* Tolm.

Affinis *Luzulae confusae*, a qua differt inflorescentia glomerulis minoribus numerosis, sepalis angustioribus et longioribus, capsulam superantibus. A *L. unalaschkensi* differt caespitibus densioribus, vaginis foliorum lucidis purpurascentibus, foliis canaliculatis rufescentibus, nec laete viridibus planis.

Species combining obvious resemblance to *L. confusa* (densely tufted growth form, setiform channelled leaves with brownish tint, intensely reddish sheaths) with a strongly branched small-headed inflorescence resembling that of *L. unalaschkensis* or typical *L. arcuata*. We confused *L. beringensis* in the field with *L. unalaschkensis kamtschadalorum* precisely on account of external resemblance of its inflorescence to that of this Kamchatka race.

In addition to the morphological differences described in the key, *L. beringensis* also differs ecologically from *L. unalaschkensis kamtschadalorum*. It is a plant of stony mountain tundras, occurring on exposed mountain ridges among utterly depauperate vegetation.

Soviet Arctic. Extreme east and southeast of Chukotka Peninsula; Ratmanov Island in the Bering Strait; lower course of the Anadyr River; Beringian coast of Koryakia (district of Bay of Korf); tundra portion of Penzhina Basin; coast of Gizhiga Bay. (Map IV–12).

Foreign Arctic. Apparently absent.

Also not found **outside the Arctic.**

8. *Luzula unalaschkensis* (Buch.) Satake in Nakai & Honda, Nov. Fl. Jap. 1, Juncac. (1938), 31 and in Journ. Jap. Bot. XIV (1938), 260; Vasilev in Bot. mat. Gerb. Bot. inst. XV (1935), 43; id., Fl. Komand. ostr. 68.

L. arcuata var. *unalaschkensis* Buchenau in Engl. Bot. Jahrb. XII (1890), 124; id. in Pflanzenreich IV, 36, 70.

L. arcuata ssp. *unalaschkensis* Gorodk. in Sched. Herb. Inst. Bot. (1935).

L. arcuata — Hultén, Fl. Al. III, 433.

Growing on dry tundra mounds among lichens, and sometimes on stony sites.

Soviet Arctic. SE Chukotka Peninsula; eastern part of Anadyr Basin; Northern Koryakia (NE part of Penzhina Basin). (Map IV–13).

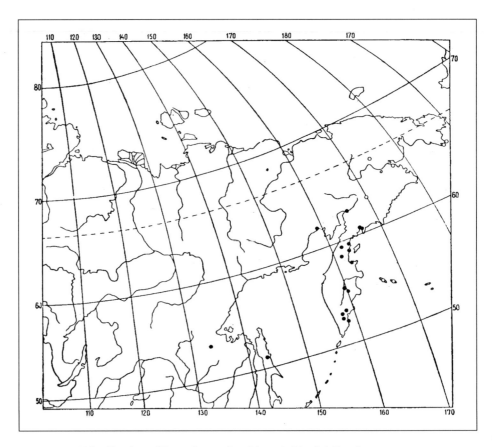

Map IV-13 Distribution of *Luzula unalaschkensis* (Buch.) Satake. *Open circles*— L. unalaschkensis s. str.; *black circles* — L. unalaschkensis ssp. kamtschadalorum (Sam.) Tolm.

    Foreign Arctic. West coast of Alaska and islands in the Bering Sea.
    Outside the Arctic. Commander and Aleutian Islands, South Alaska.

8a. *Luzula unalaschkensis* ssp. *kamtschadalorum* (Sam.) comb. n.
    L. arcuata var. *kamtschadalorum* Samuelsson in Hultén, Fl. Kamtch. 1 (1927), 223.
    L. kamtschadalorum Gorodk. — Krechetovich in Fl. SSSR III, 568 (pro minima parte).
        Plant of small swales with low herbs or mosses and low herbs in mountain tundra and in openings within shrub thickets.
    Soviet Arctic. Koryakia (coast of Gizhiga Bay, lower reaches of Penzhina, Bay of Korf). (Map IV-13).
    Foreign Arctic. Not reported.
    Outside the Arctic. Northern Kamchatka; mountains of Central and Southern Kamchatka; mountains of Sakhalin (Mount Lopatin) and possibly of Northern Japan.

9. *Luzula arcuata* (Wahlb.) Sw., Summa Veget. (1814); Lange, Consp. fl. groenl. 126 (pro parte); Gelert, Fl. arct 1, 29 (pro parte); Buchenau in Pflanzenreich IV, 36, 70; Andreyev, Mat. fl. Kanina 164; Tolmachev, Fl. Kolg. 17; id., Mat. fl. Mat. Shar 288; Perfilev, Fl. Sev. I, 138; Krechetovich in Fl. SSSR III, 569; Gröntved, Pterid. Spermatoph. Icel. 192; Hultén, Atlas 116 (462), pro parte; Shlyakov in Fl. Murm. II, 188.

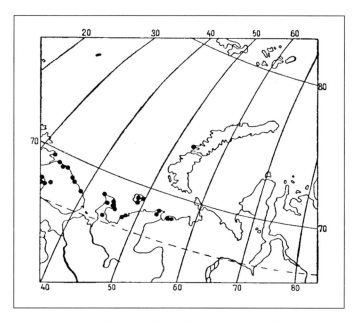

MAP IV–14 Distribution of *Luzula arcuata* (Wahlb.) Sw.

Ill.: Fl. Murm. II, pl. LX, 2.

We refer to *L. arcuata* only a minority of the arctic plants which have been identified as belonging to this species. The considerable majority of arctic "*L. arcuata*" must undoubtedly be considered to be individuals of *L. confusa* deviating from the norm to varying degree. Additionally, a certain portion of plants from the European Arctic actually possess characters intermediate between *L. confusa* and *L. arcuata* and are possibly hybrids between these species. It is less probable that these transitional forms are primary in nature, because we find them exclusively in regions relatively recently colonized by plants (after glaciations or transgressions) while within the most ancient region of development of the arctic flora we invariably find clearly expressed (and evidently hereditarily stable) *L. confusa*.

Luzula arcuata in our sense is a subarctic European species. It is possible that it also exists in Greenland especially in the south, while in more northern districts *L. confusa* grows exclusively.

In the Soviet Arctic *L. arcuata* is a plant of moderately moist mossy sites, and is encountered in openings among shrubs. Forms transitional between it and *L. confusa* occur most frequently in tundras with peat mounds.

Soviet Arctic. Murman; Kanin; Timanskaya and Malozemelskaya Tundras; western part of Bolshezemelskaya Tundra; Kolguyev Island; Novaya Zemlya (Volchikha Bay in the SW part of the North Island). Plants transitional between *L. arcuata* and *L. confusa* occur in eastern districts of the Bolshezemelskaya Tundra, on Vaygach Island and on the west coast of the South Island of Novaya Zemlya. (Map IV–14).

Foreign Arctic. Southern districts of Greenland (?); Iceland; Arctic Scandinavia.

Outside the Arctic. Mountains of Scandinavia and of the central part of the Kola Peninsula.

10. *Luzula nivalis* Laest. ex Spreng., Syst. veg. 2 (1825), 111; Gelert, Fl. arct. 1, 30; M. Porsild, Fl. Disko 62; Lynge, Vasc. pl. N.Z. 91; Tolmatchev, Contr. Fl. Vaig. 128; Tolmachev, Fl. Taym. I, 105; id., Mat. fl. Mat. Shar 288; id., Obz. fl. N.Z. 150; Hanssen & Lid, Fl. pl. Franz Josef L.; Perfilev, Fl. Sev. I, 135–136; Krechetovich in Fl. SSSR III, 568; Sørensen, Vasc. pl. E Greenl. 162; Seidenfaden & Sørensen, Vasc. pl.

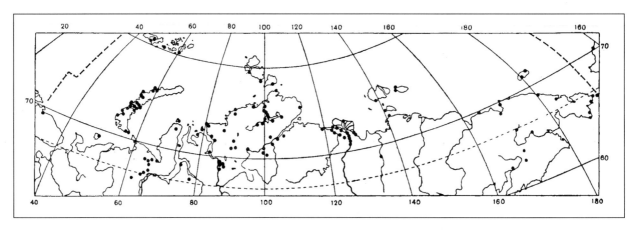

MAP IV–15 Distribution of *Luzula nivalis* Laest.

NE Greenl. 106, 180; Hultén, Fl. Al. III, 438; Tikhomirov, Fl. Zap. Taym. 31; Shlyakov in Fl. Murm. II, 186; A.E. Porsild, Vasc. pl. W Can. Arch 98; id., Ill. Fl. Arct. Arch 62; Korotkevich, Rast Sev. Zemli; Polunin, Circump. arct. fl. 119.

L. arctica Blytt — Kjellman, Phanerog. sib. Nordk. 120; Lange, Consp. fl. groenl. 127; Buchenau in Pflanzenreich IV, 36, 68; Krylov, Fl. Zap. Sib. III, 550; Hultén, Atlas 116 (461); Böcher & al., Grønl. Fl. 236.

L. hyperborea var. *minor* — Trautvetter, Fl. taim. 24.

One of the plants very characteristic of our Arctic and of the Arctic Region generally. It grows mainly in mountain tundras, but often colonizes the better moistened sites there, for example where a certain quantity of meltwater runs off in spring. Although a common plant, it does not play a substantial role in the formation of vegetational communities but always grows in small quantity as scattered tufts.

The species shows rather considerable variation in details and in the constancy of basic diagnostic characters. For example, in the district of Matochkin Shar on Novaya Zemlya we observe two forms of this species. One of them differs in its taller stature, dark glossy leaves, and relatively large but little branched (most frequently fully condensed) inflorescences. It occurs in relatively moist sites on tundra. This form may with the most justification be considered the typical *L. nivalis*. Plants with this habit may be encountered throughout the whole expanse of the wide range of this species. The other form from Novaya Zemlya grows in dry stony mountain tundras on summits of the plateau usually exceeding 200 m in height; it differs in being extremely short, in forming very dense small tufts, and in having opaque light greyish-green leaves and small but distinctly branched inflorescences. I have not been able to find this form among material of *L. nivalis* from many parts of the range. It taxonomic status is still unclear.

Soviet Arctic. West Murman (one locality!); Kolguyev Island (very rare); Polar Ural (common) and adjacent tundra; Vaygach Island; throughout Novaya Zemlya (common); Franz Josef Land; Northern Yamal; coasts of Tazovskaya Bay; mouth of River Pur; Gydanskaya Tundra; lower reaches of Yenisey and islands in Yenisey Bay; entire Taymyr Peninsula from the forest-tundra to Cape Chelyuskin (common); Severnaya Zemlya; lower reaches of the Olenek and Lena, and coasts of Buorkhaya Bay; New Siberian Islands; lower reaches of Indigirka; coasts of the East Siberian and Chukchi Seas from the mouth of the Kolyma to the Bering Strait; Wrangel Island; Beringian coast of Chukotka Peninsula; Ratmanov Island; Anadyr Basin; Northern Koryakia. (Map IV–15).

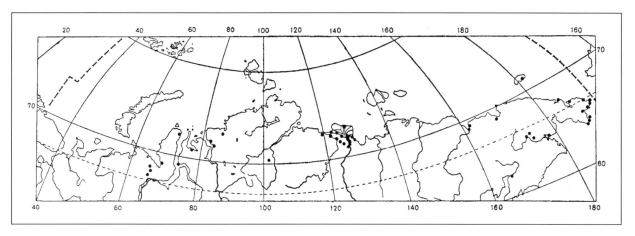

MAP IV–16 Distribution of *Luzula tundricola* Gorodk.

> **Foreign Arctic.** Coasts of Alaska (rare); arctic coast of Canada (relatively frequent); entire Canadian Arctic Archipelago north to the limits of land; Labrador; Greenland, except districts below the Arctic Circle on the west coast and below the 70th parallel on the east coast; Arctic Scandinavia (mountainous districts).
> **Outside the Arctic.** Mountains of Northern Scandinavia; Northern Ural; subarctic districts of Alaska and Western Canada.

11. ***Luzula tundricola*** Gorodk. in Not. Syst. ex Herb. Inst. Bot. XV (1935), 40; Karavayev, Konsp. fl. Yak. 74.

   L. kamtschadalorum Gorodk. — Krylov, Fl. Zap. Sib. III, 351 (non Sam.!); Krechetovich. in Fl. SSSR III, 568.

   L. arcuata f. *latifolia* Kjellman in Vega Exp. Vet. Iaktt. I (1882), 566.

   L. nivalis var. *latifolia* (Kjellm.) Sam. in Hultén, Fl. Aleut. Isl. (1937) 127; Hultén, Fl. Al. III, 566.

   L. beeringiana Gjaerevoll in Kgl. Norske Vid. Selsk. Skr. 1958, Nr. 5, 63.

   This species long remained unrecognized as such, although specimens of it were studied and variously determined by botanists long ago. In habit it combines characters of *L. nivalis* (constricted obtusish tips of the generally broad leaves) with characters reminiscent of forms with extremely divided inflorescences from the *L. arcuata* group or sometimes even *L. Wahlenbergii*. This was reflected in my provisional determinations of plants from the lower course of the Lena as probably the hybrid *L. nivalis* × *Wahlenbergii*, and later in B.N. Gorodkov's identification of plants collected by himself with *L. arcuata* var. *kamtschadalorum* Sam. described from Kamchatka. Samuelsson showed that the latter identification was incorrect, which led to subsequent description of the species in question under the new name *L. tundricola*.

   Information on herbarium labels about the conditions of growth of *L. tundricola* is extremely scanty. Judging from this, the species grows on grassy slopes, on open streamside areas of tundra, and sometimes in coastal meadows.

> **Soviet Arctic.** Polar Ural; east coast of Yamal; at the entrance of Tazovskaya Bay; Gydanskaya Tundra; coast of Yenisey Bay; lower course of Pyasina; lower reaches of the Olenek and Lena and the tundra and forest-tundra between them (the greatest concentration of localities within the range); north coast of Chukotka from the mouth of the Kolyma to the Bering Strait; Wrangel Island; SE and southern coasts of Chukotka; tundra portion of Anadyr Basin; lower reaches of Penzhina. (Map IV–16).
> **Foreign Arctic.** Not reported (but may quite possibly be found in Alaska).

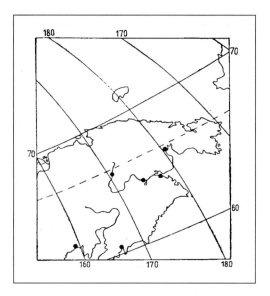

Map IV–17 Distribution of *Luzula capitata* (Miq.) Nakai.

Outside the Arctic. Northern edge of the Central Siberian Plateau in the district west of Khatanga; Bering Island.

12. **Luzula capitata** (Miq.) Nakai, Rep. Veget. Mt. Apoi (1930), 76; Krechetovich in Fl. SSSR III, 572.

    L. campestris var. *capitata* Miq. in Ann. Mus. Lugd.-Bat. 3 (1867), 165; Buchenau in Pflanzenreich IV, 36, 92.

    Plant of the moderately northern part of Pacific Asia, distributed mainly on islands. It penetrates the Arctic in the north of its range, where it is reported on grassy slopes on the sea coast and in meadows in the lower reaches of tundra streams. Further south (outside the Arctic) it locally ascends mountains.

    Soviet Arctic. South coast of Chukotka (Bay of Krest); Anadyr Basin; Koryakia (Bay of Korf and north coast of Gizhiga Bay). (Map IV–17).

    Foreign Arctic. Not reported.

    Outside the Arctic. Kamchatka, Kurile Islands, Sakhalin, Northern Japan, Korean Peninsula.

13. **Luzula multiflora** (Retz.) Lej., Fl. env. Spa 1 (1811), 169; Gelert, Fl. arct. 1, 31; Krylov, Fl. Zap. Sib. III, 555; Perfilev, Fl. Sev. I, 136; Krechetovich in Fl. SSSR III, 572; Gröntved, Pterid. Spermatoph. Icel. 193; Hultén, Fl. Al. III, 436; id., Atlas (466); Böcher & al., Grønl. Fl. 236; Polunin, Circump. arct. fl. 119.

    Juncus multiflorus Ehrh. — Retz., Fl. scand. (1795).

    Luzula campestris var. *multiflora* Celak. — Buchenau in Pflanzenreich IV, 36, 94.

    L. campestris — Gelert, Fl. arct. 1, 31; Scheutz, Pl. jeniss. 170.

    We have been forced to revert to a wide treatment of *L. multiflora* as a species, both because of the presence of transitions between its various forms and because of inadequate study of the question of its geographical races and the interrelations between them. The actual relations between the forms of this polymorphic aggregate are apparently considerably more complex than indicated in the simple scheme for subdividing it into a series of geographically vicariant races (species) given in the third volume of the "Flora of the USSR." The species *L. multiflora* (Retz.) Lej. s. l. definitely needs monographic revision in its entirety. We restrict ourselves below to considering those forms of it which grow in arctic districts of

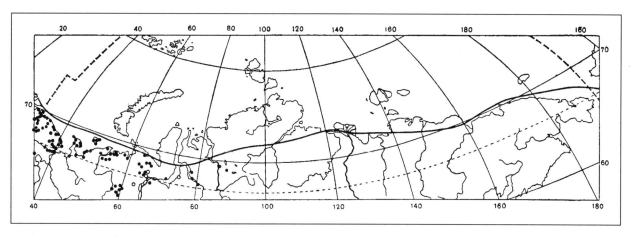

Map IV–18 Distribution of *Luzula multiflora* ssp. *frigida* (Buch.) V. Krecz. Continuous line: northern boundary of distribution of *L. multiflora* s. l.

the USSR. We can with full confidence report that three races grow in the Soviet Arctic, namely ssp. *frigida* (Buch.) Krecz., ssp. *sibirica* Krecz. and ssp. *Kjellmaniana* (Miyabe et Kudo). These are relatively easily recognizable but connected by transitional forms.

Plants agreeing with the concept of typical *L. multiflora* (*L. multiflora* Lej. s. str.) do not occur in the Arctic.

13a. Luzula multiflora ssp. *frigida* (Buch.) Krecz. in Journ. Soc. Bot. 12, No. 4 (1927); Krylov, Fl. Zap. Sib. III, 556; Andreyev, Mat. fl. Kanina 164; Tolmachev, Fl. Kolg. 17; id., Obz. fl. N.Z. 150; Perfilev, Fl. Sev. I, 136.

L. multiflora var. *frigida* — Leskov, Fl. Malozem. tundry 42.

L. frigida Sam. in Lindman, Svensk Fanerogam. fl. (1918), 161; M. Porsild, Fl. Disko 63; Devold & Scholander, Fl. pl. SE Greenl. 412; Krechetovich in Fl. SSSR III, 573; Seidenfaden & Sørensen, Vasc. pl. SE Greenl. 180; Hultén, Atlas (465); Shlyakov in Fl. Murm. II, 194; Böcher & al., Grønl. Fl. 237.

L. campestris var. *frigida* Buchenau in Öst. Bot. Zeitschr. (1898), 184; id. in Pflanzenreich IV, 36, 93.

L. multiflora — Lange, Consp. fl. groenl. 125.

L. campestris — Ruprecht, Fl. samojed. cisur. 59 (pro parte).

L. Pohleana Krecz. — Andreyev, Mat. fl. Kanina 164 (nomen nudum!).

At first glance *L. multiflora* ssp. *frigida* seems to merit treatment as a separate species, since it is quite sharply distinguished from *L. multiflora* in the narrow sense. While such is especially the case in the European North, delimiting *frigida* from the Asian forms of the same aggregate proves to be a considerably more difficult task. Thus, within the stretch from the Polar Ural to the lower reaches of the Yenisey we encounter quite clearly differentiated individuals of ssp. *frigida* and of the Siberian ssp. *sibirica*, as well as considerable numbers of plants whose referral to one or other of these races is to a considerable degree tentative. *Luzula multiflora* ssp. *sibirica* is in turn not so sharply distinguished from *L. multiflora* s. str. as is ssp. *frigida*. As we progress towards NE Asia we observe a broad zone of overlap with a considerable degree of "blurring of the boundaries" between ssp. *sibirica* and the Far Eastern race *L. multiflora* ssp. *Kjellmaniana*. The latter is in many respects very close to the Northern European ssp. *frigida*, to the extent that in cases of inadequate labelling mistakes can easily be made in determining the plants. All this compels us to refrain from treating *L. frigida* as a separate species.

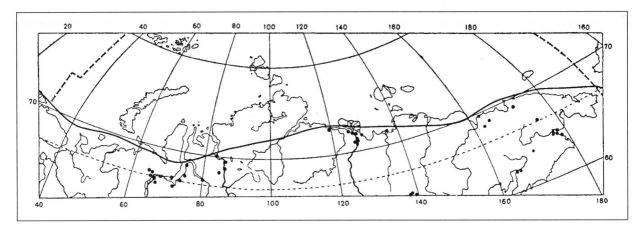

MAP IV–19 Distribution of *Luzula multiflora* ssp. *sibirica* V. Krecz. Continuous line: northern boundary of distribution of *L. multiflora* s. l.

Luzula multiflora frigida is a plant of grassy slopes and well drained meadow-like openings among shrubs in the southern part of the tundra zone. Although not playing a substantial community role, it is often reported here as a common plant. It is not reported in communities of typical tundra vegetation.

Soviet Arctic. Murman (ubiquitous); Kanin (common throughout the peninsula); Timanskaya and Malozemelskaya Tundras; Kolguyev Island (common); Bolshezemelskaya Tundra; Polar Ural; coasts of Ob Sound and Tazovskaya Bay; lower reaches of Yenisey (Dudinka district); Norilsk. Records for Southern Novaya Zemlya are doubtful. (Map IV–18).

Foreign Arctic. Greenland from the southern tip north to 72°30´ on the west coast and to 70°30´ on the east coast (old records of its having been found north of the 80th parallel are scarcely credible); Iceland; Arctic Scandinavia.

Outside the Arctic. Mountains and subarctic districts of Fennoscandia (including the Kola Peninsula); Solovetskiye Islands; Northern Ural and the adjacent zone of the Pechora Basin.

13b. *Luzula multiflora* ssp. *sibirica* V. Krecz. in Zhurn. Russk. bot. obshch. XII, No. 4 (1927).

L. multiflora ssp. *asiatica* Kryl. et Serg., Fl. Zap. Sib. III (1929), 556.

L. sibirica Krecz. in Fl. Zabayk. II (1931), 144; id. in Fl. SSSR III, 574; Karavayev, Konsp. fl. Yak. 74.

Occurring as a not uncommon plant on grassy slopes in the southern part of the tundra zone of Siberia and the Far East. In the vicinity of the Pacific Ocean, it occurs mainly in districts rather remote from the coast, near which it is normally replaced by the specifically Far Eastern ssp. *Kjellmaniana.*

Soviet Arctic. Polar Ural; lower reaches of the Ob and the coasts of Ob Sound; mouth of the Taz; lower reaches of the Yenisey (more frequent than ssp. *frigida*); lower reaches of the Olenek and Lena (common on the Lena); coast of Buorkhaya Bay; at the mouths of the Yana and Kolyma; NW districts of Chukotka; Anadyr Basin; lower reaches of Penzhina. (Map IV–19).

Foreign Arctic. Not occurring.

Outside the Arctic. Northern districts of Yakutia; mountains of Southern Siberia.

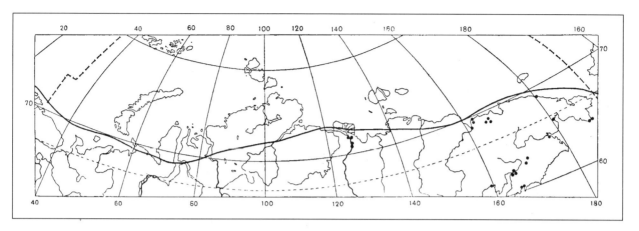

Map IV-20 Distribution of *Luzula multiflora* ssp. *Kjellmaniana* (Miyabe et Kudo) Tolm. Continuous line: northern boundary of distribution of *L. multiflora* s. l.

13c. *Luzula multiflora* ssp. *Kjellmaniana* (Miyabe et Kudo) comb. n.
  L. Kjellmaniana Miyabe et Kudo in Trans. Sapporo Nat. Hist. Soc. V (1913), 38; Krechetovich in Fl. SSSR III, 573.
  L. multiflora var. *Kjellmaniana* Sam. in Hultén, Fl. Kamtch. 1 (1927), 227.

  Widely distributed Far Eastern race of the aggregate species *L. multiflora*, common in the Far Eastern Arctic proper (east of the mouth of the Kolyma) but also occurring in Northern Yakutia. In contrast with ssp. *sibirica*, it is most common in the Far East in maritime districts, yielding place to ssp. *sibirica* in more inland parts of the country. Thus, the overlap of the ranges of the two races does not occupy such a large area as might be supposed at first glance.

  A plant of slopes with low herbs and meadows among shrub thickets.

Soviet Arctic. Lower reaches of Lena (in a restricted stretch); lower reaches of the Kolyma and tundra east of its mouth; north coast of Chukotka (Cape Schmidt); SE Chukotka Peninsula; Bay of Krest; district of the mouth of the Anadyr; Beringian coast of Koryakia (district of Bay of Korf); Penzhina Basin; coast of Gizhiga Bay. (Map IV-20).

Foreign Arctic. Probably occurring in arctic districts of Alaska.

Outside the Arctic. Kamchatka; Commander and Aleutian Islands; Alaska; Kurile Islands; coast of Sea of Okhotsk; Sakhalin; Northern Japan.

14. *Luzula sudetica* (Willd.) DC., Fl. franç. V (1805), 306; Krechetovich in Fl. SSSR III, 575; Hultén, Atlas 119 (473); Shlyakov in Fl. Murm. II, 195; Gröntved, Pterid. Spermatoph. Icel. 195; Polunin, Circump. arct. fl. 119.

  Plant of moist habitats on the shores of rivers, streams and lakes, among meadowlike vegetation.

Soviet Arctic. Murman (more or less ubiquitous). No reliable finds in districts east of the White Sea.

Foreign Arctic. Iceland; Arctic Scandinavia.

Outside the Arctic. Fennoscandia (including the south of the Kola Peninsula and Northern Karelia); mountains of Central Europe.

15. *Luzula oligantha* Sam. in Hultén, Fl. Kamtch. 1 (1927), 227; Krechetovich in Fl. SSSR III, 574.

  A species described from Kamchatka and found at a few sites in the Far Eastern Arctic. At the most northern point of its range reported to grow on riverine sands.

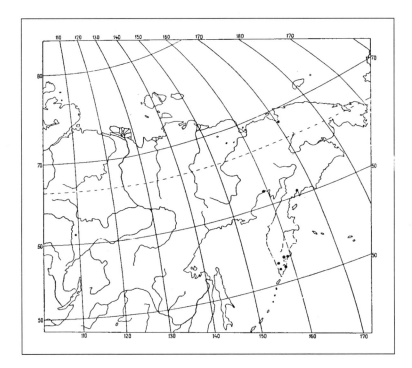

MAP IV–21 Distribution of *Luzula oligantha* Sam.

Soviet Arctic. NW Chukotka (south of the Bay of Chaun); coast of Gizhiga Bay and the Bay of Korf. (Map IV–21).
Foreign Arctic. Not occurring.
Outside the Arctic. Kamchatka.

16. ***Luzula pallescens*** (Wahlb.) Bess., Enum. pl. Volhyn. (1822), 15; Krylov, Fl. Zap. Sib. III, 554; Krechetovich in Fl. SSSR III, 576; Hultén, Atlas 118 (468); id., Amphi-atl. pl. 140 (121); Shlyakov in Fl. Murm. II, 196; Karavayev, Konsp. fl. Yak. 74; Polunin, Circump. arct. fl. 120.

Boreal species just penetrating arctic limits.

Soviet Arctic. Murman (rare: western portion of coast, south of Murmansk, and Kildin Island); lower reaches of Ob (district of Salekhard); Chubukulakh in the lower reaches of the Lena (near the 71st parallel).
Foreign Arctic. Iceland (one locality in the northwest); Arctic Scandinavia.
Outside the Arctic. Eastern half of Europe (rare in the north); temperate West Siberia; less regularly in East Siberia and the Far East; a few localities in North America (introduced from Europe?).

FAMILY XVII
Liliaceae Hall.
LILY FAMILY

EXTENSIVE FAMILY, RICHLY REPRESENTED in holarctic floras in regions with temperate or subtropical climate. In forested regions of the north possessing fewer representatives belonging mainly to widespread holarctic genera. In the Arctic only isolated species of a few genera occur, and they are mainly associated with the southern fringes of the region. The greater part of these plants possess a wide distribution outside the Arctic in the northern forest zone. As a rule, they are represented in the Arctic by the same forms that grow further south. An exception is a species of *Veratrum* widespread in the tundras of Eurasia, *V. Lobelianum*, in which the differentiation of a predominantly arctic variety is reported.

Two genera, *Tofieldia* and *Lloydia*, are represented in the Arctic by typical arctic-alpine species. All these are relatively widely distributed and rather characteristic of certain parts of the Arctic. With respect to northern distribution (within the USSR) the first place is occupied by *Lloydia serotina*, which reaches the polar limits of the mainland almost everywhere in Siberia and occurs on many arctic islands.

1. Large robust plants of a height of 25–50 cm or more (sometimes up to 1 m), with thick erect stem (basally about 1 cm thick, in large specimens up to 2 cm) and large acuminate leaves (the lower ovate, the upper oblong-lanceolate) whose surfaces are plaited and ribbed (with thick raised veins on lower surface). Inflorescence large (up to half of total height of plant), paniculate, branched to varying degree, with numerous densely arranged flowers. Flowers rather large (when open about 1.5 cm in diameter), greenish. Fruit a somewhat succulent, triangular, distally acuminate capsule. ....3. **VERATRUM L. — FALSE HELLEBORE**
– Less robust, sometimes very small plants. Stems always considerably thinner than 1 cm. Leaves narrow and linear or broad, but in the latter case flat (not plaited). Arrangement and appearance of flowers various, but not as above. ....2.

2. Leaves narrow, linear or linear-lanceolate. Fruit a capsule. ....3.
– Leaves broad, elliptical or ovate. Fruit juicy, berrylike. ....8.

3. Leaves lanceolate, widely divergent from stem, arranged in whorls. Flowers solitary or 2–3, large (perianth segments 2.5–3 cm long), dark purplish brown, nodding. Capsule erect, up to 2 cm long. ....
 ....6. **FRITILLARIA L. — FRITILLARY**

- Leaves linear or lanceolate, upright or obliquely ascending, not grouped in whorls. Flowers small (not more than 1 cm long), solitary or aggregated in terminal inflorescences. ....4.

4. Stems erect and relatively tall (up to 40 cm), succulent, rather thick (up to 5 mm in diameter), bearing dense many-flowered terminal inflorescence of almost globose shape. Until opening this is covered by broad membranous bracts. Leaves long, linear, few, flat or hollow (circular in cross-section). ....5. **ALLIUM** L. — **ONION**
- Stems slender, not succulent. Flowers few or solitary; or, if many, then forming elongate inflorescence of more or less cylindrical shape. ....5.

5. Basal leaves flat, linear-lanceolate, assembled in equitant rosette, not reaching more than half the height of the flowering stem. Stem leaves weakly developed or absent. Inflorescence many-flowered, compact, of more or less cylindrical shape. Flowers very small (not more than 3 mm long). Plants not forming bulbs. ....1. **TOFIELDIA** HUDS. — **FALSE ASPHODEL**
- Leaves few, not forming basal rosette, capable of reaching the height of the flowering stem or even exceeding it. Inflorescence loose (paniculate or racemose), or consisting of a few (2–5) flowers situated on long pedicels and forming a virtual umbel, or flowers solitary. Flowers always larger (length of perianth more than 5 mm). Plants forming bulb at base of stem. ....6.

6. Plant 20–25 cm high, with numerous flowers assembled in loose paniculate or racemose inflorescence. Pedicels slightly longer than perianth or of equal length. Perianth whitish (greenish on outside), widely spreading, with segments connate at base. Capsule obtusely triangular, oblong-ovoid, tapering distally, about 1 cm long. ....2. **ZYGADENUS** RICH. — **CAMAS**
- Plant not more than 20 cm high (often 10–20 cm, sometimes less), with a few (1–5) flowers on long slender pedicels which sometimes form an inflorescence resembling an umbel. Perianth campanulate, with its segments free from base. Capsule usually obovoid, not tapering distally. ....7.

7. Flowers 1–5, sometimes forming umbellate inflorescence. Perianth segments narrow, lanceolate, yellow (often greenish on outside). Capsule small, shorter than the withered perianth retained on it. All leaves flat, bright green. Bulb short, ovoid. ....
.......................4. **GAGEA** SALISB. — **YELLOW STAR-OF-BETHLEHEM**
- Flowers solitary or 2 (rarely 3) per stem. Perianth segments broad, oblong-elliptical, milky white with yellowish base, slightly lilac (or brownish) on outside, with conspicuous lilac (or reddish) veins, falling soon after flowering. Capsule rather large, obovoid or almost barrelshaped, with depressed ribs. Stem and leaves pale green; basal leaf filiform. Bulb long and narrow; bulb and stem base surrounded by remains of dead leaves. ....7. **LLOYDIA** SALISB. — **LLOYDIA**

8. Plant about 20 cm high, with four large oval acuminate leaves arranged in cruciate whorl on upper part of stem. Flower large and solitary, on upright pedicel; outer perianth segments ovoid-lanceolate, green, up to 2.5 cm long; inner perianth segments linear, greenish yellow, about 1.5 cm long. Fruit a black berry, about 1 cm in diameter. . . . . .10. **PARIS** L. — **HERB PARIS**
– Plant 10–20 cm high, with 2–3 alternate leaves. Flowers small (perianth segments 3–5 mm long), white, assembled in terminal racemose inflorescences. . . . . .9.

9. Leaves 3 (more rarely 2), oblong-elliptical, tapering basally, sessile, amplexicaulous, situated on lower half of stem. Inflorescence loose, with 4–10 flowers; pedicels directed obliquely upwards. . . . . .8. **SMILACINA** DESF. — **FALSE SOLOMON'S-SEAL**
– Leaves 2, ovoid, with broadly cordate base, on long petioles, situated on upper half of stem. Inflorescence rather compact, with numerous (20–30) flowers; pedicels thin, spreading. . . . . .9. **MAIANTHEMUM** WEB. IN WIGG. — **MAY LILY**

GENUS 1   Tofieldia Huds. — FALSE ASPHODEL

SMALL GENUS DISTRIBUTED in the northern temperate zone. Represented in the Arctic by two arctic-alpine species.

1. Leaves mostly with 3 nerves. Inflorescence dense, almost capitate, somewhat sparser basally. Pedicels about 0.5 mm long at start of flowering, 1–2 mm in fruit. Bract single. Flowers yellowish white with stamens of same colour as perianth; inflorescence as a whole uniformly light coloured. . . . . .1. **T. PUSILLA** (MICHX.) PERS.
– Leaves mostly with 5 nerves. Inflorescence dense at start of flowering, then elongating and becoming looser. Pedicels about 0.5 mm long at start of flowering, 1–3 mm in fruit, horizontally spreading or slightly reclinate. Bracts 2. Perianth segments pale yellow with violet tint, anthers dark violet; inflorescence as a whole nonuniformly coloured, dark, reddish. . . . . .2. **T. COCCINEA** RICHARDS.

1. Tofieldia pusilla (Michx.) Pers., Syn. pl. 1 (1805), 399; Hultén, Fl. Al. III, 447; id., Atlas 120 (476); Semenova-Tyan-Shanskaya in Fl. Murm. II, 200; A.E. Porsild, Vasc. pl. W Can. Arch. 98; id., Ill. Fl. Arct. Arch. 64; Böcher & al., Grønl. Fl. 222; Polunin, Circump. arct. fl. 123.

Tofieldia palustris Huds. — Ledebour, Fl. ross. IV, 209; Trautvetter, Pl. Sib. bor. 116; Feilden, Fl. pl. N.Z. 20; Ostenfeld, Fl. arct. 1, 32; id., Fl. Greenl. 69; Simmons, Survey Phytogeogr. 67; Macoun & Holm, Vasc. pl. 10a; Holm, Contr. morph. syn. geogr. distr. arct. pl. 69; M. Porsild, Fl. Disko 63; Krylov, Fl. Zap. Sib. III, 588; Tolmachev, Fl. Kolg. 17, 41; Andreyev, Mat. fl. Kanina 165; Devold & Scholander, Fl. pl. SE Greenl. 107; Sørensen, Vasc. pl. E Greenl. 164; Perfilev, Fl. Sev. I, 144; Kuzeneva in Fl. SSSR IV, 3; Leskov, Fl. Malozem. tundry 45; Seidenfaden & Sørensen, Vasc. pl. NE Greenl. 180; Gröntved, Pterid. Spermatoph. Icel. 195; Karavayev, Konsp. fl. Yak. 75.

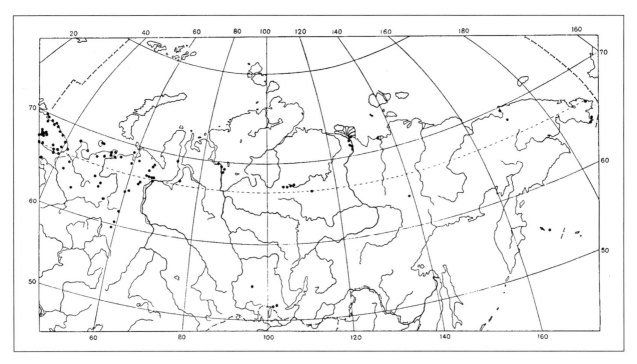

Map IV-22 Distribution of *Tofieldia pusilla* (Michx.) Pers.

Tofieldia borealis Wahlb. — Lange, Consp. fl. groenl. 122.
Narthecium pusillum Michx., Fl. Bor. Am. 1 (1803), 209.

    Common plant mainly in parts of the Arctic with relatively temperate climatic conditions. Growing mainly in moderately moist sites, on patches of bare loamy soil in patchy tundras, on slopes with mosses and low herbs, and sometimes in peat bogs. Occurring locally in great quantity, for example on the shores of the River Usa in its upper reaches, in the Malozemelskaya Tundra, and on islands in the lower reaches of the Lena. Judging by the nature of its habitats, it is associated with sites more or less protected by snow cover in wintertime.

Soviet Arctic. Murman (ubiquitous); Kanin (northern part); Timanskaya and Malozemelskaya Tundras (a more or less ubiquitous common plant); Kolguyev Island (rare, only on the eastern part of the island); Bolshezemelskaya Tundra (common at a series of sites); Dolgiy Island; Polar Ural; lower reaches of the Ob and the southern part of Yamal; lower reaches of Yenisey (north to 69°40´); Khatanga (near the forest boundary); between the Olenek and Lena Rivers; lower reaches of Lena (locally common); lower reaches of Yana; lower reaches of Kolyma; NW and eastern part of Chukotka. (Map IV–22).

Foreign Arctic. West coast of Alaska; arctic coast of Canada; Labrador; Canadian Arctic Archipelago (Victoria and Baffin Islands); Greenland from the southern extremity to 73°30´N on the west coast and to 74°30´N on the east coast; Iceland; Spitsbergen; Arctic Scandinavia.

Outside the Arctic. Mountains of Great Britain; the north and mountains of Fennoscandia; Alps; scattered localities in the forested region of the NE European part of the USSR on outcrops of limestone or gypsum (at Pinega, Soyana, Pechorskaya Pizhma and other sites); Northern Ural; extreme north of the forest zone on the Yenisey; NW Yakutia (Olenek Basin); Eastern Sayan at high elevation; forested region of Alaska and Northern Canada (in the mountains reaching the United States border); Newfoundland.

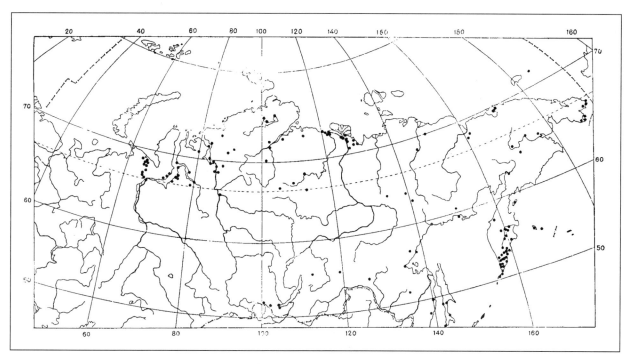

Map IV-23 Distribution of *Tofieldia coccinea* Richards.

2. Tofieldia coccinea Richards. in Franklin Narr. 1st Journ., Appendix (1823), 736; Schmidt, Fl. jeniss. 121; Kjellman, Phanerog. sib. Nordk. 129; Scheutz, Pl. jeniss. 168; Ostenfeld, Fl. arct. 1, 32; id., Fl. Greenl. 69; M. Porsild, Fl. Disko 63; Devold & Scholander, Fl. pl. SE Greenl. 107; Sørensen, Vasc. pl. E Greenl. 164; Seidenfaden & Sørensen, Vasc. pl. NE Greenl. 107, 180; Hultén, Fl. Al. III, 444; A.E. Porsild, Vasc. pl. W Can. Arch. 98; id., Ill. Fl. Arct. Arch. 64; Böcher & al., Grønl. Fl. 222; Polunin, Circump. arct. fl. 123.

Tofieldia nutans Willd. — Ledebour, Fl. ross. IV, 4; Krylov, Fl. Zap. Sib. III, 583; Kuzeneva in Fl. SSSR IV, 4; Karavayev, Konsp. fl. Yak. 75.

Characteristic plant of the Siberian and American Arctic. Judging by its overall distribution and its association with certain types of habitat, it gravitates towards more continental conditions than does *T. pusilla* although sometimes occurring side by side with the latter. *Tofieldia coccinea* grows mainly in patchy (polygonal) tundras where it colonizes both patches of bare soil and edges of the borders of closed vegetation. Sometimes occurring in moderately stony sites on tundra hillocks. It can grow and be common in areas with very modest protection by snow cover in winter. On the fringe of the Arctic it penetrates open forests carpeted with mosses and lichens. Boggy mossy sites are avoided in much of its range, but in certain districts (Anadyr and Penzhina Basins) the species is also reported from sphagnum bogs.

The ranges of *T. coccinea* and *T. pusilla* overlap to a considerable extent, but both in the Siberian and American Arctic *T. coccinea* generally extends further north. Conversely, it is not found in districts of the Atlantic Arctic with milder oceanic climate.

Soviet Arctic. Northern extremity of the Polar Ural; tundra on the Kara River; lower reaches of Ob (near Salekhard); Yamal, coasts of Ob Sound and Tazovskaya Bay, lower reaches of the Rivers Pur and Taz; Gydanskaya Tundra; lower reaches of the Yenisey from the forest-tundra north to Zverevo and Golchikha; large part of the

Pyasina Basin; central part of Taymyr (on north coast of Lake Taymyr); forest-tundra in Khatanga Basin; lower reaches of River Popigay; lower reaches of the Anabar and Olenek; between the Olenek and the Lena; lower reaches of the Lena (Bulun, Tas-Ary, Tit-Ary, etc.); coasts of Buorkhaya Bay (Tiksi Bay, Cape Eliyden); district of Bay of Chaun; Gerald Island; Chukotkan coast of the Bering Strait; Bay of Krest; Anadyr Basin; lower reaches of Penzhina. (Map IV–23).

Foreign Arctic. West coast of Alaska; arctic coast of Canada; Labrador; Canadian Arctic Archipelago (Banks, Victoria, Devon and Baffin Islands); West and East Greenland between 67°30´ and 78°N.

Outside the Arctic. Northern part of Khatanga Basin; subarctic and mountainous districts of Yakutia; mountains of Kamchatka, the Okhotsk Coast, Northern Preamuria and Transbaikalia; Sakhalin; Kurile Islands; Hokkaido; Alaska; Mackenzie River Basin; Canadian Rocky Mountains.

Plants from the most southern parts of the range in the Far East (Sakhalin, Northern Japan, Preamuria) should possibly be referred to a distinct race.

GENUS 2 — Zygadenus Rich. — CAMAS

1. Zygadenus sibiricus (L.) A. Gray in Ann. Lyc. Nat. Hist. New York IV (1837), 112; Ostenfeld, Fl. arct. 1, 33; Krylov, Fl. Zap. Sib. III, 584; Kuzeneva in Fl. SSSR IV, 9; Karavayev, Konsp. fl. Yak. 75; Polunin, Circump. arct. fl. 124.

Anticlea sibirica Kunth — Ledebour, Fl. ross. IV, 207; Trautvetter, Pl. Sib. bor. 115; id., Fl. rip. Kolym. 563.

Melanthium sibiricum L., Amoen. Acad. 2 (1758), 349.

Plant of the forest zone, just penetrating arctic limits in East Siberia. Growing in meadowlike areas and sometimes in moist patches in mossy patchy tundra.

Soviet Arctic. Lower reaches of Lena (series of localities south of the start of branching of the arms of the delta); district of Tiksi Bay. Near the arctic border on the northern slopes of the Central Siberian Plateau (in the Kheta Basin) and in the valleys of the Kheta and Khatanga.

Foreign Arctic. Not occurring.

Outside the Arctic. East Siberia from the edge of the Arctic in Yakutia, from the country between the Yenisey and the Khatanga, and from 65°30´N on the Yenisey south to the Amur Basin, Northern Mongolia, the Sayans and the Tannu-Ola. In the southwest reaching the Altay and the Kuznetskiy Alatau, but not penetrating the West Siberian Lowland. Disjunctly in the Middle Ural.

GENUS 3 Veratrum L. — FALSE HELLEBORE

Rather extensive (several tens of species) holarctic genus, most diversely represented in East Asia. A series of species are distributed in the northern temperate zone, especially in the USSR. Two species of section *Alboveratrum* penetrate arctic limits. They occur only in its climatically more temperate districts, but are locally not uncommon.

The group to which our arctic *Veratrum* species belong is centered on *V. Lobelianum* Bernh., a very widespread species in temperate Eurasia; this group undoubtedly still needs critical monographic revision with consideration of all accumulated material. Neither the monograph of Loesener (who studied only extremely limited material from the USSR) nor the treatment of this genus in the "Flora of the USSR" give a complete concept of its species composition, especially with respect to the interrelations between particular species and the variation shown by each of them separately.

The apparent ease of distinguishing certain species of *Veratrum* has to a considerable degree been attributable to their description from too limited material (sometimes single specimens), so that special characters of individuals or populations could be readily assumed to be specific characters and contrasted with the characters of other, obviously different populations. In particular, despite the fact that it has long been well known that the size of the plant and of its separate parts is variable in many species of *Veratrum*, the size of the plant has rather often been invoked as a character "facilitating" the distinction of those "species" whose qualitative differences appear insufficiently distinctive. However, the stature of the plant as a whole, the size of its leaves, and the size and degree of branching of the inflorescence are all subject to a great deal of variation in several species. The more extensive the material under study and the more diverse with respect to geographical origin and conditions of growth, the greater the amplitude of this variation appears.

Along with inadequately founded species distinctions in *Veratrum*,[1] there have apparently also been occasions when the conviction that one particular species (race) of the *V. Lobelianum* group must necessarily grow at each given locality diverted authors from analysis of the diversity of material available from the relevant districts, and led to "automatic" determinations mainly on the basis of geographical data with neglect of morphological characters. It was also inadequately considered that the features which in combination characterize individuals recognized as typical of a particular species are sometimes not correlated with one another and may be associated in certain combinations with characters considered to be attributes of another species.

For example, the distinction of the arctic race *Veratrum Mišae* (Sir.) Loes. from the boreal *V. Lobelianum* Bernh. is based on the combination of the following basic characters: small stature, correspondingly small size and weak branching of the inflorescence, absence of pubescence on the lower surface of the leaf, and narrower shape of the perianth segments. The characters which most obviously catch the eye (small stature, small size and weak branching of the inflorescence) appear to possess possible taxonomic significance only in

---

[1] For instance, *V. dolichopetalum* Loes. f. described from the Korean Peninsula is unlikely to be a species distinct from *V. oxysepalum* Turcz. described from depauperate Kamchatkan specimens; this species is in fact widespread in East Siberia and the Far East, and even on Kamchatka reaches one and a half metres in height. If *V. oxysepalum* were formally treated on the basis of Turchaninov's original diagnosis, then a considerable portion of the Kamchatka plants would also have to be referred to *V. dolichopetalum*.

selective comparison with large specimens of *V. Lobelianum* from the forest region. But observation of large numbers of *Veratrum* from neighbouring habitats will readily convince one of the extent to which these characters, directly affected by environmental influence, are unstable. They are relatively stable only in plants growing at the very edge of the range of *V. Lobelianum*, where such traits as low stature and limited inflorescence development inevitably arise under the pressure of extreme conditions of existence. Even at a short distance from the northern range boundary, the fact that these characters do not possess taxonomic significance becomes obvious. Absence of leaf pubescence is a trait of the majority of *Veratrum* plants from tundra, but plants with well developed pubescence on the lower leaf surface are also found among them. In turn, plants without leaf pubescence can be found among *V. Lobelianum* from the northern forest zone. This is reflected in Loesener's description of the glabrous-leaved *V. Lobelianum asiaticum* Loes. f., a form not differentiated geographically from "typical" *V. Lobelianum* (with leaves pubescent beneath). As for the narrower shape of the perianth segments, this can be correlated with a considerable stature of the plant, strong branching of the inflorescence and the presence of leaf pubescence. In summary, attempts to distinguish *V. Mišae* from *V. Lobelianum* have proved unsuccessful, and we have been compelled to recognize the former as one of the forms of *V. Lobelianum* unlikely to deserve even the rank of subspecies.

On the other hand, we cannot agree with the treatment of the East Siberian-Kamchatkan *V. oxysepalum* Turcz. as a subspecies of *V. Lobelianum* (or *V. album* L.). This is inconsistent with the considerable overlap of the ranges of these species with retention of their morphological identity in districts where they grow together. Moreover, we are not inclined to associate *V. oxysepalum* especially closely with the tundra *V. Lobelianum* var. *Mišae*, as was done (for example) in the "Flora of the USSR." Far northern specimens of these species are in fact similar to one another in such characters as the small size and generally weak branching of the inflorescence, short stature, and absence of leaf pubescence. But all these characters are also inconstant in *V. oxysepalum*: both the stature of the plant and the size and degree of branching of the inflorescence increase considerably in more southern parts of its range (including Kamchatka, whence the species was originally described); plants with the leaves pubescent beneath also occur there. Thus, we are faced with a significant expression of parallel variation in the two species. At the same time, when we compare specimens of *V. Lobelianum* (including *V. Mišae*) and *V. oxysepalum* which are similar in overall size and in inflorescence parameters, we are convinced that the inflorescence of the latter species is more extended and that its branches are shorter in comparison with the main rachis and more appressed to it. The inflorescence as a whole is narrower and more condensed in *V. oxysepalum*, broader and somewhat diffuse in *V. Lobelianum*. The flowers of the species under comparison are also distinctly different: in *V. Lobelianum* they open widely due to the perianth segments being considerably bent outwards; in *V. oxysepalum* the perianth segments are not only narrower but also almost straight, due to which the flower as a whole has an obconical shape. The two species also differ significantly in the shape and size of the bracts. Consideration of all these differences, which are not the direct result of local conditions but retained throughout the extensive ranges of the two species, allows us to state that these species are much better differentiated than is the case with *V. Lobelianum* and *V. Mišae*. The fact that the morphological boundary between these two species is not blurred in the partial overlap of their ranges fully supports this interpretation.

1. Perianth campanulate or obconical, with *more or less straight, obliquely divergent, narrow, lanceolate*, acuminate segments with denticulate margins, mostly about 10 mm long. Inflorescence *narrow, almost ropelike*, unbranched or weakly branched *with the branches closely appressed to the main rachis and considerably shorter than it*. Bracts oblong-lanceolate, *elongate*, on lower part of inflorescence of same length as flowers or exceeding them. Leaves glabrous on both surfaces or slightly pubescent beneath, the lower elliptical or oblong-elliptical, the upper lanceolate, gradually tapering apically and usually all acuminate. . . . .
. . . . .1. **V. OXYSEPALUM** TURCZ.
- Perianth *widely open, with outwardly bent, oblong-ovate or ovate* segments which strongly taper basally, mostly 7–8 mm long. Bracts ovate or oblong-ovate, *always considerably shorter than flowers*. Lower leaves broadly ovate, often obtusish. . . . . .2.

2. Inflorescence abbreviated, *weakly branched, narrow*. Leaves *glabrous on both surfaces*. . . . . .2a. **V. LOBELIANUM** VAR. **MIŠAE** SIR.
- Inflorescence larger, *relatively broad, with well developed, obliquely divergent branches*. Leaves *more or less pubescent beneath* with small pale hairs (inconspicuous if pubescence weakly developed). . . . .
. . . . .2. **V. LOBELIANUM** BERNH., S. STR.

   1. ***Veratrum oxysepalum*** Turcz. in Bull. Soc. Nat. Mosc. (1840) No. 1, 79; Loesener in Verh. Bot. Ver. Brandenb. 68 (1926), 135; id. in Fedde, Rep. spec. nov. XXIV (1928), 65; Kuzeneva in Fl. SSSR IV, 14; Karavayev, Konsp. fl. Yak. 75.
   V. album ssp. *oxysepalum* (Turcz.) Hultén, Fl. Kamtch. 1, 233; id., Fl. Al. III, 450.
   V. album var. *viridis* Rgl. — Trautvetter, Fl. rip. Kolym. 69; id., Fl. Tschuktsch. 38; id., Syll. pl. Sib. bor.-or. 55.
   V. album L. — Ostenfeld, Fl. arct. I, 33 (pro parte); Komarov, Fl. Kamch. I, 291.
   V. Mišae (Sir.) Loes. — Karavayev, Konsp. fl. Yak. 75.
   ? = *V. dolichopetalum* Loes. f. in Fedde, Rep. spec. nov. XXIV (1928), 65.
   Among willow bushes, in meadows in river valleys, and on more or less protected slopes in the East Siberian and Far Eastern Arctic. Reported as a scarce plant in the majority of districts where it occurs.
   Soviet Arctic. Lower reaches of the Yenisey (vicinity of Dudinka, Rychkov, 18 VII 1909); lower reaches of the Lena (numerous localities; Bulun, Ayakit, Kumakh-Surt, Tit-Ary, etc.); Kharaulakh River; coasts of Buorkhaya Bay (Tiksi Bay, Cape Eliyden); lower reaches of the Kolyma (Pokhodskoye, Panteleyevka, Nizhne-Kolymsk); Chukotka (district of Bay of Chaun, Lawrence Bay); Anadyr Basin (numerous localities, apparently ubiquitous); coast of Koryakia (Bay of Korf); Penzhina Basin and coasts of Penzhina Bay. (Map IV–24).
   Foreign Arctic. Extreme west of Alaska (Seward Peninsula).
   Outside the Arctic. Far East of Asia from the arctic border south to Preamuria, the northern half of Sakhalin and the Korean Peninsula (?); Kolyma Basin; Yakutia. In

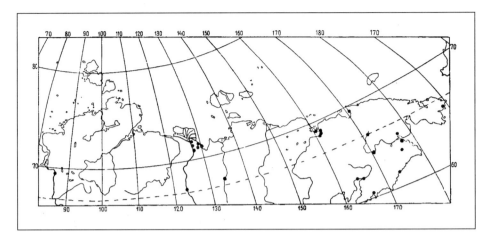

MAP IV-24 Distribution of *Veratrum oxysepalum* Turcz.

the west isolated localities in the Yenisey Basin (in addition to the Dudinka locality stated above, found especially on Mount Yenashimskiy Polkan on Yenisey Ridge).

2. ***Veratrum Lobelianum*** Bernh. in Schrad. N. Journ. 2 (1807), 356; Loesener in Verh. Bot. Ver. Brandenb. 68, 131; Ruprecht, Fl. samojed. cisur. 57; Krylov, Fl. Zap. Sib. III, 547; Perfilev, Fl. Sev. I, 144; Kuzeneva in Fl. SSSR IV, 13; id. in Fl. Murm. II, 201; Leskov, Fl. Malozem. tundry 45.
    V. album ssp. *Lobelianum* — Hultén, Atlas 120 (478).
    V. album L. — Schrenk, Enum. pl. 529; Schmidt, Fl. jeniss. 121; Scheutz, Pl. jeniss. 168; Ostenfeld, Fl. arct. 1, 33 (pro maxima parte).
    V. album floribus viridibus — Ledebour, Fl. ross. IV, 208.

2a. ***Veratrum Lobelianum*** var. ***Mišae*** Sirjaev in Acta Bot. Bohem. 2 (1923), 41; Loesener in Verh. Bot. Ver. Brandenb. 68, 132; Perfilev, Fl. Sev. I, 145.
    V. Mišae (Sir.) Loes. f. in Fedde, Rep. spec. nov. XXIV (1928), 65; Andreyev, Mat. fl. Kanina 165; Kuzeneva in Fl. SSSR III, 14; id. in Fl. Murm. II, 204; Karavayev, Konsp. fl. Yak. 75 (pro parte).
    V. Mischae — Krylov, Fl. Zap. Sib. III, 588.
    V. Lobelianum Bernh. — Tolmachev, Mat. fl. yevr. arkt. ostr. 464; id., Fl. pober. Karsk. morya 189.
    V. album var. *Lobelianum* — Tolmatchev, Contr. Fl. Vaig. 128.

    Widely distributed boreal species, occurring from northern and eastern districts of the European part of the USSR as far as Prebaikalia and sporadically further to the northeast. It penetrates the Arctic along the entire expanse from extreme NE Norway to the Bering Sea, and is common in the European and West Siberian sectors (including the lower reaches of the Yenisey and apparently the whole of the country between the Yenisey and the Khatanga). Further east it occurs more rarely than *V. oxysepalum*. In many districts from which the latter is reported, *V. Lobelianum* is certainly absent. On the northern fringe of its distribution it is mainly represented by the arctic variety, var. *Mišae*, or by forms transitional to this.

    Growing among willow shrubs and in open meadows, on slopes and in river valleys, often encountered in considerable quantity. In districts with greater human activity, this species sometimes becomes very abundant because it is unpalatable to cattle and reindeer. Accordingly, a profusion of *V. Lobelianum* is characteristic of the vicinity of settlements in the forest-tundra and southern tundra zones, as well as of sites where reindeer herds spend the spring.

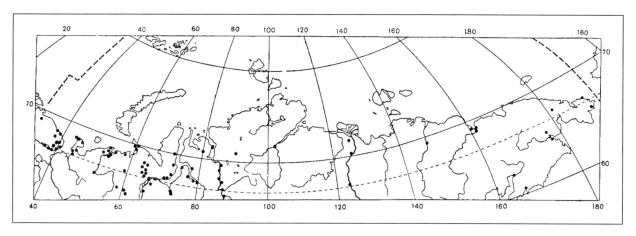

MAP IV–25 Distribution of *Veratrum Lobelianum* Bernh.

In southern parts of the tundra zone this species reaches one metre in height; further north (and in more exposed habitats) it becomes shorter, but continues to stand out from surrounding vegetation through its relatively large size. On this account it is always prominent wherever it occurs and is rarely overlooked by observers. Its role in the formation of the herb layer is readily overestimated on visual appraisal.

Soviet Arctic. Murman (not uncommon, but in small quantity; mainly the typical form, var. *Mišae* only in the extreme east); Kanin (mainly var. *Mišae*, but the typical form also occurs in the northern part of the peninsula); coast of Cheshkaya Bay (transitional forms and var. *Mišae*); Malozemelskaya Tundra (mainly the typical form; often reported as a common plant); entire Bolshezemelskaya Tundra (typical form predominating; often very common and growing in considerable quantity where suitable habitats are available); Polar Ural (mainly var. *Mišae*); Pay-Khoy (var. *Mišae* and transitional forms); the south of Vaygach Island (var. *Mišae*); lower reaches of Ob, Yamal north to 70°, coasts of Ob Sound and Tazovskaya Bay (var. *Mišae* and transitional forms); Gydanskaya Tundra (var. *Mišae*); lower reaches of the Yenisey (Dudinka and Tolstyy Nos, mainly var. *Mišae*; reaching Zverevo on the left bank of the Yenisey, on the right bank not reaching its mouth); southern half of Pyasina Basin (var. *Mišae*); Khatanga (var. *Mišae*); isolated localities between the Olenek and Lena Rivers (var. *Mišae*); lower reaches of the Kolyma (Pokhodskoye and Panteleikha, var. *Mišae*); Northern Chukotka (River Amguyema and Cape Serdtse-Kamen, var. *Mišae*); district of the mouth of the Anadyr (var. *Mišae*); coast of Koryakia (typical form); Penzhina Basin (typical form). (Map IV–25).

Foreign Arctic. Arctic Scandinavia (Eastern Finland).

Outside the Arctic. Central Europe (mainly in the mountains); European territory of the USSR, except NW districts and the steppe portion; Caucasus; West and Central Siberia, as far as the frontiers of Mongolia. The distribution in East Siberia, where other closely related species occur, is still unclarified.

GENUS 4   Gagea Salisb. — YELLOW STAR-OF-BETHLEHEM

EXTENSIVE GENUS, richly represented in the Mediterranean Region, in the USSR found mainly in the Caucasus and Central Asia. Not characteristic of the northern forest zone. One or two species penetrate districts on the fringe of the Arctic.

1. Bulb globose-ovoid, *with numerous small bulblets at base.* Stem 10–20 cm high. Basal leaf elongate, flat, *broad* (4–9 mm wide), strongly tapering at base, *of about same length as stem.* Bracts *very dissimilar,* one rather broadly lanceolate, the other almost filiform. Flowers 1–5 (in the Arctic 1–2), on long pedicels. Perianth segments lanceolate, acutish, yellow, *with green or brown stripes on outside,* 12–15 mm long. . . . . .1. **G. GRANULOSA** TURCZ.
– Bulb ovoid, *with a very small second bulb.* Height of stem together with flowers about 10 cm, after flowering up to 13–14 cm. Basal leaf elongate, 12–20 cm long, *considerably exceeding inflorescence, hollow* basally (1–2 mm in diameter), flat and broader on distal half. Bracts *more or less similar,* narrowly lanceolate. Flowers 2–3, on long slender pedicels. Perianth segments about 12 mm long, narrowly elliptical, obtusish, yellow, *almost lacking greenish tint on outside.* . . . . . . .2. **G. SAMOJEDORUM** GROSSH.

   1. ***Gagea granulosa*** Turcz. in Bull. Soc. Nat. Mosc. XXVII, 3 (1854), 112; Krylov, Fl. Zap. Sib. III, 496; Perfilev, Fl. Sev. I, 147; Grossheim in Fl. SSSR IV, 71.
     G. lutea — Schrenk, Enum. pl. 528.
        Occurring among shrubs and in deciduous forests of the forest zone, penetrating the forest-tundra.
    Soviet Arctic. Southern Kanin (forest-tundra) in forest on the south-facing slope of the valley of the Semzha River (Detlaf, 2 VI 1914). Perfilev's (l. c.) statement that *G. granulosa* has been found in the Bolshezemelskaya Tundra (with citation of Schrenk) was apparently based on a misunderstanding. Schrenk records "*G. lutea*" for the River Rochuga situated between the Mezen and the Pechora. There are no records of the finding of "*G. lutea*" in tundras beyond the Pechora in that work, nor are there any herbarium specimens from there.
    Foreign Arctic. Not occurring.
    Outside the Arctic. The northeast of the European part of the USSR, West Siberia (south of the 60th parallel), and the south of Central Siberia.

   2. ***Gagea samojedorum*** Grossh. in Fl. SSSR IV (1935), 84 and 736.
    ? *G. pusilla* Schult. — Perfilev, Fl. Sev. I, 147.
        Species described from the Prepolar and Northern Urals (Mount Sablya, upper reaches of River Sertynya in the Lyapin Basin, Mount Telpos-Iz), closely related to the Central European *G. fistulosa* Ker-Gawl. In the "Flora of the USSR" there is a record of *G. samojedorum* also occurring in "Arctic Europe." But there are no definite records for any particular localities. The herbarium of the Botanical Institute in Leningrad does not contain any material of this species from the European tundras. The record of its occurrence in the Arctic given in the "Flora of the USSR" is apparently erroneous.

◆ GENUS 5   **Allium** L. — ONION

EXTREMELY EXTENSIVE GENUS, especially richly represented in the floras of Southern Europe, Western and Central Asia. A considerable number of species occur in Southern Siberia. Two species also distributed in the northern forest zone occur within the Arctic (mainly the Asian Arctic). One of these is widespread and locally very common in the more temperate parts of the Soviet Arctic; the other is a rare plant.

1. Inflorescence *semiglobose or globose*. Pedicels *1½–2 times as long as perianth*. Perianth segments 4–5 mm long, pink with conspicuous purple midvein. Stamen filaments of about *same length as perianth or slightly longer*. Leaves *flattish, stiff*, shorter than stem. Bulb *elongate* (6–8 cm long), with brown reticulate coats. Growing mainly on rocky sites and steep slopes, sometimes on gravelbars. . . . . 1. **A. STRICTUM** SCHRAD.
– Inflorescence of *irregularly globose* shape (sometimes slightly acuminate apically), *very dense*. Pedicels *⅓ - ½ as long as perianth*. Perianth segments 7–15 mm long, lilaceous pink (or very rarely white) with darker midvein. Stamen filaments *half as long as perianth*. Leaves cylindrical or semicylindrical, *hollow, succulent*, shorter than or equalling length of stem. Bulb *oblong-ovoid*, with brown, almost leathery coats with longitudinal fibres. Plant of meadows and meadowlike areas of tundra, riparian sandbars and gravelbars, etc. . . . . .2. **A. SCHOENOPRASUM** L. — **CHIVES**

*1. **Allium strictum** Schrad., Hort. Goett. 7 (1809), 1; Ledebour, Fl. ross. IV, 178; Trautvetter, Pl. Sib. bor. 115; id., Fl. rip.Kolym. 562; Ostenfeld, Fl. arct. 1, 35; Vvedenskiy in Fl. SSSR IV, 151; Hultén, Atlas 123 (489); Karavayev, Konsp. fl. Yak. 75.

A. lineare var. *strictum* — Krylov, Fl. Zap. Sib. III, 626.

A. lineare L. — Schmidt, Fl. jeniss. 121.

Occurring in small quantity on rocky slopes, stony tundra hillocks, and the slopes of hills, sometimes also in groves and shrubbery in valleys; near the southern fringe of the Arctic Region, mainly within Yakutia and in the extreme northeast of Asia.

Soviet Arctic. Prepolar Ural in the upper reaches of the Manya River (a tributary of the Lyapin) and of the Voykar River (Gorodkov, 1915, 1925); Selyakino in the lower reaches of the Yenisey (isolated locality; recorded by Schmidt from collections in 1866, herbarium specimen not preserved); lower reaches of the Lena (series of sites, the most northern being Kumakh-Surt); lower reaches of Kolyma; Bay of Chaun; lower course of the Belaya River, a tributary of the Anadyr; coast of Gizhiga Bay; coast of Bay of Korf. (Map IV–26).

Foreign Arctic. Not occurring.

Outside the Arctic. The east of the European part of the USSR, the south of West Siberia and of the Yenisey Basin, the greater part of East Siberia (in Northern Yakutia in the upper part of the Olenek Basin and on the Yana and Kolyma), Northern Mongolia, the Far East. Disjunctly on the shores of the northern part of Lake Ladoga and in Central Europe.

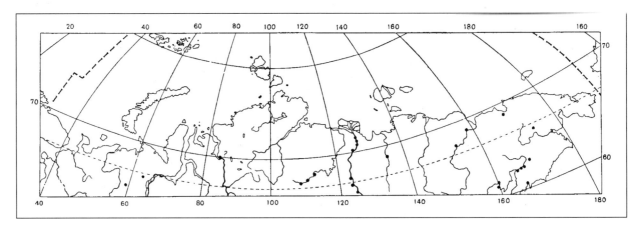

MAP IV–26 Distribution of *Allium strictum* Schrad.

2. Allium schoenoprasum L., Sp. pl. (1753), 301; Ruprecht, Fl. samojed. cisur. 57; Schrenk, Enum. pl. 528; Trautvetter, Pl. Sib. bor. 114; id., Fl. rip. Kolym. 562; Schmidt, Fl. jeniss. 121; Ostenfeld, Fl. arct. 1, 35; Tolmatchev, Contr. Fl. Vaig. 128; Tolmachev, Fl. Kolg. 41; Perfilev, Mat. fl. N.Z. Kolg. 53; id., Fl. Sev. I, 148; Vvedenskiy in Fl. SSSR IV, 190; Karavayev, Konsp. fl. Yak. 76; Polunin, Circump. arct. fl. 120.

A. schoenoprasum β *sibiricum* — Scheutz, Pl. jeniss. 167.

A. schoenoprasum var. *sibiricum* (L.) Hartm. — Hultén, Fl. Al. III, 453; id., Atlas 122 (486).

A. sibiricum L., Mant. 11 (1771), 562; Ledebour, Fl. ross. IV, 167; Andreyev, Mat. fl. Kanina 165; Perfilev, Fl. Sev. I, 148; Leskov, Fl. Malozem. tundry 46; Semenova-Tyan-Shanskaya in Fl. Murm. II, 207.

Ill.: Fl. Murm. II, pl. LXVI.

Common plant of tundra meadows, mainly in river valleys, especially in stretches immediately adjacent to shores (outside tundra proper). Often occurring abundantly on low terraces of sand or sand and gravel on the shores of rivers and floodplain lakes, where it grows in association with *Polygonum bistorta*, *Pyrethrum bipinnatum*, *Hedysarum arcticum*, various grasses, *Equisetum arvense*, etc. Sometimes, for example at certain sites in the upper reaches of the Usa (Bolshezemelskaya Tundra), on the Lukovaya Channel in the lower reaches of the Yenisey, and on the shores of Boganidskoye Lake in the forest-tundra beyond the Yenisey, it is so numerous that it becomes one of the dominant plants in small riparian meadows. It is also often encountered on gravelbars flooded at high water. Sometimes growing on loamy or clayey-stony slopes, but less numerous under these conditions. In mountains (for example, the Urals) it sometimes occurs along streams flowing through rock debris.

Confined to climatically more temperate parts of the tundra zone, but rather widely distributed there and advancing a considerable distance beyond the furthest northern outposts of woody vegetation. In the European Arctic and in Siberia east of the Yenisey this species is more or less uniformly and apparently continuously distributed. The lack of records of its occurrence in the tundras of NE Yakutia may easily be the result of the very low level of botanical investigation of this area. On the other hand, the almost complete absence of *A. schoenoprasum* from the expanse between the eastern foothills of the Urals and the Yenisey Valley despite the obvious availability of favourable habitats is apparently a real fact. It is natural to compare this with the absence of the species also from moderately

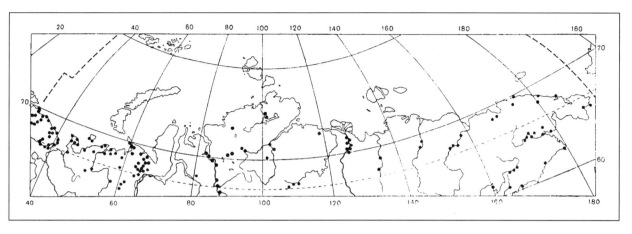

Map IV–27 Distribution of *Allium schoenoprasum* L. (Correction: Localities in moderately northern districts of the European part of the USSR are not shown).

northern parts of the West Siberian Lowland, within which it is found only on the eastern fringe.

Soviet Arctic. Murman (more or less ubiquitous); Kanin (common, north to Kanina Nos); Timanskaya and Malozemelskaya Tundras (more or less ubiquitous), Southern Kolguyev Island (not reported from the north); Bolshezemelskaya Tundra (everywhere, mainly in river valleys); Pay-Khoy, southern part of Vaygach Island; Polar Ural east to the upper reaches of the Shchuchya River; mouth of the River Taz; lower reaches of the Yenisey north to Tolstyy Nos and Nikandrovskiy Island; Pyasina River Basin (reaching north to the mouth of the Tareya River); Southern Taymyr tundra and forest-tundra; Central Taymyr (NW edge of Lake Taymyr and at the outflow from it of the Lower Taymyra River, at about the 75th parallel, the most northern localities for the species); lower reaches of Olenek; lower reaches of the Lena as far as the start of the delta; Chukotka from the Bay of Chaun to the Bering Strait; Wrangel Island; Anadyr Basin (ubiquitous); coasts of Gizhiga and Penzhina Bays; Penzhina River Basin; Bay of Korf. (Map IV–27).

Foreign Arctic. West coast of Alaska; Arctic Scandinavia (inside Norway).

Outside the Arctic. Forested part of Kola Peninsula, the coasts of the White Sea, the basins of the Onega, Severnaya Dvina and Pechora, and near the Urals in Perm Oblast. In Siberia from the Yenisey Valley, the River Vakh in the Ob Basin and the western foothills of the Altay in the west to the Kolyma Basin, Kamchatka and northern districts of Sakhalin Island in the east. South to Northern Mongolia. NW North America. In a considerable part of Europe represented by a form possibly not identical with ours. *Allium schoenoprasum* is especially recorded for Southern Scandinavia and generally on the coasts of the southern half of the Baltic Sea. It is reported from Great Britain and Ireland, from the middle zone of the European part of the USSR, and from the mountains of Central Europe. Additionally, it occurs in the Caucasus and the mountains of Central Asia (Tien Shan, Himalayas). In Southern Siberia and the Far East it is partly replaced by *A. Ledebourianum* Roem. et Schult. and *A. Maximowiczii* Rgl., while in Eastern North America "*A. schoenoprasum* var. *laurentianum*" occurs (separated by a considerable gap from the Alaskan portion of the range of the species).

The question of the delimitation of *Allium schoenoprasum* as a species and of which plants should properly bear this name and which the name *A. sibiricum* L. has not previously been settled. But now that *A. schoenoprasum* is generally treated in a wide sense (apparently with good reason), the separation from it of Siberian-Northern European plants under the name *A. sibiricum* is scarcely justifiable. It should not be overlooked that in describing *A. schoenoprasum* Linnaeus (1753) named as its origin both the island of Öland (Oelandia) and Siberia, thus not drawing a distinction between Swedish and Siberian plants. Eighteen years later (1771) he described *A. sibiricum* not as a form removed from *A. schoenoprasum*, but as a truly new species. What was the reason for this? If we refer to the description, then the phrase "Petala . . . alba, carine virescente" leaves no doubt that the author was dealing with the *albinistic form* of *A. schoenoprasum* and not with plants differing from it in any racial characters. At the present time, white forms of *A. schoenoprasum* can be found both in Siberia and in Northern Europe surrounded by typical plants with bright lilaceous pink flowers; the latter always absolutely predominate. To separate the white forms from the typical form under a separate species name would, of course, have no foundation. Any other treatment of "*A. sibiricum*" is inadmissable, as arbitrarily deviating from the characterization given by Linnaeus for this inappropriately described "species."

Allium schoenoprasum is undoubtedly a polymorphic species. In Scandinavia two geographically differentiated forms can apparently be contrasted with one another to some degree: a Baltic form (*A. schoenoprasum* s. str. in Hultén's sense) and a Norwegian–Murman form (*A. sch.* var. *sibiricum* Hartm., non L.). The situation is more complex in the Soviet North and in Siberia. Various botanists have turned their attention to the fact that the plants occurring here are nonuniform in certain respects. It is no accident that I.A. Perfilev was inclined to recognize both the species of chives described by Linnaeus as growing in the former Northern Territory. The peculiarity of *A. schoenoprasum* from the mountains of Central Asia was investigated by A.I. Vvedenskiy (in the "Flora of the USSR"), who had extensive knowledge of this species.

In the north of the USSR, *A. schoenoprasum* (incl. *A. sibiricum*) generally exhibits considerable individual variation. It is perhaps possible to detect incipient racial differentiation of this species in certain differences.

TABLE IV-1 Variation in certain parameters in *Allium schoenoprasum* from the European North and the lower reaches of the Yenisey

Parameter measured	Arctic Europe			Lower reaches of Yenisey
	Murman	Kanin	Bolshezemelskaya Tundra	
Height of plant in flower (cm)	28–41(–52)	21–45	16–36	20–40
Diameter of inflorescence (cm)	2.5–4(–4.5)	2–4	2.5–3.5	3–5.5
Length of perianth (mm)	8–10	8–12	8–11(–14)	9–15
Ratio of perianth length to pedicel length	3:2 (2:1–1:1)	2:1–1:1	3:2 (2:1–5:4)	2:1 (3:1–3:2)

Note: Values in brackets refer to isolated deviant plants.

In particular, plants from the lower reaches of the Yenisey and from the tundra and forest-tundra of Southern Taymyr differ (at the same height) in being somewhat more robust in comparison with European plants from the tundras near the Pechora or from the Kola Peninsula. Their inflorescences are usually larger and their perianth length greater both absolutely and relatively (most frequently twice as long as the pedicel length). In European plants the perianth is generally shorter and frequently only 1½ times as long as the pedicel, and sometimes these lengths are almost equal. The data in the above table give some idea of the parameter differences characterizing Northern European and Lower Yenisean *A. schoenoprasum*.

The bulk of Lower Yenisean plants also differ from Northern European plants in having more numerous flowers.

Plants from the Pyasina and Khatanga Basins do not differ at all obviously from those from the Lower Yenisey. Similar plants also occur in Arctic Yakutia. Here, however, they are accompanied by another form that is very different at first glance. This is a relatively low-growing plant (usually 14–22 cm high), possessing slender (up to 2 mm in diameter) stems and not very large inflorescences with a diameter of 2.5–3 cm (although appearing relatively larger than in Northern European plants). Their perianth length is 8–12 mm, with pedicels ½ - ⅔ as long. This form is rather widespread in Northern Yakutia (lower reaches of the Lena within the forest-tundra, upper part of the Olenek Basin), and occurs in the mountain country of the Verkhoyansk Range and here and there in the Aldan Basin. We find the same form in the basin of the Podkamennaya Tunguska on the Chunya River. Geographically it is *not separated* from the large form of *A. schoenoprasum*, but is apparently associated with somewhat different habitat conditions (stony slopes which are probably poorer with respect to soil and moisture).

Considerable variation is shown by *A. schoenoprasum* also in Southern Siberia. But we have not been able to detect differences which could serve as the basis for a general contrast between plants from Southern and Northern Siberia.

GENUS 6   Fritillaria L. — FRITILLARY

1. ***Fritillaria kamtschatcensis*** (L.) Ker-Gawl. in Botan.-Magaz. XXX (1809), tabl. 1216; Ledebour, Fl. ross. IV, 147; Komarov, Fl. Kamch. I, 300; Lozino-Lozinskaya in Fl. SSSR IV, 318; Hultén, Fl. Al. III, 454.
   Lilium kamtschatcense L., Sp. pl. (1753), 303.

   Plant of larch forests and sparse shrub thickets in maritime districts of the Far East of Asia and NW North America. Just penetrating the Arctic in the extreme north of its range, where it grows among shrubbery in valleys.

   Soviet Arctic. Coast of Koryakia at the Bay of Korf. Reported by Komarov for the Anadyr Basin, but there is no relevant material in the herbarium of the Botanical Institute of the USSR Academy of Sciences.

   Foreign Arctic. There are two old records of occurrence on the west coast of Alaska, but these have not been confirmed by subsequent collections or observations.

   Outside the Arctic. Coastal zone of the Asian mainland from the southern borders of the USSR to the north coast of the Sea of Okhotsk (Magadan district); Kamchatka; Northern Japan, Sakhalin, Kurile Islands, Commander and Aleutian Islands; Pacific coast of Alaska and Canada, extreme NW United States.

GENUS 7   Lloydia Salisb. — LLOYDIA

SMALL GENUS native to Central and Eastern Asia. One species of the genus, the most widely known *L. serotina*, is distributed throughout the Siberian Arctic and crosses into NW North America and marginally into NE Europe. The same species is widespread in mountains of the temperate zone.

1. ***Lloydia serotina*** (L.) Rchb., Fl. germ. exc. (1830), 102; Ledebour, Fl. ross. IV, 144; Trautvetter, Fl. taim. 24; id., Pl. Sib. bor. 114; id., Syll. pl. Sib. bor.-or. 535; Schrenk, Enum. pl. 528; Schmidt, Fl. jeniss. 120; Kjellman, Phanerog. sib. Nordk. 129; Scheutz, Pl. jeniss. 167; Ostenfeld, Fl. arct. 1, 35; Tolmatchev, Contr. Fl. Vaig. 128; Krylov, Fl. Zap. Sib. III, 642; Perfilev, Fl. Sev. I, 149; Tolmachev & Pyatkov, Obz. rast. Diksona 157; Tolmachev, Fl. Taym. I, 105; Komarov in Fl. SSSR IV, 369; Hultén, Fl. Al. III, 456; Tikhomirov, Fl. Zap. Taym. 31; Karavayev, Konsp. fl. Yak. 76; Polunin, Circump. arct. fl. 121.

Bulbocodium serotinum L., Sp. pl. (1753), 294.
Anthericum serotinum L., Sp. pl. ed. II (1762), 444.
Ornithogalum altaicum Laxm. (1774); *O. striatum* Willd. (1799); *Lloydia alpina* Salisb. (1812); *Nectarobothrium striatum* Ledb. (1830); *N. Redowskianum* Schl. et Cham. (1831).

Widespread arctic-alpine plant. Often growing in relatively dry polygonal tundras, where it colonizes poorly vegetated patches of loamy or loamy-stony soil. Occurring on sites moderately snow-covered in winter (but not where snow is absent), often in association with *Dryas octopetala* (or *D. punctata*) which together with other flowering plants and mosses forms the borders of closed vegetation delimiting the unvegetated "patches." *Lloydia serotina* also occurs on more or less protected slopes, but avoids the more densely vegetated portions of them. On the southern fringe of the arctic part of its range, it is native to steep, discontinuously vegetated slopes and often grows near rocks. It is a common plant in many arctic districts.

In Arctic Siberia this species possesses a continuous distribution both in mountainous and lowland districts. On the southern fringe of the tundra zone it occurs only in more or less mountainous or (at least locally) rocky districts. The species becomes more common as one advances northwards. It is most characteristic of the arctic tundra subzone, where it occurs practically everywhere.

With respect to the European Arctic, this species only penetrates the eastern fringe in the zone of the foothills of the Polar Ural and in Pay-Khoy (together with Vaygach Island, but apparently not reaching its northern extremity). East of the Bering Strait the species is widespread in Alaska, but does not penetrate further east than the Mackenzie Delta.

Lloydia serotina is well distinguished from all other arctic plants in appearance and structure, but locally shows rather considerable variation in details. It is relatively more stable in the western part of its range (Polar Ural, district of Yugorskiy Shar, NW Siberia), where almost all plants produce only a single flower. It is more variable in Taymyr and Arctic Yakutia, where plants with two or sometimes even three flowers often occur together with single-flowered plants. Plants with two or three flowers are more robustly developed overall. The same phenomenon is observed in the mountains of Southern Siberia and Central Asia.

Soviet Arctic. Eastern fringe of Bolshezemelskaya Tundra (upper reaches of River Usa); Polar Ural; Pay-Khoy; south half of Vaygach Island; Yamal; coasts of Ob Sound; Gydanskaya Tundra; lower reaches of Yenisey; coasts and islands of Yenisey Bay; Taymyr north to the NW coast, the mouth of the Lower Taymyra River and Pronchishcheva Bay; arctic coast of Yakutia; New Siberian Islands; throughout

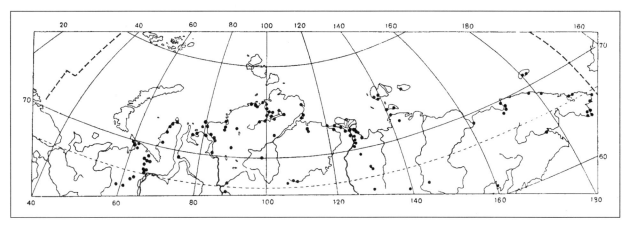

MAP IV–28 Distribution of *Lloydia serotina* (L.) Rchb.

Chukotka; Wrangel Island; Koryakia. (Map IV–28).

Foreign Arctic. Islands of the Bering Sea, west and north coasts of Alaska; coast of Canada from the Alaskan frontier to the Mackenzie Delta.

Outside the Arctic. Mountains of North Wales; Urals; Alps, Carpathians, Northern Balkans, northern slopes of the Great Caucasus; hills of the northern part of the Central Siberian Plateau; Verkhoyansk-Kolymsk mountain country; Kamchatka; high mountains of the Stanovoy Range, Northern Preamuria, the Northern Sikhote-Alin, Sakhalin, and Northern Japan; mountains of Prebaikalia and Northern Mongolia, the Sayans and the Altay; mountains of Soviet Central Asia (Dzhungarskiy Alatau, Tien Shan, the Gissarskiy-Zeravshanskiy Ranges, Pamir) and mountains of the Chinese part of Central Asia; Himalayas (the Himalayan and part of the Western Chinese plants possibly belong to the closely related species *L. Mairei*); mountains of Western North America from Alaska and Yukon south to Oregon, Nevada and New Mexico (with gaps).

GENUS 8 Smilacina Desf. — FALSE SOLOMON'S-SEAL

1. ***Smilacina trifolia*** (L.) Desf. in Ann. Mus. Paris IX (1807), 52; Ledebour, Fl. ross. IV, 128; Kuzeneva in Fl. SSSR IV, 452; Karavayev, Konsp. fl. Yak. 75.

Plant of mossy forests and sphagnum bogs, widespread in the forest zone of East Siberia, the Far East and North America. It penetrates the Arctic in extreme NE Asia, where it grows mainly in sphagnum bogs in riverine lowlands.

Soviet Arctic. Anadyr Basin (considerable number of localities); Penzhina Basin.

Foreign Arctic. No reliable records; possibly found in the arctic part of Canada.

Outside the Arctic. East Siberia from the middle course of the Lower Tunguska, the upper reaches of the Olenek, the middle part of the Indigirka Basin (reported below 67°40´N) and Srednekolymsk in the north as far as the SW part of Krasnoyarsk Kray, Preangaria, forested districts of Transbaikalia and the Amur Basin in the south. Penetrating NE China, Sakhalin and Northern Japan (but not on the Kurile Islands or Kamchatka). Taiga zone of Canada.

GENUS 9   Maianthemum Web. in Wigg. — MAY LILY

1. **Maianthemum bifolium** (L.) F.W. Schmidt, Fl. Boem. Cent. IV (1794), 55; Krylov, Fl. Zap. Sib. III, 651; Perfilev, Fl. Sev. I, 149; Fedchenko in Fl. SSSR IV, 453; Leskov, Fl. Malozem. tundry 46; Hultén, Atlas 126 (502); Semenova-Tyan-Shanskaya in Fl. Murm. II, 208; Tolmachev, Ist. temnokhv. taygi ris. 22 (map); Karavayev, Konsp. fl. Yak. 77.
   Convallaria bifolia L., Sp. pl. (1753), 316.
   Smilacina bifolia Desf. — Ruprecht, Fl. samojed. cisur. 57; Ledebour, Fl. ross. IV, 127; Schrenk, Enum. pl. 528.

   Plant characteristic of spruce-fir forests of the temperate north (like other species of the oligotypic genus *Maianthemum*). Also occurring in birch woods. Penetrating the forest-tundra in Northern Europe.

   Soviet Arctic. Occasional in the forest-tundra of the Kola Peninsula, Southern Kanin, and on the Pesha near Cheshskaya Bay. Near the forest boundary in the Pechora Basin and the lower reaches of the Ob (vicinity of Salekhard, valleys of the Poluy and Nadym). Further east the range boundary shifts further south. Records for the extreme northwest of the Kola Peninsula (Rybachiy Peninsula) are doubtful.

   Foreign Arctic. Arctic Scandinavia.

   Outside the Arctic. Widespread in the forests of Northern and Central Europe, the forest zone of the European part of the USSR, Siberia and the Far East, east all the way to Sakhalin and the central part of Kamchatka, south as far as the Upper Dnepr region, the Middle Volga, the upper reaches of the Ob and Yenisey, forested mountains of Northern Mongolia, southern parts of the Amur Basin, and the north of the Korean Peninsula. Disjunctly in the Caucasus.

GENUS 10   Paris L. — HERB PARIS

1. **Paris quadrifolia** L., Sp. pl. (1753), 367; Ledebour, Fl. ross. IV, 120; Krylov, Fl. Zap. Sib. III, 657; Perfilev, Fl. Sev. I, 150; Knorring in Fl. SSSR IV, 469; Grøntved, Pterid. Spermatoph. Icel. 196; Hultén, Atlas 127 (506); Semenova-Tyan-Shanskaya in Fl. Murm. II, 212.

   Forest plant of Europe and Siberia, just reaching arctic boundaries in the west of its range.

   Soviet Arctic. Murman (Pechenga district, Southern Rybachiy Peninsula, forest-tundra on Kola Bay, lower reaches of the Ponoy). East of the White Sea nowhere reaching the polar limit of forests.

   Foreign Arctic. Iceland, Arctic Scandinavia.

   Outside the Arctic. Widespread in forested regions of Europe, reaching latitudes a little north of 65° in Arkhangelsk Oblast and the Komi ASSR; in Siberia reaching 61°30´–62°N in the Ob Basin, crossing the 65th parallel on the Yenisey, and reaching slightly north of 60° in the Lena Basin. In the south reaching the Altay, the Sayans and forested districts of Northern Mongolia.

FAMILY XVIII
Iridaceae Lindl.
IRIS FAMILY

EXTENSIVE FAMILY, most richly represented in the floras of extratropical Southern Africa and the Mediterranean Region. Possessing very few representatives at relatively high latitudes. One of the species of the widespread genus *Iris* penetrates the Arctic to a limited extent.

GENUS 1 — Iris L. — IRIS

1. ***Iris setosa*** Pall. ex Link in Spreng., Schrad. & Link, Jahrb. Gewächsk. 1, 3 (1820), 71; Ledebour, Fl. ross. IV, 96; Trautvetter, Pl. Sib. bor. 114; id., Fl. rip. Kolym. 68; Fedchenko in Fl. SSSR IV, 520; Hultén, Fl. Al. III, 464; Karavayev, Konsp. fl. Yak. 77. — *I. sibirica* L. — Ostenfeld, Fl. arct. 1, 36.

    Ill.: Fl. SSSR IV, pl. XXXII, 6.

    Occurring on moist open sites on river shores, and sometimes on banks in protected parts of the sea coast.

    Soviet Arctic. Lower reaches of the Kolyma (Pokhodskoye, the furthest northern point of the range); Anadyr Basin; Penzhina Basin and coasts of Penzhina and Gizhiga Bays; Bay of Korf. (Map IV–29).

    Foreign Arctic. West coast of Alaska.

    Outside the Arctic. Widespread in the Far East and in Yakutia. The northern range boundary proceeds from the sea coast in the district of the mouth of the Anadyr almost to the mouth of the Kolyma (Pokhodskoye), then extends westwards with slight shift to the south; it intersects the Indigirka near Abyy (slightly north of 68°) and the Lena near the mouth of the Nazhim River (near the Arctic Circle). Then the boundary passes north of the greater part of the Vilyuy Basin and extends as far as the upper reaches of that river, where (according to observations by I.D. Kildyushevskiy in 1958) *I. setosa* is common. In Maack's collection there is a specimen collected somewhere between the upper reaches of the Vilyuy and the Olenek (without precise locality label). *Iris setosa* apparently nowhere crosses the Lena-Yenisey watershed. From the upper reaches of the Vilyuy the boundary of its range assumes a southeastward direction, intersects the Lena above Olekminsk, and passes into Preamuria along the western edge of the Zeya River Basin. The species is common in Central Yakutia, as well as in Central Preamuria. South of the Amur the range boundary passes through the northeastern provinces of China to the southern border of the Ussuri River Basin, then enters the northern part of the Korean Peninsula where it reaches the Sea of Japan. On the coasts of the Pacific Ocean *I. setosa* is apparently distributed continuously from the shores of Anadyr Bay to the north of the Korean Peninsula. It is widespread on Kamchatka, the Commander and Kurile Islands, throughout Sakhalin, on Hokkaido and in the northern part of Honshu. In North America the distribution of *I. setosa* is restricted to the Aleutian Islands and Alaska. In Eastern North America the closely related species *I. canadensis* is distributed along the sea coast from Labrador and

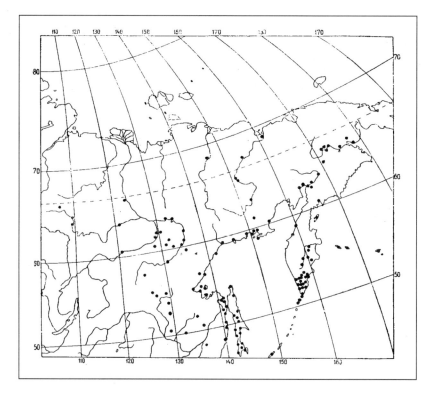

MAP IV–29 Distribution of *Iris setosa* Pall.

Newfoundland to the State of Maine; this is treated by Hultén as a race of *I. setosa* (*I. setosa* ssp. *canadensis*).

A form occurring in inland districts of Alaska described under the name *I. setosa* var. *interior* E. Anders. in Hultén's opinion merits treatment as the subspecies (geographical race) *I. setosa* ssp. *interior* (E. Anders.) Hult. In addition to Alaskan (noncoastal!) plants, Hultén also refers to this taxon of plants from the Penzhina River (from the collections of Gorodkov and Tikhomirov). The characters of var. *interior* are also shown by some specimens from the Kolyma Basin. But it is doubtful whether it is possible to separate typical *I. setosa* and var. *interior* with the status of geographical races in Soviet territory.

FAMILY XIX
Orchidaceae Lindl.
ORCHID FAMILY

EXTENSIVE FAMILY including hundreds of genera and many thousands of species. Distributed in all parts of the World, but most richly represented at equatorial latitudes, where it sometimes occupies the leading position in species numbers in the composition of floras of moist forested regions. At temperate latitudes the number of species is far fewer (for example, only 120 are recorded for the entire territory of the USSR), and they are almost completely confined to forested regions. The number of species distinctly increases in regions with stronger expression of oceanic climatic traits (increased atmospheric humidity, mild winter cold whose influence on plants may be further mitigated by deep snow cover). In the northern forest zone members of this family are mainly concentrated in its more southern parts, especially in districts closer to the Atlantic or Pacific Oceans (Northern Europe, the Far East of Asia, and the Pacific part of North America).

This family is in general not characteristic of the Arctic. It does not include specifically arctic species. The species occurring within the Arctic are confined to its more temperate parts and have the main part of their ranges situated in the temperate zone. Only two species, *Coeloglossum viride* and *Corallorhiza trifida*, are relatively characteristic of the Soviet Arctic. But they do not penetrate high arctic districts. The majority of other species have a very restricted distribution in the Arctic. Some of them occur in the Soviet Arctic only in Murman (*Cypripedium calceolus*, *Chamaeorchis alpina*, etc.); usually, these are somewhat more widespread in Arctic Fennoscandia beyond our frontier. A few species also occur, as well as in Murman, on the fringe of the Arctic in the northeast of the European part of the USSR (*Listera cordata*, *Gymnadenia conopsea*, *Leucorchis albida*). Finally, certain specifically Far Eastern species (*Coeloglossum bracteatum*, *Lysiella obtusata* ssp. *oligantha*) penetrate arctic limits in the extreme northeast of Asia. In the Siberian Arctic proper only the two species mentioned above as generally rather characteristic of the Arctic occur.

1. Flowers *solitary* (or rarely 2 on same plant), *large* (perianth segments up to 3 cm long, narrow, reddish brown; lip almost ovoid in outline, pouch-shaped, yellow, up to 3 cm long). Leaves elliptical, alternate. . . . . .1. **CYPRIPEDIUM** L. — **LADY'S SLIPPER**
   - Flowers *aggregated at tip of stem into more or less oblong inflorescence*, considerably smaller, their shape otherwise. . . . . .2.

2. *Plant leafless*, stem bearing only membranous sheaths. Entire plant pale, *yellowish brown, devoid of green colouration.* Rhizome divided into *short*

obtusish branchlets, in shape reminiscent of young reindeer antlers or coral branches. Flowers few, *small*, *white*; inflorescence rather loose. . . . . .2. **CORALLORHIZA** HALL. — **CORALROOT**

– *Green plants with normally developed leaves.* Rhizome (if present) without "antlerlike" ("corallike") branches. . . . . .3.

3. *Two broadly ovate leaves situated near middle of stem, opposite one another at the same level.* Inflorescence loose, with few (4–12) very small pale flowers. Ovary inflated, straight, on twisted pedicel. Small plant (up to 15 cm high) with erect pale stem. . . . . .3. **LISTERA** R. BR. — **TWAYBLADE**

– Leaves more or less *tapered at base, better developed on lower half of stem* or concentrated there, *alternate*; or only a *single* leaf situated at base of stem. . . . . .4.

4. *Single* oblong-lanceolate leaf, strongly tapered basally, situated *at base* of stem and reaching half to two-thirds of its height. Inflorescence loose, with few (3–4–6) flowers. Flowers white, small. Ovary *sessile, twisted*. Roots cordlike. Entire plant small, 5–8 cm high. . . . . .8. **LYSIELLA** RYDB. — **BLUNT-LEAVED ORCHID**

– Leaves *not solitary, alternate*. . . . . .5.

5. Ovary *straight, on twisted pedicel*. Inflorescence *loose, elongate*, basically a *one-sided* raceme. Flowers rather large, *nodding, purplish*. Perianth segments broad, ovate, 6–8 mm long. Lip with median constriction, reflexed at tip. Leaves rather numerous and relatively crowded, elliptical or ovate, acuminate at tip, rather stiff. . . .3a. **EPIPACTIS** SW. — **HELLEBORINE**

– Ovary *twisted, sessile*. Inflorescence otherwise in details of shape, often symmetrical. Flowers whitish, greenish or violet, *not purplish*. . . . . .6.

6. Flowers *white, yellowish white or yellowish green.* .7.

– Flowers *violet*, often with darker pattern on relatively light background. . . . . .10.

7. Plant with *long slender rhizome*. Leaves slender, weakly coloured, with conspicuous fine network of veins. Inflorescence *one-sided*, with small white or yellowish flowers. Lip strongly concave, with reflexed beak at tip. . . . . .4. **GOODYERA** R. BR. — **RATTLESNAKE PLANTAIN**

– Plants not developing rhizomes, with *tuberously swollen roots*. Inflorescence *more or less symmetrical*. Leaves bright green, sometimes with dark spots. . . . . .8.

8. Leaves *narrow, linear*, somewhat twisted, *reaching height of stem*. Flowers small, greenish yellow, nodding. Lip obtuse, *not divided into lobes. Very small plant*, reaching 4–8 cm (sometimes 10 cm) high. . . . . .5. **CHAMAEORCHIS** L.C. RICH. — **ALPINE ORCHID**

– Leaves *oblong or elliptical* (not linear), *considerably shorter than stem.* Flowers standing obliquely upright. Lip *trilobed* (or sometimes middle lobe not developed). Plant 8–20 cm high. . . . . .9.

9. Flowers *yellowish green*. Bracts *large, lanceolate, the lower considerably longer than the flowers*. Lower leaves *elliptical*, upper narrower. Lip *flat, extended like a ribbon*, 5–6 mm long, divided at tip into three lobes of which the middle is shorter than the lateral or not developed at all. ....
..............................6. **COELOGLOSSUM** HARTM. — **FROG ORCHID**

– Flowers *white, small* (perianth with ovary about 5 mm long), aggregated in *dense many-flowered oblong* inflorescence (with 20 or more flowers). All bracts *shorter than flowers* (lower bracts longer than ovary, upper shorter). Lower leaves *oblong*, upper lanceolate. Lip slightly longer than rest of perianth segments, with its middle lobe longer than the lateral. ....
........................7. **LEUCORCHIS** E. MEY. — **SMALL WHITE ORCHID**

10. Leaves *lanceolate, somewhat folded lengthwise, uniformly shining green*. Inflorescence *narrow*, 3–6 cm long, cylindrical, *dense and many-flowered*. Lower bracts longer than flowers, upper shorter. Flowers lilaceous pink. Perianth segments up to 5 mm long, *the lateral spreading horizontally*. Lip trilobed, about 5 mm long and wide. *Spur slender, curved like a sickle, twice as long as lip and usually longer than ovary.* ....
..............................9. **GYMNADENIA** R. BR. — **REIN ORCHID**

– Leaves with complex *pattern of dark spots on green background* (often disappearing or becoming scarcely evident after drying). Inflorescence *relatively short, loose*, ovoid. Most frequently 10–15 flowers (sometimes less than 10 or up to 25). Bracts lanceolate, the lower of about same length as flowers. Flowers light or bright violet, *with darker pattern on relatively light background, rather large*. Perianth segments up to 8 mm long, *the lateral directed obliquely upwards*. Lip 8–10 mm long and of the same or greater width, trilobed, slightly undulate on margin. *Spur straight, shorter than ovary and slightly shorter than lip.* ....
................10. **DACTYLORCHIS** (KLINGE) VERMLN. — **SPOTTED ORCHID**

◆ GENUS 1   Cypripedium L. — LADY'S SLIPPER

1. Cypripedium calceolus L., Sp. pl. (1753), 951; Ledebour, Fl. ross. IV, 86; Krylov, Fl. Zap. Sib. III, 679; Perfilev, Fl. Sev. I, 153; Nevskiy in Fl. SSSR IV, 598; Hultén, Atlas 129 (512); Orlova in Fl. Murm. II, 216; Karavayev, Konsp. fl. Yak. 78.

Ill.: Fl. Murm. II, pl. LXIX.

Forest plant widespread in temperate Eurasia. Penetrating arctic limits only in Northern Fennoscandia.

Soviet Arctic. One locality known in the extreme northwest of Murmansk Oblast, on the Norwegian frontier (Hultén, l. c.).

Foreign Arctic. Arctic Scandinavia (within Norway).

Outside the Arctic. Northern and Central Europe, southern part of the forest zone of Siberia and the Far East. In the north of the European part of the USSR reaching almost 68°N in the central part of the Kola Peninsula; reaching about 65°N in Arkhangelsk Oblast, 60°N in the Urals, 59°N in the basins of the Ob and Yenisey, and 62°N in Yakutia. Ranging south to the Balkan Peninsula, Crimea, the forest-steppe of West Siberia, the Altay, Northern Mongolia, Northern China, and Japan.

GENUS 2   Corallorhiza Hall. — CORALROOT

1. **Corallorhiza trifida** Châtel., Sp. inaug. Corall. (1760), 8; M. Porsild, Fl. Disko 65; Ostenfeld, Fl. Greenl. 69; Nevskiy in Fl. SSSR IV, 608; Gröntved, Pterid. Spermatoph. Icel. 197; Hultén, Fl. Al. III, 492; id., Atlas 141 (558); Orlova in Fl. Murm. II, 220; Böcher & al., Grønl. Fl. 227; Karavayev, Konsp. fl. Yak. 78; Polunin, Circump. arct. fl. 126.
   C. innata R. Br. — Trautvetter, Fl. boganid. 146; Scheutz, Pl. jeniss. 164; Lange, Consp. fl. groenl. 120; Ostenfeld, Fl. arct. 1, 37; Perfilev, Fl. Sev. I, 159; Leskov, Fl. Malozem. tundry 46.
   C. neottia Scop. — Krylov, Fl. Zap. Sib. III, 717.
   Ophrys corallorhiza L., Sp. pl. (1753), 945.
   Ill.: Fl. Murm. II, pl. LXXI.

   Plant with a distribution of the boreal type, penetrating the tundra zone along a considerable stretch of the northern edge of its range. Growing among shrubs, sometimes also on sparsely vegetated open slopes, at sites well protected by snow cover in winter and adequately moistened in summer (although well drained), on soil more or less enriched with organic substances. Continuously vegetated sites are avoided. Normally occurring solitarily or in small groups, but not a rarity. Readily overlooked during itinerant investigations, and evidently much more common than established by the data so far gathered.

   Soviet Arctic. Murman (relatively frequent, but not ubiquitous?); Northern Malozemelskaya Tundra (apparently rare); Bolshezemelskaya Tundra (right bank of the Lower Pechora, the Vangurey Fan, Varandey Island, Vorkuta, upper reaches of River Usa; locally rather common); lower reaches of Ob; Yamal (to 70°N); coasts of Ob Sound and Tazovskaya Bay; lower reaches of the Yenisey north at least to Dudinka; Taymyr-Khatanga forest-tundra (Petrov Krest on the Boganida River at about 71°30´N); lower reaches of the Lena north to 71°05´; Penzhina River Basin; Bay of Korf. (Map IV–30).

   Foreign Arctic. West coast of Alaska; West Greenland north to Disko Island; Iceland; Arctic Scandinavia.

   Outside the Arctic. Forested region of Europe (south to the Pyrenees, Yugoslavia and Bulgaria); European part of the USSR from the edges of the tundra as far south as the 50th parallel in the west and the 55th parallel in the east; disjunctly in the mountains of Crimea and the Caucasus; greater part of Siberia south to forest-steppe districts. Absent from the mountains of Southern Siberia, but occurring disjunctly from the main range in the mountainous district of Central Kazakhstan and in the NE part of the Tien Shan, also in extreme NW India. In the Far East found in Northern Preamuria, on the Okhotsk Coast and in Kamchatka (but absent from Sakhalin, the Kurile Islands and Japan). Widespread in subarctic and temperate forested districts of North America.

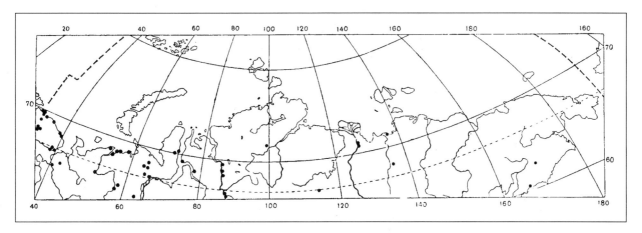

MAP IV-30 Distribution of *Corallorhiza trifida* Châtel.

🍂 GENUS 3   **Listera** R. Br. — **TWAYBLADE**

1. ***Listera cordata*** (L.) R. Br. in Ait., Hort. Kew. ed. 2, V (1813), 201; Ledebour, Fl. ross. IV, 80; Lange, Consp. fl. Groenl. 120; Ostenfeld, Fl. arct. 1, 37; id., Fl. Greenl. 69; M. Porsild, Fl. Disko 64; Krylov, Fl. Zap. Sib. III, 709; Perfilev, Fl. Sev. I, 158; Nevskiy in Fl. SSSR IV, 613; Seidenfaden & Sørensen, Vasc. pl. NE Greenl. 180; Gröntved, Pterid. Spermatoph. Icel. 149; Hultén, Atlas 137 (540); id., Amphi-atl. pl. 252 (233); Orlova in Fl. Murm. II, 220; Böcher & al., Grønl. Fl. 224; Karavayev, Konsp. fl. Yak. 79; Polunin, Circump. arct. fl. 129.

L. cordata var. *nephrophylla* (Rydb.) Hult. — Hultén, Fl. Al. III, 488.

Ophrys cordata L., Sp. pl. (1753), 946.

Ill.: Fl. Murm. II, pl. LXXII.

Forest plant, associated mainly with spruce-fir forests of the temperate north and the mountains of more southern districts. Penetrating the Arctic in Northern Europe.

Soviet Arctic. Murman (Pechenga, Rybachiy Peninsula, district of Kola Bay, Iokanga Basin, Terskiy Shore); Northern Malozemelskaya Tundra (west of Dvugolovaya Hill, mossy willow stand; Igoshina, 1930).

Foreign Arctic. Labrador (on the border of the Arctic Region); Greenland (on the west coast north to Disko Island; extreme southeast); Iceland; Arctic Scandinavia.

Outside the Arctic. Northern Europe south to Southern England, Northern Germany, the middle course of the Dnepr, and forested districts of the Lower Volga. Disjunctly in the mountains of Central and parts of Southern Europe, Crimea and the Caucasus. Reported in the northeast of the European part of the USSR from the Ukhta River Basin, the upper reaches of the River Shchugor and Mount Sablya. In Siberia distributed north to 61°15´ on the Ob, to 63°30´ on the Yenisey, and to approximately 62°N in the Lena Basin. The southern range boundary runs close to the southern edge of closed forests, excluding the Altay and Sayans. In the Far East occurring in Northern Preamuria, the northern part of Primorskiy Kray, Kamchatka, Sakhalin and Japan (var. *japonica* Hara). In the forested region of North America represented by the variety var. *nephrophylla* (Rydb.) Hult.

GENUS 3A — Epipactis Sw. — HELLEBORINE

1. ***Epipactic atrorubens*** (Hoffm.) Schultes, Oest. Fl. ed. 2, 1 (1814), 58; Ledebour, Fl. ross. IV, 83; Hultén, Atlas 137 (543); Orlova in Fl. Murm. II, 222.
    E. rubiginosa Crantz. — Nevskiy in Fl. SSSR IV, 626.
    E. atropurpurea Raf. — Krylov, Fl. Zap. Sib. III, 704; Perfilev, Fl. Sev. IV, 157.
    Ill.: Fl. Murm. II, pl. LXXIII.

    Plant of Central and Northern Europe, in Norway penetrating the arctic part of the country. Reported in the Northern USSR from the central part of the Kola Peninsula (the Khibins) and from the Adakskoye Gorge on the River Usa, but apparently nowhere reaching the limits of the tundra zone.

    Records given on map 79 in the "Flora of Murmansk Oblast" that this species has been found on the east coast of the Kola Peninsula north of the Ponoy and in the district of Kola Bay are clearly inconsistent with statements in the text of the same Flora. Apparently the relevant localities refer to *Goodyera repens*.

GENUS 4 — Goodyera R. Br. — RATTLESNAKE PLANTAIN

1. ***Goodyera repens*** (L.) R. Br. in Ait., Hort. Kew. ed. 2, V (1813), 198; Ledebour, Fl. ross. IV, 86; Krylov, Fl. Zap. Sib. III, 712; Perfilev, Fl. Sev. I, 159; Nevskiy in Fl. SSSR IV, 639; Hultén, Atlas 140 (554); Orlova in Fl. Murm. II, 226; Tolmachev, Ist. temnokhv. taygi 68 (map 18); Karavayev, Konsp. fl. Yak. 79.
    G. repens var. *ophioides* Fern. — Hultén, Fl. Al. III, 490.
    Satyrium repens L., Sp. pl. (1753) 945.
    Ill.: Fl. Murm. II, pl. LXXIV.

    One of the characteristic plants of coniferous forests of the temperate north, just penetrating arctic limits in Fennoscandia.

    Soviet Arctic. Murman (Pechenga, Terskiy Shore north of the mouth of the Ponoy).
    Foreign Arctic. Arctic Scandinavia.
    Outside the Arctic. Northern Europe and northern parts of Central Europe; mountains of Central and Southern Europe. Forest zone of the European part of the USSR north to the forest-tundra on the Kola Peninsula and to 65°–65°30´N east of the White Sea. Disjunctly in the mountains of Crimea, the Caucasus and Asia Minor. Forests of Siberia north to 65°, south to the Altay, the Sayans and the mountains of Northern Mongolia. Disjunctly in the Tien Shan and Himalayas. North of its continuous Siberian range it has recently been found on the River Khatanga just below 72°N (most northern locality for the species!) near the polar limit of forests. In the Far East found in Northern Preamuria, on the coasts of the Sea of Japan and of the southern part of the Sea of Okhotsk, on Sakhalin and southern islands in the Kurile Chain, and in Northern Japan. Central part of Kamchatka. Mountains of Central Japan, the Korean Peninsula and Northern China. Represented in the forested region of North America by var. *ophioides* Fern.

◆ GENUS 5   Chamaeorchis L.C. Rich. — ALPINE ORCHID

1. Chamaeorchis alpina (L.) Rich. in Mém. Mus. Par. IV (1818), 57; Nevskiy in Fl. SSSR IV, 644; Hultén, Atlas 134 (529); Orlova in Fl. Murm. II, 228.
Ophrys alpina L., Sp. pl. (1753), 948.

European mountain plant, penetrating arctic limits in Northern Fennoscandia.

Soviet Arctic. Murman (single known locality on the Rybachiy Peninsula).
Foreign Arctic. Arctic Scandinavia.
Outside the Arctic. Mountains of Scandinavia and Central Europe.

◆ GENUS 6   Coeloglossum Hartm. — FROG ORCHID

1. Bracts *relatively short*, on lower flowers equalling, slightly longer than (1¼–1½ times as long as) or shorter than flower, on upper flowers always shorter. Lowest one or two leaves situated *almost at base of stem*. Inflorescence loose, *short* during flowering, *with few flowers* (from 3 to 16, most often 5–10 at least in the Arctic). Lip trilobed, *with middle lobe usually obtusish, sometimes weakly developed*. . . . . . 1. **C. VIRIDE** (L.) HARTM.
−   Bracts *large, on lower flowers 2–3 times as long as flower*, further up 1½ times as long, the uppermost sometimes of equal length. Lower leaves *crowded and inserted distinctly above the base of the stem*, whose lower part only bears membranous sheaths. Stem erect, firm. Inflorescence *with relatively numerous flowers* (from 11–12 to 25–30). Lip either trilobed (often *with narrow acute middle lobe*), *or with middle lobe weakly developed* or entirely absent (i.e., lip bilobed). . . . . .2. **C. BRACTEATUM** (MÜHLB.) PARL.

1. Coeloglossum viride (L.) Hartm., Handb. Scand. Fl. ed. 1 (1820), 329; Ruprecht, Fl. samojed. cisur. 57; Kjellman, Phanerog. as. K. Ber. Str. 555; Scheutz, Pl. jeniss. 165; Krylov, Fl. Zap. Sib. III, 691; Tolmachev, Mat. fl. yevr. arct. ostr. 464; Perfilev, Fl. Sev. I, 156; Nevskiy in Fl. SSSR IV, 647; Leskov, Fl. Malozem. tundry 46; Gröntved, Pterid. Spermatoph. Icel. 197; Hultén, Fl. Al. III, 472; id., Atlas 135 (535); Orlova in Fl. Murm. II, 228; Karavayev, Konsp. fl. Yak. 79.
Peristylus viridis Lindl. — Ledebour, Fl. ross. IV, 72; Trautvetter, Pl. Sib. bor. 535; id., Syll. pl. Sib. bor.-or. 55; Kurtz, Fl. Tschuktsch. 475.
Habenaria viridis R. Br. — Ostenfeld, Fl. arct. 1, 38.
Satyrium viride L., Sp. pl. (1753), 914; Polunin, Circump. arct. fl. 129.
Peristylus islandicus Lindl., Orchid. (1835), 297.
Coeloglossum islandicum — Komarov, Fl. Kamch. I, 313.
Ill.: Fl. Murm. II, pl. LXXV.

Rather common plant but usually occurring in small quantity, found on protected grassy slopes well covered by snow in winter. Also encountered among willow shrubbery. Avoiding surplus moisture. In the Lena Valley also reported from level portions of dwarf shrub-moss tundra (on sandy substrate). Occurring only in moderately northern parts of the Arctic Region in Northern Europe and NE Asia. Apparently not penetrating arctic limits at all over the wide stretch from the eastern slopes of the Urals to the lower reaches of the Lena.

Soviet Arctic. Murman (more or less ubiquitous); Kanin, Timanskaya and Malozemelskaya Tundras (common); Kolguyev Island; greater part of

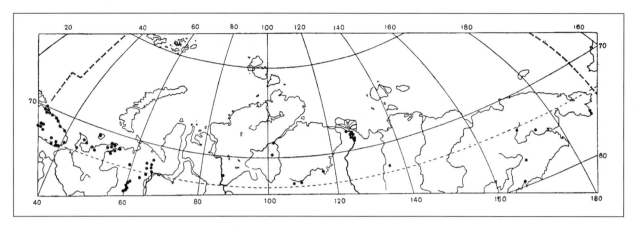

Map IV–31 Distribution of *Coeloglossum viride* (L.) Hartm. (Correction: Localities in moderately northern districts of the European part of the USSR are not shown).

Bolshezemelskaya Tundra (except for the maritime zone proper); Polar Ural. Absent from the arctic part of the West Siberian Lowland and on the Yenisey reported as a rarity at Plakhino settlement (about 68°N) within the forest zone; found in the lower reaches of the Kheta River in the furthest north of the forest zone (almost 72°N), but only in the southern part of the Olenek Basin. Lower reaches of the Lena (many localities, locally frequent; reaching 72°10–15´N; common on Tit-Ary Island just below 72°N); the southeast of the Chukotka Peninsula (Konyyam Bay, Arakamchechen Island, Chaplino); Anadyr Basin (occasional); Penzhina Basin; Bay of Korf. (Map IV–31).

Foreign Arctic. Extreme west of Alaska (Seward Peninsula); Iceland; Arctic Scandinavia.

Outside the Arctic. Northern and Central Europe; mountains of Southern Europe (as far south as Southern Spain, Italy and Greece); entire forest zone of the European part of the USSR; mountains of Crimea, the Caucasus and adjacent parts of Asia Minor; southern half of West Siberia (from the forest-steppe to 60–61°N); the Altay and the mountains of Central Asia. Frequent in the southern part of East Siberia (generally not ranging northwards beyond 60°N, southwards as far as Northern Mongolia); extremely rare north of the 60th parallel. In the east reaching the Okhotsk Coast (Ayan) and Central Preamuria. On Kamchatka found in relatively northern districts, only in the mountains. In North America only in Western Alaska (see above) and Newfoundland.

Coeloglossum viride is not uniformly distributed within its range. It is still unclear whether its distribution in extreme NE Asia is continuous or fragmented. Although common in the lower reaches of the Lena and in extreme SE Chukotka, it has been found only at a few sites in the generally rather well investigated Anadyr Basin and has so far not been reported even once from the Kolyma Basin. There are also no reliable records of the presence of this species on the Okhotsk Coast north of Ayan. For the middle zone of Yakutia there is so far only a single record (Namskiy Ulus on the middle course of the Lena). It cannot be excluded that the localities for this species in the East Siberian Subarctic are separated from the main range. Also noteworthy is the restricted distribution of *C. viride* in the West Siberian Lowland, where it has not been reported for the whole expanse north of 60–61°.

A low-growing northern form of *C. viride* perhaps merits treatment as a distinct variety, *C. viride* var. *islandicum* M. Schulze (= *Peristylus islandicus* Lindl.).

2. ***Coeloglossum bracteatum*** (Mühlb.) Parl., Fl. Ital. (1858) 409; Schlechter in Fedde, Rep. sp. nov. XVI (1920), 374.

C. viride β *bracteatum* Rchb. f. — Komarov, Fl. Kamch. I, 312.

C. viride ssp. *bracteatum* — Hultén, Fl. Al. III, 472.

C. viride var. *bracteatum* Richt. — Hultén, Fl. Kamtch. 1, 260; id., Fl. Aleut. 137; Vorobev, Mat. fl. Kuril. 30.

Peristylus bracteatus Ledb., Fl. ross. IV, 71.

Orchis bracteata Mühlb. in Willd., Sp. pl. 4 (1805), 34.

Soviet Arctic. Southeast of Chukotka Peninsula (vicinity of Chaplino Hot Springs; among mounds in valley, in area heated by warm waters; V. Gavrilyuk, 26 VI 1957).

Foreign Arctic. Not reported.

Outside the Arctic. Kamchatka (common), Kurile Islands, Sakhalin (rare), Northern Japan, Northern Sikhote-Alin, Central Preamuria (rare); Central China; Aleutian Islands; forest region of North America from the south coast of Alaska to the southern fringe of Hudson Bay and Nova Scotia, south to the states of Washington, Colorado, Ohio, Pennsylvania and North Carolina.

The single specimen of this species collected in the Arctic belongs to the variety described by Fernald as var. *interjecta* Fern., whose essential character is complete nondevelopment of the middle lobe of the perianth lip. With respect to the general appearance of the plant, the position of the leaves, the robustness of the stem, the dense many-flowered inflorescence and large bracts, the Chukotkan *C. bracteatum differs sharply* from *C. viride* occurring in the same district, of which a good series of specimens was also collected by V.A. Gavrilyuk. The latter plants are very similar to specimens of *C. viride* from the Anadyr and Penzhina Basins, as well as to the alpine Kamchatka plants which V.L. Komarov determined as *Coeloglossum islandicum* with emphasis on the striking differences between them and *C. bracteatum* (= *C. viride* β *bracteatum*).

Despite the considerable variation both in the European-Siberian *C. viride* and in the Far Eastern-North American *C. bracteatum*, their distinction as species on the basis of a combination of characters does not present substantial difficulties. We will fail in this only so long as we try to rely on some striking diagnostic character (for example, the size of the bracts) taken in isolation. S.A. Nevskiy's (Fl. SSSR IV, page 648) record of typical *C. viride* growing "next" to *C. bracteatum*, though true per se, does not refute their real existence as species. On the contrary, the very fact that the two forms can grow in one and the same district (as is the case, for instance, in the southeast of the Chukotka Peninsula) and do not lose their morphological distinctness in such cases supports their recognition as separate species. We note, moreover, that the ranges of *C. viride* and *C. bracteatum* are in general not only distinct but partly even separate from one another, as occurs (for example) in Alaska (see Hultén, l. c., maps 378a and 378b). The areas where the ranges of the two species overlap (SE Chukotka Peninsula, Central Kamchatka, part of Central Preamuria) are relatively insignificant.

GENUS 7 — Leucorchis E. Mey. — SMALL WHITE ORCHID

1. **Leucorchis albida** (L.) E. Mey. ex Schur, Enum. Pl. Trans. (1866), 645; Nevskiy in Fl. SSSR IV, 649; Grøntved, Pterid. Spermatoph. Icel. 199; Hultén, Atlas 134 (530); id., Amphi-atl. pl. 116 (233); Orlova in Fl. Murm. II, 230; Polunin, Circump. arct. fl. 127; Böcher & al., Grønl. Fl. 226.
   Habenaria albida R. Br. — Lange, Consp. fl. groenl. 118; Ostenfeld, Fl. arct. 1, 38; id., Fl. Greenl. 69; M. Porsild, Fl. Disko 64.
   Gymnadenia albida L.C. Rich. — Krylov, Fl. Zap. Sib. III, 695; Perfilev, Fl. Sev. I, 157.
   Peristylus albidus Lindl. — Ledebour, Fl. ross. IV, 73.
   Habenaria straminea Fern. — Devold & Scholander, Fl. pl. SE Greenl. 151; Seidenfaden & Sørensen, Vasc. pl. NE Greenl. 180.
   Satyrium albidum L., Sp. pl. (1753), 944.
   Ill.: Fl. Murm. II, pl. LXXVI.

   North Atlantic (predominantly European) plant, occurring in a series of districts of the Atlantic Arctic.

   Soviet Arctic. Murman (Rybachiy Peninsula); Prepolar Ural (Khoyl Pass in the upper reaches of the Voykar).

   Foreign Arctic. Greenland north to 69°30′ on the west coast and to 66° on the east coast; Iceland; Arctic Scandinavia.

   Outside the Arctic. Scandinavia (mainly in the mountains), the north of Central and Atlantic Europe, mountains of Central and parts of Southern Europe. Central part of Kola Peninsula; the northeast of the European part of the USSR (within Arkhangelsk Oblast and the northern part of the Komi ASSR). Records for Central Russia (the former Vladimir and Penza Provinces) require checking. In North America known from the west coast of Labrador (in the forest region) and from Newfoundland.

GENUS 8 — Lysiella Rydb. — BLUNT-LEAVED ORCHID

1. **Lysiella obtusata** (Pursh) Britt. & Rydb. ssp. **oligantha** (Turcz.) comb. n.
   Platanthera obtusata ssp. *oligantha* (Turcz.) Hultén, Fl. Al. III, 481.
   Platanthera oligantha Turcz., Fl. baic.-dahur. 11, 2 (1856), 182; Hultén, Atlas 136 (538).
   Lysiella oligantha (Turcz.) Nevski in Fl. SSSR IV, 663; Vasilev, Fl. Komandr. ostr. 79; Karavayev, Konsp. fl. Yak. 79.
   Platanthera obtusata Lindl. — Scheutz, Pl. jeniss. 165.
   Habenaria obtusata Rich. — Ostenfeld, Fl. arct. 1, 38 (pro parte); Polunin, Circump. arct. fl. 128.
   Platanthera parvula Schlecht. — Hultén, Fl. Kamtch. 1, 265.

   Rare forest plant native mainly to East Siberia and the Far East. Occurring both in coniferous and in deciduous valley forests (formed by willow, poplar or *Chosenia*). Penetrating arctic limits in extreme NE Asia and the mountains of Scandinavia.

   Soviet Arctic. Near the mouth of the Anadyr River (at Anadyr settlement, formerly Novo-Mariinskoye; Tyulina, 1931; Vasilev, 1932). Near the arctic boundary at Pokhvalnoye settlement on the lower course of the River Indigirka (69°30′N; Sheludyakova, 1936).

   Foreign Arctic. Mountains of Scandinavia between 68° and 70°N (ssp. *oligantha*!); west coast of Alaska (*L. obtusata* s. str.).

   Outside the Arctic. Subspecies *oligantha*: Eastern Sayan (Nukhu-Daban); Southern and Western Prebaikalia (River Goryachkinskaya at its former Balagansk outlet;

Slyudyanka village, Goremyka Valley and Cape Onguren on the shores of Baykal; Arshak village and the vicinity of Kyakhta on the Mongolian frontier); Aldan Valley between Tommot and the mouth of the Uchur River; Ayan; Kamchatka (rare); Bering Island. Completely disjunct from its main range near the entry of the Lower Tunguska into the Yenisey (Arnell, 1876).

Lysiella obtusata s. str.: forested region of North America from Alaska to Newfoundland, south to the northern states of the USA.

GENUS 9 — Gymnadenia R. Br. — REIN ORCHID

1. ***Gymnadenia conopsea*** (L.) R. Br. in Ait., Hort. Kew. ed. 2, V (1813), 191; Ledebour, Fl. ross. IV, 64; Krylov, Fl. Zap. Sib. III, 693; Perfilev, Fl. Sev. I, 157; Nevskiy in Fl. SSSR IV, 668; Hultén, Atlas 135 (532); Orlova in Fl. Murm. II, 234; Karavayev, Konsp. fl. Yak. 79.

Orchis conopsea L., Sp. pl. (1753), 942.

Ill.: Fl. Murm. II, pl. LXXVII.

Plant of sparse forests, forest glades and moist meadows of the temperate zone of Eurasia. Just penetrating arctic limits in Northern Europe.

Soviet Arctic. Murman (from the Norwegian frontier to Teriberka); Southern Kanin (valley of Malaya Nes River; C. Grigorev, 1914).

Foreign Arctic. Arctic Scandinavia.

Outside the Arctic. Greater part of Western and Central Europe; mountains of Southern Europe; entire forested part of the European territory of the USSR (north to the Kola Peninsula inclusively; in Arkhangelsk Oblast only exceptionally reaching the forest-tundra). Mountains of Crimea, the Caucasus and Northern Iran. In Siberia from the southern fringe approximately to Surgut, Yeniseysk and Yakutsk. Mountains of Central Kazakhstan and Northern Mongolia. Preamuria, Primorskiy Kray, Sakhalin, Northern China and Northern Japan.

GENUS 10 — Dactylorchis (Klinge) Vermln. — SPOTTED ORCHID

1. Lower leaves *oblong-elliptical, flat, tapered basally*. Flowers *light lilac*, their lip with fine darker pattern on light background. . . . .
. . . . 1. **D. MACULATA** (L.) VERMLN.
– Leaves *linear-lanceolate, somewhat folded lengthwise*, straight or falcate-recurved. Flowers *violet-purple*, their lip with darker violet pattern. . . . .
. . . . .2. **D. TRAUNSTEINERI** (SAUT.) VERMLN.

1. ***Dactylorchis maculata*** (L.) Vermln., Studies on Dactylorchids (1947), 68; Löve & Löve, Consp. Icel. Fl. 152.

Orchis maculata L., Sp. pl. (1753), 942; Ruprecht, Fl. samojed. cisur. 57; Ledebour, Fl. ross. IV, 58; Schrenk, Enum. pl. 528; Krylov, Fl. Zap. Sib. III, 688; Perfilev, Fl. Sev. I, 155; Nevskiy in Fl. SSSR IV, 703; Gröntved, Pterid. Spermatoph. Icel. 201; Hultén, Atlas 131 (517, 517a); Orlova in Fl. Murm. II, 233.

Ill.: Fl. Murm. II, pl. LXXVIII.

Plant of mossy forests of the temperate north, penetrating the Arctic in the extreme north of Europe. Growing in the tundra zone among shrubs or in islets of birch krummholz, on moss-covered substrate. Not uncommon, but usually occurring in small quantity.

Soviet Arctic. Murman (more or less ubiquitous). Reliable localities from the tundra zone east of the White Sea are unknown to me. Perfilev reports *Orchis maculata* for the western part of the Bolshezemelskaya Tundra, apparently on the basis of collections from the upper reaches of the River Pay-Yakha (Chernaya) made by V.N. Andreyev and Z. Savkina in the autumn of 1930 and provisionally (in the field) identified as *Orchis maculata*. These plants in fact represent *Coeloglossum viride*. The reliable localities for *D. maculata* closest to tundra in the northeast of the European part of the USSR are the mouth of the Mezen (Ruprecht) and the River Peza (Schrenk).

Foreign Arctic. Iceland, Arctic Scandinavia.

Outside the Arctic. Fennoscandia; the north of West and Central Europe, and the north of the forest zone of the European part of the USSR including the entire Kola Peninsula, the forested north of Arkhangelsk Oblast, and the Pechoran part of the Ural Region (north to the Rivers Synya and Kosyu). In Siberia recorded for the West Siberian Lowland north to the upper reaches of the River Pur (63°30´N), and on the Yenisey north to the mouth of the Podkammenaya Tunguska (61°30´N). Nevskiy (Fl. SSSR IV, pages 704–707) refers all records of the finding of *Orchis maculata* in Siberia and in the east of the European part of the USSR to *O. Fuchsii* Druce (or, as seemed to him more probable, to a closely related undescribed species). However, typical *D. maculata* is apparently distributed further to the east than is stated in the "Flora of the USSR." At any rate, a series of plants from the Ural Region belong to typical *D. maculata*.

2. ***Dactylorchis Traunsteineri*** (Saut.) Vermln., Studies on Dactylorchids (1947), 66.
Orchis Traunsteineri Saut. ex Rchb. f., Fl. Germ. Exs. (1830), 140; Krylov, Fl. Zap. Sib. III, 687; Perfilev, Fl. Sev. 1, 155; Nevskiy in Fl. SSSR IV, 711; Hultén, Atlas 133 (527); Orlova in Fl. Murm. II, 237.
O. curvifolia Nym. — Ledebour, Fl. ross. IV, 55.

In peat bogs and marshy meadows in the forest zone, also in the south of the tundra zone in Northern Europe.

Soviet Arctic. Murman (isolated localities in the district of Polyarnyy village on Kola Bay and in East Murman east of Teriberka); Kanin (recorded by Perfilev; I have seen no specimens).

Foreign Arctic. Arctic Scandinavia (rare, only within Norway).

Outside the Arctic. Northern and Central Europe, the north of the European part of the USSR, and the south of the taiga zone of West Siberia.

Records of the finding of *D.* (*Orchis*) *incarnata* (L.) Vermln. (or *O. sambucina* L.) in the Soviet Arctic are obviously erroneous. Data on the distribution of *D. sambucina* in Murman inserted on map 83 in the "Flora of Murmansk Oblast" (volume II) in fact refer to *D. maculata*.

APPENDIX I

Summary of Data on the Geographical Distribution of Vascular Plants of the Soviet Arctic

SUPPLEMENTARY TO THE information provided in the text of this work, we present below a tabular summary of data on the geographical distribution of the species of plants treated in the present volume of this Flora (see Tables 3–4). This summary is given on the basis of the preliminary subdivision of the Soviet Arctic into districts published briefly in our article in the "Botanicheskiy Zhurnal" (1956) (see Map I–1). Along with data on the distribution of plants in these districts, we also include brief information on their distribution in foreign arctic districts. For the convenience of readers we repeat here the characterization of the districts adopted, supplementing it with corresponding characterization of districts of the Foreign Arctic.

1. *Murman District:* from the Norwegian frontier to the neck of the White Sea.

2. *Kanin-Pechora District:* Kanin Peninsula, Malozemelskaya and Timanskaya Tundras, lower reaches of River Pechora, Bolshezemelskaya Tundra (except Ural foothills and Pay-Khoy), Kolguyev Island.

3. *Polar Ural District:* Polar Ural, upper reaches of River Usa, district south of Baydaratskaya Bay.

4. *Yugorskiy District:* Pay-Khoy, Vaygach Island.

5. *Novaya Zemlya.*

6. *Franz Josef Land.*

7. *Ob-Tazovskiy District:* Yamal Peninsula, tundra from the mouth of the Ob to the Taz, Gydanskaya Tundra (except left bank of Yenisey); islands north of Yamal and the Gydanskaya Tundra.

8. *Yenisey District:* lower reaches of Yenisey with the strip watered by tributaries of the Yenisey; shores of the very narrow part of Yenisey Bay; Sibiryakov Island.

9. *Taymyr District:* Taymyr Peninsula from the Yenisey-Pyasina watershed to the Khatanga River; Severnaya Zemlya and other islands north of Taymyr.

10. *Anabar-Olenek District:* from the Khatanga River to the lower reaches of the Olenek inclusively; Begichev and Preobrazheniye Islands.

11. *Lena District:* delta and lower course of Lena River (west to the watershed with the Olenek, east to Buorkhaya Bay).

12. *Yana-Kolyma District:* from Buorkhaya Bay to the eastern boundary of the Yakut ASSR; New Siberian Islands, De Long Islands.

13. *North Chukotka District:* Chukotka north of the Anadyr watershed, east to Kolyuchin Bay; Ayon Island, Wrangel Island.

14. *Beringian Chukotka District:* Chukotka east of Kolyuchin Bay and Krest Bay, with adjacent small islands.

15. *Anadyr District:* Anadyr River Basin and neighbouring areas.

16. *Koryak District:* Koryak Range and adjacent coast; district of Penzhina River, Bay of Penzhina and Bay of Gizhiga.

17. *Arctic part of Alaska.*

18. *Arctic mainland of Canada west of Hudson Bay.*

19. *Arctic part of Labrador Peninsula* (including Arctic Quebec).

20. *Canadian Arctic Archipelago.*

21. *North-West Greenland:* districts of Greenland north of 80° on the east coast and 74° on the west coast.

22. *South-West Greenland:* west coast of Greenland from 74°N to its southern extremity.

23. *East Greenland:* east coast of Greenland from 80°N to its southern extremity.

24. *Iceland* (with Jan Mayen).

25. *Svalbard* (Spitsbergen, Bear Island).

26. *Arctic Scandinavia.*

TABLE 3

The Distribution of Vascular Plants of the Soviet Arctic

Cyperaceae

KEY

+ indicates the existence of reliable records of the presence of the species in the relevant district;

• indicates the absence of such records.

	SOVIET ARCTIC																FOREIGN ARCTIC								
	Murman	Kanin-Pechora	Polar Ural	Yugorskiy	Novaya Zemlya	Franz Josef Land	Ob-Tazovskiy	Yenisey	Taymyr	Anabar-Olenek	Lena	Yana-Kolyma	North Chukotka	Beringian Chukotka	Anadyr	Koryak	Arctic Alaska	NW Canada	Labrador	Canadian Archipelago	NW Greenland	SW Greenland	East Greenland	Iceland	Svalbard	Arctic Scandinavia
XIV. CYPERACEAE
Eriophorum angustifolium	+	+	+	+	+	•	+	+	+	+	+	+	+	+	+	+	+	+	+	+	+	+	+	+	+	+
Eriophorum Komarovii	•	•	•	•	•	•	•	•	•	•	•	•	•	+	+	?	•	•	•	•	•	•	•	•	•	•
Eriophorum gracile	+	•	•	•	•	•	+	•	•	•	?	?	•	•	+	•	•	•	•	•	•	•	•	•	•	+
Eriophorum russeolum	+	+	+	+	+	•	+	+	+	+	+	•	+	+	+	•	+	+	+	+	+	+	+	•	+	+
Eriophorum medium	+	•	+	•	•	•	+	+	+	+	+	+	+	+	+	•	+	+	+	+	+	•	+	•	+	+
Eriophorum Scheuchzeri	+	+	+	+	+	+	+	+	+	+	+	+	+	+	+	+	+	+	+	+	+	+	+	+	+	+
Eriophorum callitrix	•	•	•	?	•	•	+	+	•	•	+	+	+	+	+	•	+	+	+	+	+	•	+	•	+	•
Eriophorum brachyantherum	+	+	+	+	+	•	+	+	+	•	+	+	•	+	+	•	+	+	•	•	•	•	•	•	•	•
Eriophorum vaginatum	+	+	+	+	+	•	+	+	+	+	+	+	+	+	+	•	+	+	+	•	•	•	+	•	•	+
Trichophorum caespitosum	+	+	•	•	•	•	•	•	•	•	•	•	+	+	+	•	•	+	+	•	•	+	+	+	+	+
Trichophorum alpinum	+	+	•	•	•	•	•	+	•	•	•	•	•	•	•	•	•	•	•	•	•	•	•	•	•	+
Scirpus Maximowiczii	•	•	•	•	•	•	•	•	•	•	•	+	•	+	•	•	•	•	•	•	•	•	•	•	•	•
Eleocharis acicularis	+	•	•	•	•	•	+	•	•	•	+	•	•	•	•	•	+	+	+	•	+	•	+	•	+	+
Eleocharis quinqueflora	+	•	•	•	•	•	•	•	•	•	•	•	•	•	•	•	•	•	•	•	•	•	•	•	•	+
Eleocharis palustris	+	+	•	•	•	•	•	•	•	•	•	•	•	•	•	•	•	•	•	•	•	•	+	•	+	•
Eleocharis intersita	•	•	•	•	•	•	•	•	•	•	+	•	•	•	+	+	•	•	•	•	•	•	•	•	•	•
Eleocharis uniglumis	+	+	•	•	•	•	•	•	•	•	•	•	•	•	+	•	•	•	•	•	•	•	+	•	+	•
Kobresia sibirica	•	•	+	•	•	•	•	+	+	+	•	•	+	?	•	+	+	+	•	+	•	•	•	•	•	•
Kobresia filifolia	•	•	•	•	•	•	•	•	•	•	•	•	•	+	•	•	•	•	•	•	•	•	•	•	•	•
Kobresia Bellardii	+	•	•	•	•	•	•	+	+	+	•	+	+	+	+	•	+	+	+	+	+	+	+	+	•	+
Kobresia simpliciuscula	•	•	•	•	•	•	•	+	•	•	•	•	+	•	•	•	+	+	+	+	•	•	+	•	+	+
Carex obtusata	•	+	•	•	•	•	+	•	+	•	+	•	•	•	•	•	+	+	•	•	•	•	•	•	•	•
Carex rupestris	+	+	+	+	+	•	•	+	+	+	+	+	+	+	+	•	+	+	+	+	+	+	+	+	+	+
Carex micropoda	•	•	•	•	•	•	•	•	•	•	•	•	•	+	?	•	•	•	•	•	•	•	•	•	•	•
Carex pauciflora	+	•	•	•	•	•	•	•	•	•	•	•	•	•	•	•	•	•	•	•	•	•	•	•	•	+
Carex microglochin	+	•	•	•	•	•	•	•	•	•	•	•	•	•	•	•	+	+	+	+	•	+	+	+	•	+
Carex scirpoidea	•	•	•	•	•	•	•	•	•	•	•	•	•	+	+	•	+	+	+	+	+	+	+	•	•	•
Carex anthoxanthea	•	•	•	•	•	•	•	•	•	•	•	•	•	•	+	•	•	•	•	•	•	•	•	•	•	•
Carex Hepburnii	•	•	•	•	•	•	•	•	•	•	•	•	+	•	•	•	+	+	+	+	+	•	+	•	+	•
Carex capitata	+	•	+	•	•	•	•	+	+	+	+	•	+	+	+	•	•	•	+	•	•	+	+	+	•	+
Carex arctogena	+	•	•	•	•	•	•	•	•	•	•	•	•	•	•	•	•	•	•	+	•	+	+	•	•	+
Carex diandra	+	+	•	•	•	•	+	•	•	•	•	•	•	•	•	•	+	+	•	•	•	•	•	+	•	+
Carex pallida	•	•	•	•	•	•	•	•	•	•	+	•	•	+	•	•	•	•	•	•	•	•	•	•	•	•
Carex chordorrhiza	+	+	+	•	•	•	+	+	+	+	+	+	+	+	+	•	+	+	+	•	•	•	+	•	•	+
Carex duriuscula	•	•	•	•	•	•	•	•	•	•	•	+	•	•	•	•	•	•	•	•	•	•	•	•	•	•
Carex maritima	+	+	•	+	+	•	+	+	+	+	•	+	+	•	+	•	+	+	+	+	+	+	+	•	•	+
Carex leporina	+	•	•	•	•	•	•	•	•	•	•	•	•	•	•	•	•	•	•	•	•	•	•	•	•	•
Carex dioica	+	+	•	•	•	•	•	•	•	•	•	•	•	•	•	•	•	•	•	•	•	•	•	+	•	+

TABLE 3

	SOVIET ARCTIC																FOREIGN ARCTIC								
	Murman	Kanin-Pechora	Polar Ural	Yugorskiy	Novaya Zemlya	Franz Josef Land	Ob-Tazovskiy	Yenisey	Taymyr	Anabar-Olenek	Lena	Yana-Kolyma	North Chukotka	Beringian Chukotka	Anadyr	Koryak	Arctic Alaska	NW Canada	Labrador	Canadian Archipelago	NW Greenland	SW Greenland	East Greenland	Iceland	Svalbard	Arctic Scandinavia
Carex gynocrates	•	•	•	•	•	•	•	•	•	•	+	•	•	+	+	+	+	+	•	+	•	+	•	•	•	•
Carex Redowskiana	•	+	+	+	•	•	+	+	+	+	+	•	•	•	•	•	•	•	•	•	•	•	•	•	•	+
Carex parallela	+	+	•	?	+	•	•	•	•	•	•	•	•	•	•	•	•	•	•	•	•	•	+	•	+	+
Carex Kreczetoviczii	•	•	•	•	•	•	•	•	•	•	•	•	+	•	•	•	•	•	•	•	•	•	•	•	•	•
Carex brunnescens	+	+	+	+	•	•	+	+	•	•	•	•	•	•	•	•	+	+	•	•	•	+	+	+	•	+
Carex Mackenziei	+	+	•	•	•	•	•	•	•	•	•	+	+	•	+	•	+	+	•	•	•	+	+	+	•	+
Carex canescens	+	+	+	•	•	•	+	+	•	•	•	•	+	•	•	•	+	+	•	•	•	+	+	+	•	+
Carex lapponica	+	+	+	•	•	•	+	•	•	•	•	•	•	+	+	•	•	•	•	•	•	•	•	•	•	+
Carex bonanzensis	•	•	•	•	•	•	•	•	•	+	•	•	+	+	•	•	•	•	•	•	•	•	•	•	•	•
Carex glareosa	+	+	•	+	+	•	•	•	•	+	+	•	+	+	+	+	+	+	+	+	•	+	+	+	+	+
Carex ursina	•	•	•	+	+	+	•	•	•	+	+	+	+	+	+	•	+	+	•	+	•	•	+	•	+	•
Carex loliacea	+	•	•	•	•	•	•	+	•	•	•	•	•	•	+	•	+	•	•	•	•	•	•	•	•	+
Carex tenuiflora	+	+	+	•	•	•	+	+	•	•	+	•	+	+	+	•	+	•	•	•	•	•	•	•	•	+
Carex heleonastes	+	+	•	•	•	•	•	•	•	•	•	•	•	•	•	•	•	•	•	•	•	•	•	+	•	+
Carex amblyorhyncha	•	•	+	•	•	•	+	+	+	+	+	•	•	•	•	•	+	+	+	+	•	+	+	•	+	•
Carex tripartita	+	+	+	+	+	•	+	+	+	+	+	+	+	+	+	+	+	+	+	+	•	+	+	+	+	+
Carex pribylovensis	•	•	•	•	•	•	•	•	•	•	•	•	•	+	•	•	•	•	•	•	•	•	•	•	•	•
Carex disperma	+	•	•	•	•	•	+	•	•	•	•	•	•	•	•	•	+	+	•	•	•	•	•	•	•	•
Carex acuta	+	+	+	•	•	•	+	•	•	•	•	•	•	•	•	•	•	•	•	•	•	•	•	•	•	+
Carex nigra	+	+	•	•	•	•	•	•	•	•	•	•	•	•	•	•	•	•	•	•	•	+	+	?	•	+
Carex wiluica	+	+	+	•	•	•	+	+	•	•	•	•	•	•	•	•	•	•	•	•	•	•	•	•	•	?
Carex appendiculata	•	•	•	•	•	•	+	+	+	•	•	+	+	•	+	•	•	•	•	•	•	•	•	•	•	•
Carex caespitosa	+	+	+	•	•	•	+	+	•	•	•	•	•	•	•	•	•	•	•	•	•	•	•	•	•	+
Carex Schmidtii	•	•	•	•	•	•	•	•	•	•	•	•	•	+	+	•	•	•	•	•	•	•	•	•	•	•
Carex aquatilis	+	+	+	+	•	•	+	+	+	•	•	•	•	•	•	•	•	•	•	•	•	•	+	•	•	+
Carex stans	+	+	+	+	+	•	+	+	+	+	+	+	+	+	+	+	+	+	+	+	+	+	+	•	•	•
Carex Bigelowii	+	•	•	•	•	•	•	•	•	•	•	•	•	•	•	•	•	+	+	+	•	+	+	+	+	+
Carex ensifolia ssp. arctisibirica	?	+	+	+	+	•	+	+	+	+	+	•	•	•	•	•	•	•	•	•	•	•	+	•	•	+
Carex rigidioides	•	•	•	•	•	•	•	•	•	+	+	•	•	•	•	•	•	•	•	•	•	•	•	•	•	•
Carex lugens	•	•	•	•	•	•	•	•	•	•	+	+	+	+	+	•	+	•	•	•	•	•	•	•	•	•
Carex Soczavaeana	•	•	•	•	•	•	•	•	•	•	•	•	•	+	•	+	+	•	•	•	•	•	•	•	•	•
Carex eleusinoides	•	•	•	•	•	•	•	•	•	+	+	+	•	+	+	•	+	•	•	•	•	•	•	•	•	•
Carex bicolor	•	+	+	•	•	•	•	•	•	•	•	+	•	+	+	+	+	+	•	+	+	?	•	+
Carex paleacea	+	•	•	•	•	•	•	•	•	•	•	•	•	•	•	•	•	•	•	•	•	•	•	•	•	+
Carex cryptocarpa	•	•	•	•	•	•	•	•	•	•	•	•	•	+	+	+	+	•	•	•	•	+	+	•	•	•
Carex recta	+	•	•	•	•	•	•	•	•	•	•	•	•	•	•	•	•	•	•	•	•	+	+	+	•	+
Carex salina	+	•	•	•	•	•	•	•	•	•	•	•	•	•	•	•	•	•	•	•	•	?	?	+	•	+
Carex subspathacea	+	+	•	+	+	•	•	+	•	+	+	+	+	+	•	+	+	+	+	•	+	+	•	?	+
Carex limosa	+	+	+	•	•	•	+	+	•	•	+	•	+	+	+	•	+	+	•	•	•	•	+	•	+

Species	Murman	Kanin-Pechora	Polar Ural	Yugorskiy	Novaya Zemlya	Franz Josef Land	Ob-Tazovskiy	Yenisey	Taymyr	Anabar-Olenek	Lena	Yana-Kolyma	North Chukotka	Beringian Chukotka	Anadyr	Koryak	Arctic Alaska	NW Canada	Labrador	Canadian Archipelago	NW Greenland	SW Greenland	East Greenland	Iceland	Svalbard	Arctic Scandinavia
Carex magellanica	+	+	+	·	·	·	+	·	·	·	·	·	·	·	·	·	+	+	·	·	·	+	+	+	·	+
Carex rariflora	+	+	+	+	+	·	+	+	·	+	+	+	+	+	+	+	+	+	+	+	·	+	+	+	·	+
Carex atrofusca	+	·	+	·	·	·	·	·	+	+	·	·	+	+	+	+	+	+	+	+	+	·	+	+	·	+
Carex misandra	·	·	+	+	+	·	·	+	+	+	+	+	+	+	+	+	+	+	+	+	+	+	+	·	+	+
Carex macrogyna	·	·	·	·	·	·	·	·	+	+	+	·	·	·	·	·	·	·	·	·	·	·	·	·	·	·
Carex Ktausipalii	·	·	·	·	·	·	·	·	·	·	·	·	·	+	·	·	·	·	·	·	·	·	·	·	·	·
Carex atrata	+	·	·	·	·	·	·	·	·	·	·	·	·	·	·	·	·	·	·	·	·	+	+	+	·	+
Carex perfusca	·	·	·	·	·	·	+	·	·	·	·	·	·	·	·	·	·	·	·	·	·	·	·	·	·	·
Carex Buxbaumii	+	·	·	·	·	·	·	·	·	·	·	·	·	·	·	·	·	·	·	·	·	+	+	·	·	+
Carex adelostoma	+	·	·	·	·	·	·	·	·	·	·	·	·	·	·	·	·	·	·	·	·	·	·	+	·	+
Carex angarae	+	+	+	·	·	·	+	+	·	·	+	+	·	+	+	+	+	·	·	·	·	·	·	·	·	?
Carex norvegica	+	+	+	·	·	·	·	·	·	·	+	·	+	+	+	+	·	·	+	·	·	+	+	+	·	+
Carex sabulosa	·	·	·	·	·	·	·	·	·	·	+	·	·	·	·	·	·	·	·	·	·	·	·	·	·	·
Carex Gmelinii	·	·	·	·	·	·	·	·	·	·	·	·	+	+	+	+	·	·	·	·	·	·	·	·	·	·
Carex holostoma	+	·	·	·	·	·	·	+	·	·	·	+	·	+	+	+	+	+	+	·	·	+	·	+	·	+
Carex podocarpa	·	·	·	·	·	·	·	·	·	+	·	+	·	+	+	+	·	·	·	·	·	·	·	·	·	·
Carex Tolmiei	·	·	·	·	·	·	·	·	·	·	·	·	+	+	+	+	+	·	·	·	·	·	·	·	·	·
Carex macrochaeta	·	·	·	·	·	·	·	·	·	·	·	·	·	+	·	+	·	·	·	·	·	·	·	·	·	·
Carex stylosa	·	·	·	·	·	·	·	·	·	·	·	·	·	+	·	·	+	·	·	·	·	·	+	+	·	·
Carex melanocarpa	·	·	+	+	·	·	·	+	+	+	+	+	+	+	+	·	·	·	·	·	·	·	·	·	·	·
Carex ericetorum	+	+	·	·	·	·	·	·	·	·	·	·	·	·	·	·	·	·	·	·	·	·	·	·	·	·
Carex Vanheurckii	·	·	·	·	·	·	·	·	·	·	·	·	·	+	·	·	·	·	·	·	·	·	·	·	·	·
Carex globularis	+	+	+	·	·	·	+	+	+	+	+	·	·	+	+	·	·	·	·	·	·	·	·	·	·	+
Carex sabynensis	·	+	+	·	·	·	·	+	·	·	·	·	·	·	·	·	·	·	·	·	·	·	·	·	·	·
Carex Trautvetteriana	·	·	·	·	·	·	·	·	+	+	·	·	·	·	·	·	·	·	·	·	·	·	·	·	·	·
Carex livida	+	·	·	·	·	·	·	·	·	·	·	·	·	·	·	·	·	+	·	·	·	·	·	·	+	+
Carex vaginata	+	+	+	+	·	·	+	+	+	+	+	+	+	·	+	+	+	+	+	·	·	·	+	+	·	+
Carex falcata	·	·	·	·	·	·	·	·	·	·	·	·	·	+	+	·	·	·	·	·	·	·	·	·	·	·
Carex pediformis	·	·	·	·	·	·	·	·	·	·	+	·	·	+	·	·	·	·	·	·	·	·	·	·	·	·
Carex supina ssp. spaniocarpa	·	·	·	·	·	·	+	·	+	·	·	·	+	·	+	·	+	·	·	·	·	·	+	+	·	·
Carex glacialis	+	+	+	·	·	·	·	+	+	+	·	+	·	+	+	+	+	+	+	·	·	+	+	+	·	+
Carex flava	+	·	·	·	·	·	·	·	·	·	·	·	·	·	·	·	·	·	·	·	·	·	·	·	+	+
Carex Oederi	+	·	·	·	·	·	·	·	·	·	·	·	·	·	·	·	·	·	·	·	·	·	·	·	+	+
Carex Williamsii	·	·	+	·	·	·	+	·	·	+	·	+	+	+	+	+	+	+	·	·	·	·	·	·	·	·
Carex capillaris	+	+	+	·	·	·	+	·	+	+	·	·	·	·	·	·	?	?	?	?	·	·	+	+	+	+
Carex fuscidula	·	+	+	·	·	·	+	+	+	+	+	+	+	+	+	+	?	?	+	·	·	+	·	+	·	·
Carex Krausei	·	·	·	·	·	·	·	+	·	·	+	·	·	·	·	·	·	·	+	·	·	·	+	+	+	·
Carex Ledebouriana	·	+	+	·	·	·	+	+	·	+	·	·	+	+	·	·	·	·	·	·	·	·	·	·	·	·
Carex lasiocarpa	+	+	·	·	·	·	·	·	·	·	·	·	·	·	·	·	·	·	·	·	·	·	·	·	·	+

TABLE 3 — 221

	\multicolumn{16}{c	}{SOVIET ARCTIC}	\multicolumn{10}{c	}{FOREIGN ARCTIC}
	Murman	Kanin-Pechora	Polar Ural	Yugorskiy	Novaya Zemlya	Franz Josef Land	Ob-Tazovskiy	Yenisey	Taymyr	Anabar-Olenek	Lena	Yana-Kolyma	North Chukotka	Beringian Chukotka	Anadyr	Koryak	Arctic Alaska	NW Canada	Labrador	Canadian Archipelago	NW Greenland	SW Greenland	East Greenland	Iceland	Svalbard	Arctic Scandinavia
Carex rhynchophysa	•	+	•	•	•	•	+	•	•	•	•	•	•	•	+	+	•	•	•	•	•	•	•	•	•	+
Carex rostrata	+	+	+	•	•	•	+	+	•	•	+	+	•	•	+	+	+	•	•	•	•	+	+	+	•	+
Carex rotundata	+	+	+	+	+	•	+	+	•	+	+	+	+	+	+	+	+	+	•	•	•	•	•	•	•	+
Carex membranacea	•	•	•	•	•	•	•	•	•	•	•	•	+	+	+	+	+	+	+	+	•	•	•	•	•	•
Carex vesicaria	+	+	•	•	•	•	•	•	•	•	•	•	•	•	•	•	•	•	•	•	•	•	•	•	•	+
Carex vesicata	•	•	•	•	•	•	•	•	•	•	•	•	•	•	+	+	•	•	•	•	•	•	•	•	•	•
Carex sordida	•	•	•	•	•	•	•	•	•	•	•	•	•	•	+	+	•	•	•	•	•	•	•	•	•	•
Carex saxatilis ssp. saxatilis	+	+	•	•	•	•	•	•	•	•	•	•	•	•	•	•	•	+	+	•	•	+	+	+	+	+
Carex saxatilis ssp. laxa	•	•	+	+	+	•	+	+	+	+	+	+	+	+	+	+	+	+	•	+	•	•	•	•	•	•

TABLE 4

The Distribution of Vascular Plants of the Soviet Arctic

Lemnaceae–Orchidaceae

KEY	
+	indicates the existence of reliable records of the presence of the species in the relevant district;
•	indicates the absence of such records.

	\multicolumn{16}{c}{SOVIET ARCTIC}	\multicolumn{10}{c}{FOREIGN ARCTIC}
	Murman	Kanin-Pechora	Polar Ural	Yugorskiy	Novaya Zemlya	Franz Josef Land	Ob-Tazovskiy	Yenisey	Taymyr	Anabar-Olenek	Lena	Yana-Kolyma	North Chukotka	Beringian Chukotka	Anadyr	Koryak	Arctic Alaska	NW Canada	Labrador	Canadian Archipelago	NW Greenland	SW Greenland	East Greenland	Iceland	Svalbard	Arctic Scandinavia
XV. LEMNACEAE																										
Lemna minor	·	+	·	·	·	·	·	·	·	·	·	·	·	·	·	·	·	·	·	·	·	?	·	·	·	·
Lemna trisulca	·	+	·	·	·	·	+	+	·	·	·	·	·	·	·	·	·	·	·	·	·	·	·	·	·	·
XVI. JUNCACEAE																										
Juncus bufonius	+	+	·	·	·	·	+	·	·	·	·	·	·	·	+	·	·	·	·	·	·	+	·	+	·	+
Juncus trifidus	+	+	+	·	·	·	·	+	·	·	·	·	·	·	·	·	·	·	+	·	+	+	+	+	·	+
Juncus compressus	·	·	·	·	·	·	·	·	·	·	·	·	·	·	·	·	·	·	·	·	·	·	·	·	·	·
Juncus Gerardii	+	·	·	·	·	·	·	·	·	·	·	·	·	·	·	·	·	·	·	·	·	·	·	·	·	+
Juncus Gerardii ssp. *atrofuscus*	+	+	·	·	·	·	·	·	·	·	·	·	·	·	·	·	·	·	·	·	·	·	·	·	·	·
Juncus nodulosus	+	+	·	·	·	·	+	+	·	·	·	·	·	·	·	·	·	·	·	·	·	·	+	·	+	+
Juncus articulatus	·	·	·	·	·	·	·	·	+	·	+	·	·	·	·	·	·	·	·	·	·	·	·	+	·	+
Juncus biglumis	+	+	+	+	+	+	+	+	+	+	+	+	+	+	+	+	+	+	+	+	+	+	+	+	+	+
Juncus stygius	+	·	·	·	·	·	·	·	·	·	·	·	·	·	·	·	·	·	·	·	·	·	·	·	·	+
Juncus triglumis	+	+	+	+	+	·	+	+	+	+	+	·	·	·	·	·	·	·	·	·	·	·	·	+	+	+
Juncus triglumis ssp. *albescens*	·	·	·	·	·	·	+	+	·	·	+	·	+	+	+	+	+	+	+	+	+	·	·	·	·	·
Juncus castaneus	+	+	+	+	+	·	+	+	+	+	+	+	·	+	+	+	+	+	+	+	+	+	+	+	+	+
Juncus leucochlamys	·	·	·	·	·	·	+	+	·	+	·	·	·	+	·	·	·	·	·	·	·	·	·	·	·	·
Juncus filiformis	+	+	+	·	·	·	+	+	·	·	·	·	·	·	+	+	·	·	·	·	·	·	·	·	·	+
Juncus brachyspathus	·	·	·	·	·	·	+	·	·	·	+	·	·	·	·	·	·	·	·	·	·	·	·	·	·	·
Juncus beringensis	·	·	·	·	·	·	·	·	·	·	·	·	·	+	·	·	·	·	·	·	·	·	·	·	·	·
Juncus arcticus	+	+	+	·	·	·	+	+	+	+	·	·	·	·	·	·	+	+	+	·	+	+	+	+	·	+
Juncus arcticus ssp. *alaskanus*	·	+	+	·	·	·	+	·	·	·	+	+	·	+	+	·	+	+	·	·	·	·	·	·	·	·
Juncus Haenkei	·	·	·	·	·	·	·	·	·	·	·	·	+	+	+	·	·	·	·	·	·	·	·	·	·	·
Juncus balticus	+	·	·	·	·	·	·	·	·	·	·	·	·	·	·	·	·	·	·	·	·	·	·	+	·	+
Luzula rufescens	·	·	·	·	·	·	·	·	·	·	+	+	·	·	+	+	·	·	·	·	·	·	·	·	·	·
Luzula pilosa	+	+	·	·	·	·	·	·	·	·	·	·	·	·	·	·	·	·	·	·	·	·	·	·	·	+
Luzula parviflora	+	+	+	·	·	·	+	+	+	·	+	·	·	+	+	·	·	·	·	·	·	·	·	·	·	+
Luzula parviflora ssp. *melanocarpa*	·	·	·	·	·	·	·	·	·	·	·	·	·	+	+	+	+	+	·	+	·	·	+	+	·	·
Luzula Wahlenbergii	+	+	+	+	+	·	+	+	+	+	+	+	+	+	+	+	+	+	+	+	·	·	+	·	+	+
Luzula spicata	+	+	+	·	·	·	·	·	·	·	·	·	·	·	·	·	·	+	+	·	+	+	+	+	·	+
Luzula confusa	+	+	+	+	+	+	+	+	+	+	+	+	+	+	+	+	+	+	+	+	+	+	+	+	+	+
Luzula beringensis	·	·	·	·	·	·	·	·	·	·	·	·	·	+	·	·	·	·	·	·	·	·	·	·	·	·
Luzula unalaschkensis	·	·	·	·	·	·	·	·	·	·	·	·	·	+	+	·	·	·	·	·	·	·	·	·	·	·
Luzula unal. ssp. *kamtschadalorum*	·	·	·	·	·	·	·	·	·	·	·	·	·	+	·	·	·	·	·	·	·	·	·	·	·	·
Luzula arcuata	+	+	·	·	+	·	·	·	·	·	·	·	·	·	·	·	·	·	·	·	·	?	?	+	·	+
Luzula nivalis	+	+	+	+	+	+	+	+	+	+	+	+	+	+	+	+	+	+	+	+	+	+	+	+	+	+
Luzula tundricola	·	·	+	·	·	+	·	+	+	+	+	+	+	+	+	·	·	·	·	·	·	·	·	·	·	·
Luzula capitata	·	·	·	·	·	·	·	·	·	·	·	·	·	+	+	+	·	·	·	·	·	·	·	·	·	·

	Soviet Arctic																Foreign Arctic									
	Murman	Kanin-Pechora	Polar Ural	Yugorskiy	Novaya Zemlya	Franz Josef Land	Ob-Tazovskiy	Yenisey	Taymyr	Anabar-Olenek	Lena	Yana-Kolyma	North Chukotka	Beringian Chukotka	Anadyr	Koryak	Arctic Alaska	NW Canada	Labrador	Canadian Archipelago	NW Greenland	SW Greenland	East Greenland	Iceland	Svalbard	Arctic Scandinavia
Luzula multiflora ssp. *frigida*	+	+	+	+	•	•	+	+	•	•	•	•	•	•	•	•	•	•	•	•	•	+	+	+	•	+
Luzula multiflora ssp. *sibirica*	•	•	+	•	•	•	+	+	+	+	+	+	+	•	+	+	•	•	•	•	•	•	•	•	•	•
Luzula multiflora ssp. *kjellmaniana*	•	•	•	•	•	•	•	•	•	+	+	+	+	+	+	+	•	•	•	•	•	•	•	•	•	•
Luzula sudetica	+	•	•	•	•	•	•	•	•	•	•	•	•	•	•	•	•	•	•	•	•	•	•	+	•	+
Luzula oligantha	•	•	•	•	•	•	•	•	•	•	•	•	+	•	+	•	•	•	•	•	•	•	•	•	•	•
Luzula pallescens	+	•	•	•	•	•	+	•	•	•	+	•	•	•	•	•	•	•	•	•	•	•	•	+	•	+
XVII. LILIACEAE																										
Tofieldia pusilla	+	+	+	•	•	•	+	+	+	•	+	+	+	+	•	+	+	+	+	+	•	+	+	+	+	+
Tofieldia coccinea	•	•	+	•	•	•	+	+	+	+	+	+	+	+	+	+	+	+	+	+	+	+	+	•	•	•
Zygadenus sibiricus	•	•	•	•	•	•	•	•	+	•	+	•	•	•	•	•	•	•	•	•	•	•	•	•	•	•
Veratrum oxysepalum	•	•	•	•	•	•	•	+	+	•	+	+	+	+	+	+	+	•	•	•	•	•	•	•	•	•
Veratrum Lobelianum	+	+	+	+	•	•	+	+	+	•	+	+	•	•	•	•	•	•	•	•	•	•	•	•	•	+
Gagea granulosa	•	+	•	•	•	•	•	•	•	•	•	•	•	•	•	•	•	•	•	•	•	•	•	•	•	•
Gagea samojedorum	•	?	•	•	•	•	•	•	•	•	•	•	•	•	•	•	•	•	•	•	•	•	•	•	•	•
Allium strictum	•	•	+	•	•	•	+	+	•	•	+	+	•	•	+	•	•	•	•	•	•	•	•	•	•	•
Allium schoenoprasum	+	+	+	+	•	•	+	+	+	+	+	+	+	+	+	+	•	•	•	•	•	•	•	•	•	+
Lloydia serotina	•	•	+	+	•	•	+	+	+	+	+	+	+	+	+	+	+	+	•	•	•	•	•	•	•	•
Smilacina trifolia	•	•	•	•	•	•	•	•	•	•	•	•	•	+	+	•	•	•	•	•	•	•	•	•	•	•
Maianthemum bifolium	+	+	•	•	•	•	+	•	•	•	•	•	•	•	•	•	•	•	•	•	•	•	•	•	•	+
Paris quadrifolia	+	•	•	•	•	•	•	•	•	•	•	•	•	•	•	•	•	•	•	•	•	•	•	+	•	+
XVIII. IRIDACEAE																										
Iris setosa	•	•	•	•	•	•	•	•	•	•	•	+	•	+	+	+	•	•	•	•	•	•	•	•	•	•
XIX. ORCHIDACEAE																										
Cypripedium calceolus	+	•	•	•	•	•	•	•	•	•	•	•	•	•	•	•	•	•	•	•	•	•	•	•	•	+
Corallorhiza trifida	+	+	+	•	•	•	+	+	+	•	+	+	•	•	+	•	+	•	•	•	•	+	•	+	•	+
Listera cordata	+	+	•	•	•	•	•	•	•	•	•	•	•	•	•	•	•	•	+	•	•	+	+	+	•	+
Epipactis atrorubens	?	•	•	•	•	•	•	•	•	•	•	•	•	•	•	•	•	•	•	•	•	•	•	•	•	+
Goodyera repens	+	+	•	•	•	•	•	•	+	•	•	•	•	•	•	•	•	•	•	•	•	•	•	•	•	+
Chamaeorchis alpina	+	•	•	•	•	•	•	•	•	•	•	•	•	•	•	•	•	•	•	•	•	•	•	•	•	+
Coeloglossum viride	+	+	+	•	•	•	+	+	•	•	+	•	+	+	+	•	•	•	•	•	•	•	•	+	•	+
Coeloglossum bracteatum	•	•	•	•	•	•	•	•	•	•	•	•	•	•	+	•	•	•	•	•	•	•	•	•	•	•
Leucorchis albida	+	•	+	•	•	•	•	•	•	•	•	•	•	•	•	•	•	•	•	•	•	+	+	•	•	+
Lysiella obtusata	•	•	•	•	•	•	•	•	•	•	•	+	•	+	+	•	•	•	•	•	•	•	•	•	•	+
Gymnadenia conopsea	+	+	•	•	•	•	•	•	•	•	•	•	•	•	•	•	•	•	•	•	•	•	•	•	•	+
Dactylorchis maculata	+	•	•	•	•	•	•	•	•	•	•	•	•	•	•	•	•	•	•	•	•	•	•	+	•	+
Dactylorchis Traunsteineri	+	+	•	•	•	•	•	•	•	•	•	•	•	•	•	•	•	•	•	•	•	•	•	•	•	+

Index of Plant Names

Allium L. 182, 193
   Ledebourianum Roem. et Schult. 195
   lineare L. 193
      var. *strictum* 193
   Maximowiczii Rgl. 195
   schoenoprasum L. 193, 194, **195**
      var. *laurentianum* 195
      var. *sibiricum* (L.) Hartm. 194
      β *sibiricum* 194
   sibiricum L. 194
   strictum Schrad. 193, **194**
Anthericum serotinum L. 198
Anticlea sibirica Kunth 186

Bulbocodium serotinum L. 198

Carex L. 2, 30
   aa Kom. 80
   accrescens Ohwi 66
   acuta L. 44, 83
      var. *appendiculata* Trautv. et C.A. Mey. 86
   adelostoma V. Krecz 41, 111
   akanensis Franch. 136
   algida Turcz. 122
   alpina Sw. 112
      var. β Trev. 114
      β *inferalpina* Wahlb. 111
   amblyolepis Trautv. 119
   amblyorhyncha V. Krecz. 37, **81**
   ampullacea Good. 131
   amurensis Kük. 136
   angarae Steud. 42, **111**
   anthoxanthea Presl. 32, 34, 62
      var. *leiocarpa* (C.A. Mey.) Kük. 62
   appendiculata (Trautv. et C.A. Mey.) Kük. 46, **86**
   approximata auct. 118
   aquatilis Wahlb. 45, 88
      var. β et γ Trev. 89
   arakamensis C.B. Clarke 116
   arctogena H. Smith 33, 64
   aterrima Hoppe 110
   atrata L. 41, 109
      var. *aterrima* (Hoppe) Hartm. 110
      ssp. *perfusca* (V. Krecz.) T. Koyama 110
   atrofusca Schkuhr 50, 51, **106**
   behringensis C.B. Clarke 115
   Bellardii All. 28
   beringiana Cham. 117
   bicolor Bell. 39, 98, **99**
   Bigelowii Torr. 46, 92
   Boecheriana A. et D. Löve et Raymond 129
   bonanzensis Britt. 38, 76
   brachylepis Turcz. 111
   brachyphylla Turcz. 118
   Brenneri Christ 120
   brunnescens (Pers.) Poir. 39, 72, **74**
   bucculenta V. Krecz. 67
   bullata β *laevirostris* Blytt 130
   burejana Meinsh. 136
   Buxbaumii Wahlb. 41, 110
      var. *alpicola* Hartm. 111
   caespitosa L. 48, **87**
      β Trev. 98
      ssp. *wiluica* Kryl. 84
   Cajanderi Kük 76
   camptotropa V. Krecz. 67
   canescens L. 39, 74, **76**
      β *alpestris* Trev. 73
      var. *subloliacea* Laest. 75
   capillaris L. 52, 126, **127**
      β Trev 129
      ssp. *chlorostachys* A. Löve 127
      var. *Ledebouriana* (C.A. Mey.) Kük. 129
      var. *nana* f. *Krausei* (Boeck.) Kük. 128
   capitata L. 33, **63**
      f. *arctogena* Raymond 64
   caucasica ssp. *perfusca* (V. Krecz.) T. Koyama 110
   Chamissonis Meinsh. 129
   chlorostachys Stev. 127
   chordorrhiza Ehrh. 35, **66**
   compacta R.Br. 134
   concolor R.Br. 92

Numbers in bold indicate pages with maps.

Cordouei Lévl. 120
cryptocarpa C.A. Mey. 43, 100
curta Good. 74
Davalliana auct. 71
demissa Hornem. 126
diandra Schrank 35, **64**, 65
dioica L. 32, **70**
　　β Trev. 72
　　γ Trev. 70
　　ssp. *asiatica* Gorodk. 70
　　ssp. *gynocrates* (Wormsk.) Hult. 70
　　β *parallela* Laest. 72
　　γ *parallela* Ostenf. 72
　　δ *Redowskiana* Ostenf. 71
discolor Nyl. 101
disperma Dew. 35, 83
drymophila var. *abbreviata* (Kük.) Ohwi 136
　　var. *akanensis* (Franch.) Kük. 136
　　var. *glabrata* (Turcz.) Ohwi 136
duriuscula C.A. Mey. 36, 67
ebracteata Trautv. 121
eleocharis L.H. Bailey 67
eleusinoides Turcz. ex Kunth 39, **98**
ensifolia (Turcz.) V. Krecz. ssp.
　　arctisibirica Jurtz. 46, 93, **94**
eriandrolepis Lévl. 120
ericetorum Pall. 54, 118
　　ssp. *baicalensis* Gorodk. 118
　　ssp. *melanocarpa* (Cham.) Kük. 118
　　var. *strictifolia* Kryl. 118
excurrens Cham. 117
falcata Turcz. 50, 123
ferruginea var. β Trev. 108
filiformis Good. 130
fuliginosa var. β *misandra* (R.Br.) O.F. Lang 107
flava L. 52, 125
　　δ Trev. 126
frigida var β Trev. 107
fusca Bell. 84
fuscidula V. Krecz. 52, 127, **128**
fusco-cuprea (Kük.) V. Krecz. 103
fusco-vaginata auct. 83
glacialis Mackenz. 56, **124**, 125
glareosa Wahlb. 37, 76, **78**
　　var. *amphigena* Fern. 77
globularis L. 53, **120**
Gmelinii Hook. et Arn. 41, 56, 114
Goodenoughii Gay 84
Gorodkovii V. Krecz. 109
gracilis Curt. 83
Grahami Boott 137

gynocrates Wormsk. 32, **70**
Halleri Gunn. 112
heleonastes Ehrh. 37, 80
Hepburnii Boott 33, 62
hirta var. γ Trev. 136
　　var. *glabrata* Turcz. 136
holostoma Drej. 56, 114
hyperborea Drej. 92
incurva Lightf. 67
　　var. β *setina* Christ 67
incurviformis Mackenz. 69
inflata auct. 131
inornata Turcz. 118
irrigua (Wahlb.) Smith 104
jucunda V. Krecz. 67
juncella (Fries) Th. Fries 84
juncifolia All. 69
kattegatensis Fries 101
kokrinensis Porsild 98
koraginensis Meinsh. 115
Krausei Boeck. 40, 52, **128**
　　ssp. *Krausei* 129
　　ssp. *Porsildiana* (Polunin) A. et D. Löve 129
Kreczetoviczii Egor. 38, 72
Ktausipalii Meinsh. 51, 109
Lachenalii Schkuhr 81
laeviculmis auct. 72
laevirostris (Blytt) Fries 130
lagopina Wahlb. 81
　　var. *pribylovensis* (J.M. Macoun) Kük. 83
lanceata Dew. 101
lapponica O.F. Lang 38, **75**, 77
lasiocarpa Ehrh. 49, 53, 130
　　ssp. *americana* (Fern.) Hult. 130
　　ssp. *occultans* (Franch.) Hult. 130
Ledebouriana C.A. Mey. 51, **129**
leiocarpa C.A. Mey. 62
lenaensis Kük. 129
leporina L. 36, 69
limosa L. 58, 103
　　var. *fusco-cuprea* Kük. 103
　　β *irrigua* Wahlb. 104
　　δ *livida* Wahlb. 121
　　γ *rariflora* Wahlb. 105
lineolata Cham. 87
livida (Wahlb.) Willd. 50, 121
loliacea L. 38, 78
loliacea auct. 83
lugens H.T. Holm. 46, 95, **96**
Lyngbyei Hornem. 100
　　ssp. *cryptocarpa* (C.A. Mey.) Hult. 100

Mackenziei V. Krecz. 38, 73, **75**
macilentha Fries 79
macrochaeta C.A. Mey. 56, 117
macrogyna Turcz. 50, 51, 108, **109**
magellanica Lam. 40, 58, 103, **104**
    ssp. *irrigua* (Wahlb.) Hult. 104
marina Dew. 76
maritima Gunn. 36, 67, **68**
    ssp. *maritima* 68
    ssp. *setina* (Christ) Egor. 68
maritima O.F. Muell. 99
Maximowiczii Fr. Schmidt 87
media R.Br. 111
Medwedewii Lesk. 110
melanantha var. *sabulosa* Kük. 114
melanocarpa Cham. 54, **118**
melanostoma Fisch. ex V. Krecz. 116
melozitnensis Porsild 132
membranacea Hook. 55, **134**
membranopacta L.H. Bailey 134
microglochin Wahlb. 34, 61
micropoda C.A. Mey. 33, 60
mimula V. Krecz. 112
misandra R.Br. 40, **107**
mitsuriokensis Lévl. et Vaniot 120
nardina Fries 62
    var. *atriceps* Kük 62
    var. *Hepburnii* (Boott) Kük. 62
nesophila H.T. Holm 116
nigra (L.) Reichard 44, 84
nigritella Drej. 117
norvegica Retz. 42, 112, **113**
    ssp. *conicorostrata* Kalela 114
    ssp. *inferalpina* (Wahlb.) Hult. 112
    ssp. *inserrulata* Kalela 113
norvegica Willd. 73
Novograblenovii Kom. 126
obliqua Turcz. 120
obtusata Liljebl. 34, 58
Oederi Retz. 52, 126
oligantha Boott 61
orthocaula V. Krecz. 67
orthostachys var. *hirtaeformis* Maxim. 136
oxyleuca V. Krecz. 106
paleacea Wahlb. 43, 99
pallida C.A. Mey. 35, **65**
panicea β *sparsiflora* Wahlb. 121
paralia V. Krecz. 99
parallela (Laest.) Sommerf. 32, 72, **73**
pauciflora Lightf. 34, 60
paupercula Michx. 104
pauxilla V. Krecz. 115

pedata Wahlb. 125
pediformis C.A. Mey. 49, 123
pedunculifera Kom. 100
pennsylvanica var. *amblyolepis* (Trautv. et Mey.) Kük. 119
perfusca V. Krecz. 41, 110
perglobosa Mackenz. 69
Peshemskyi Malysch. 126
petraea Wahlb. 59
petricosa Dew. 108
philocrena V. Krecz. 126
physocarpa Presl. 138
physochlaena H.T. Holm 134
pilulifera auct. 119
pisiformis var. *ebracteata* (Trautv.) Kük. 121
    var. *subebracteata* Kük. 120
podocarpa R.Br. 57, **115**
polygama Schkuhr 110
pribylovensis J.M. Macoun 37, 83
pribylovensis auct. 74
procerula V. Krecz. 138
psychroluta V. Krecz. 68
pulla Good. 137
    var. *laxa* Trautv. 138
    f. *pedunculata* Kjellmann 138
    var. *sibirica* Christ 138
pyrenaica Wahlb. ssp. *micropoda* (C.A. Mey.) Hult 60
quasivaginata C.B. Clarke 121
rariflora (Wahlb.) Smith 57, **105**
recta Boott 43, 101
Redowskiana C.A. Mey. 32, 71, **72**
reptabunda V. Krecz. 69
rhizina Blytt 123
rhomalea Mackenz. 137
rhynchophysa C.A. Mey. 55, 130
Riabushinskii Kom. 100
rigida Good. 92, 94
    var. *concolor* (R.Br.) Kük. 92, 93
    ssp. *inferalpina* (Laest.) Gorodk. 92, 93
    var. *inferalpina* Laest. 92
    f. *infuscata* auct. 93, 95
    ssp. *rigidioides* Gorodk. 95
    var. *typica* Kryl. 93
rigidioides (Gorodk.) V. Krecz. 46, 95
rostrata Stokes 55, **131**
    var. *borealis* Hartm. 134
    ssp. *rotundata* (Wahlb.) Kük. 132
    var. *utriculata* (Boott) L.H. Bailey 131
rotundata Wahlb. 55, 132, **133**
    f. *Sommieri* Christ 132

rubra Lévl. et Vaniot 87
rupestris Bell. 34, 58, **59**
sabulosa Turcz. 40, 56, 114
sabynensis Less. 49, 54, 120
Sadae Lévl. et Vaniot 120
sajanensis V. Krecz. 69
salina Wahlb. 43, 101
    var. *kattegatensis* Almq. 101
    var. *subspathacea* (Wormsk.) Ostenf. 102
saxatilis L. s. l. 42, 136, **137**
    var. *compacta* Dew. 134
    ssp. laxa (Trautv.) Kalela **137**, 138
    var. *laxa* (Trautv.) Ohwi 138
    var. *major* Olney 138
    var. *rhomalea* Fern. 137
    ssp. *saxatilis* 137
Schmidtii Meinsh. 48, 87, **88**
scirpoidea Michx. 32, **61**
setina (Christ) V. Krecz. 67
simpliciuscula Wahlb. 29
Soczavaeana Gorodk. 47, 96, **97**
sordida Heurck et Muell. Arg. 48, 136
spaniocarpa Steud. 124
sparsiflora (Wahlb.) Steud. 121
    var. *falcata* (Turcz.) Kük. 123
    var. *Petersii* Kük. 123
stans Drej. 45, 89, **91**
stenantha auct. 109
    var. *taisetsuensis* Akiyama 109
stenolepis Less. 133
stenophylla γ Trev. 67
stilbophaea V. Krecz. 106
stylosa C.A. Mey. 57, 117
    var. *nigritella* (Drej.) Fern. 117
subspathacea Wormsk. 44, **102**
subspathacea auct. 90
suifunensis Kom. 100
supina Wahlb. ssp. *spaniocarpa* (Steud.) Hult. 56, 124
tenella Schkuhr 83
tenuiflora Wahlb. 36, 79, **80**
teretiuscula Good. 65
Tolmiei Boott 57, 116, **117**
    var. *invisa* (Bailey) Kük. 115
    var. *nigella* (Boott) Kük. 115
transmarina V. Krecz. 67
Trautvetteriana Kom. 49, 54, 121
trinervis auct. 101
tripartita All. 37, 81, **82**
tristis auct. 108
tumidocarpa Anderss. 126
umbrosa ssp. *sabynensis* (Less.) Kük. 120
ursina Dew. 32, 36, 78, **79**
ustulata Wahlb. 106
    var. γ *macrogyna* Regel 108
utriculata auct. 131, 132
vaginata Tausch 50, 121, **122**
Vanheurckii Muell. Arg. 53, **119**
vesicaria L. 54, 134, **135**
    ssp. *saxatilis* (L.) Kük. 137
    ssp. *saxatilis* var. *compacta* Dew. 134
    ssp. *saxatilis* var. *physocarpa* (Presl) Kük. 138
    var. *tenuistachya* Kük. 135
    f. *tenuistachya* (Kük.) T. Koyama 135
vesicata Meinsh. 55, 135
viridula Michx. 126
vitilis Fries 73
vulgaris Fries 84
    *juncella Fries 84
    *zonata Nyl. 84
Williamsii Britt. 39, 52, 126
wiluica Meinsh. 47, 84, **85**
    ssp. *europaea* Gorodk. 84
Chamaeorchis L.C. Rich. 204, 209
    alpina (L.) Rich. 209
Coeloglossum Hartm. 205, 209
    bracteatum (Mühlb.) Parl. 209, 211
    islandicum (Lindl.) 209
    viride (L.) Hartm. 209, **210**
    ssp. *bracteatum* Hult. 211
    var. *bracteatum* Richt. 211
    β *bracteatum* Rchb. f. 211
Convallaria bifolia L. 200
Corallorhiza Hall. 204, 206
    neottia Scop. 206
    trifida Châtel. 206, **207**
Cypripedium L. 203, 205
    calceolus L. 205

Dactylorchis (Klinge) Vermln. 205, 213
    incarnata (L.) Vermln. 214
    maculata (L.) Vermln. 213
    sambucina (L.) 214
    Traunsteineri (Saut.) Vermln. 213, 214

Eleocharis R.Br. 2, 22
    acicularis (L.) Roem. et Schult. 22
    eupalustris Lindb. 23

eu-uniglumis Zinserl. 24
intersita Zinserl. 22, 24
palustris (L.) R.Br. 22, 23
pauciflora (Lightf.) Link 23
quinqueflora (Hartm.) Schwarz 22, 23
uniglumis (Link) Schult. 22, 24
Elyna Bellardii (All.) C. Koch 28
  filicifolia Turcz. 27
  schoenoides auct. 26
  sibirica Turcz. 26
Epipactis Sw. 204, 208
  atropurpurea Raf. 208
  atrorubens (Hoffm.) Schultes 208
  rubiginosa Crantz 208
Eriophorum L. 2, 3
  alpinum L. 20
  altaicum Meinsh. 14
  angustifolium Honck. 4, 5, **6**
    var. *elatior* Koch 7
    var. *majus* Schultz 7
    ssp. *scabriusculum* Hult. 8
    ssp. *subarcticum* (V. Vassil.) Hult. 5
    β *triste* Th. Fries 5
    ssp. *triste* (Th. Fries) Hult. 5
  asiaticum V. Vassil. 9
  brachyantherum Trautv. et Mey. 4, 15, **16**
  callitrix Cham. ex C.A. Mey. 4, 14, **15**
  capitatum Host. 13
  Chamissonis C.A. Mey. 9, 11
  coreanum Palla 9
  gracile Koch 3, 8
    ssp. *asiaticum* auct. 9
    ssp. *coreanum* Hult. 9
  humile Turcz. 18
  intercedens Lindb. 11
  intermedium Bast. 9
  intermedium Cham. 9, 11
  japonicum Maxim. 21
  Komarovii V. Vassil. 4, 8
  latifolium Hoppe 21
  mandshuricum Meinsh. 11
  Maximowiczii Beetle 21
  medium Anderss. 4, 11, **12**
  opacum Fern. 16
  polystachyum L. 5
  russeolum Fries 4, 9, **10**
    var. *majus* Somm. 11
  Scheuchzeri Hoppe 4, 13, **14**
  spissum Fern. 17, 18
  subarcticum V. Vassil. 5
  triste (Th. Fr.) Löve et Hadač 5

vaginatum L. 4, **17**
  ssp. *spissum* (Fern.) Hult. 18
  var. *opacum* Björnstr. 16
  var. *brachyantherum* Kryl. 16

Fritillaria L. 181, 197
  kamtschatcensis (L.) Ker-Gawl. 197

Gagea Salisb. 182, 192
  fistulosa Ker-Gawl. 192
  granulosa Turcz. 192
  lutea 192
  pusilla Schult. 192
  samojedorum Grossh. 192
Goodyera R.Br. 204, 208
  repens (L.) R.Br. 208
    var. *ophioides* Fern. 208
Gymnadenia R.Br. 205, 213
  albida L.C. Rich. 212
  conopsea (L.) R.Br. 213

Habenaria albida R.Br. 212
  obtusata Rich. 212
  straminea Fern. 212
  viridis R.Br. 209

Iris L. 201
  canadensis 201
  setosa Pall. 201, **202**
    ssp. *canadensis* 202
    ssp. *interior* (E. Ands.) Hult. 202
    var. *interior* E. Ands. 202
  sibirica L. 201

Juncus L. 143
  albescens Fernald 153
  alpinus Vill., s.l. 145, 150
    ssp. *nodulosus* Lindm. 150
    var. *alpestris* 150
    var. *rariflorus* 150
  ambiguus Guss. 147
  articulatus L. 145, 150
  arcticus Willd., s. str. 147, **158**
    ssp. *alaskanus* Hult. 147, **158**, 159
  atrofuscus Rupr. 150
  balticus Willd. 147, 159

var. *alaskanus* Porsild 159
var. *europaeus* Engelm. 159
var. *Haenkei* Buch. 159
var. *sitchensis* (Engelm.) 159
beringensis Buch. 147, 157
biglumis L. 146, 150, **151**
brachyspathus Maxim. 147, 157
bufonius L. 144, 147
castaneus Sm. 146, **154**
compressus Jacq. 145, 149
filiformis L. 146, 156, 157
Gerardii Lois., s, str. 145, 149
  ssp. *atrofuscus* (Rupr.) Tolm. 145, 150
  var. *atrifuscus* Trautv. 150
Haenkei E. Mey. 147, 159
himalensis Klotzsch 155
lampocarpus Ehrh. 145, 150
leucochlamys Zing. 155
  var. *borealis* Tolm. 146, 155, **156**
macrantherus Krecz. et Gontsch. 155
nastanthus V. Krecz. et Gontsch. 148
nodulosus Wahlb. 145, 150
Schischkini Kryl. et Sumn. 153
sphacellatus Buch. 155
spicatus L. 169
stygius L. 146, 152
triceps Rostk. 155
trifidus L. 144, **148**
triglumis L., s. str. 146, 152, **153**
  ssp. *albescens* (Lge.) Hult. **153**
  var. *albescens* Lge. 153

Kobresia Willd. 2, 25
  arctica A.E. Porsild 26
  Bellardii (All.) Degland 26, 27, **29**
  bipartita Dalla Torre 29
  capillifolia (Decne.) C.B. Clarke var.
    filifolia (Turcz.) Kük. 27
  caricina Willd. 29
  filifolia (Turcz.) C.B. Clarke 26, 27, **28**
  Hepburnii (Boott) Ivanova 62
  hyperborea A.E. Porsild 26
  myosuroides (Vill.) Fiori et Paol. 28
  schoenoides auct. 26
  scirpina Willd. 28
  sibirica Turcz. 26, **27**
  simpliciuscula (Wahlb.) Mackenz. 26, 29, **30**

Lemna L. 141
  minor L. 141
  trisulca L. 141, 142

Leucorchis E. Mey. 205, 212
  albida (L.) Mey. 212
Lilium kamtschatcense L. 197
Listera R.Br. 204, 207
  cordata (L.) R.Br. 207
    var. *japonica* Hara 207
    var. *nephrophylla* (Rydb.) Hult. 207
Lloydia Salisb. 182, 198
  alpina Salisb. 198
  Mairei 199
  serotina (L.) Rchb. 198, **199**
Luzula DC. 143, 160
  arctica Blytt 174
  arcuata (Wahlb.) Sw. 163, 172, **173**
    ssp. *unalaschkensis* Gorodk. 171
    var. *confusa* 169
    var. *kamtschadalorum* Sam. 172
    var. *unalaschkensis* 171
    var. α et var. η 169
    f. *latifolia* Kjellman 175
  beringensis Tolm. 163, 170, **171**
  beeringiana Gjaerevoll 175
  campestris Rpr. var. *capitata* Miq. 176
    var. *frigida* Buch. 177
    var. *multiflora* Celak. 176
  capitata (Miq.) Nakai 164, **176**
  confusa Lindb. 163, 169, **170**
  frigida Sam. 177
  hyperborea R.Br. 169
    var. *major* 169
    var. *minor* 174
  kamtschadalorum Gorodk. 172, 175
  melanocarpa (Michx.) Desv. 167
  multiflora (Retz.) Lej., s.l. 164, 176
    ssp. *asiatica* Kryl. et Serg. 178
    ssp. *frigida* (Buch.) V. Krecz. 165, **177**
    ssp. *Kjellmaniana* (Miy. et Kudo) Tolm. 165, **179**
    ssp. *sibirica* V. Krecz. 165, **178**
    var. *frigida* Buch. 177
    var. *Kjellmaniana* Sam. 179
  nivalis Laest. 163, 173, **174**
    var. *latifolia* (Kjellm.) Sam. 175
  oligantha Sam. 164, 179, **180**
  pallescens (Wahlb.) Bess. 164, 180
  parviflora (Ehrh.) Desv., s. str. 162, **166**
    ssp. *melanocarpa* (Michx.) Tolm. 162, **166**, 167
    var. *melanocarpa* 167
  pilosa (L.) Willd. 162, 165
  Pohleana Krecz. 177
  rufescens Fisch. 162, 165
  sibirica Krecz. 178
  spadicea var. *Kunthii* E. Mey. 167

  var. *Wahlenbergii* 167
  melanocarpa E. Meyer 167
  ϵ *parviflora* 166
 spicata (L.) DC. 162, **168**
 sudetica (Willd.) DC. 164, 179
 tundricola Gorodk. 164, **175**
 unalaschkensis (Buch.) Satake, s. str.
  163, 171, **172**
  ssp. *kamtschadalorum* (Sam.) Tolm.
   163, **172**
 Wahlenbergii Rupr. 162, **167**
Lysiella Rydb. 204, 212
 obtusata (Pursh) ssp. *oligantha* (Turcz.)
  Tolm. 212
 oligantha (Turcz.) Nevski 212

Maianthemum Web. 183, 200
 bifolium (L.) F.W. Schmidt 200
Melanthium sibiricum L. 186

Narthecium pusillum Michx. 184
Nectarobothrium Redowskianum Schl. et
 Cham. 198
 striatum Ledb. 198

Ophrys alpina L. 209
 corallorhiza L. 206
 cordata L. 207
Orchis bracteata Mühlb. 211
 conopsea L. 213
 curvifolia Nym. 214
 Fuchsii Druce 214
 maculata L. 213
 sambucina L. 214
 Traunsteineri Saut. 214
Ornithogalum altaicum Laxm. 198
 striatum Willd. 198

Paris L. 183, 200
 quadrifolia L. 200
Peristylus albidus Lindl. 212
 bracteatus Ledb. 211
 islandicus Lindl. 209
 viridis Lindl. 209
Platanthera obtusata Lindl. 212
 parvula Schlecht. 212

Satyrium albidum L. 212
 repens L. 208
Scirpus L. 2, 20
 acicularis L. 23

 bracteatus Bigel. 19
 caespitosus L. 19
  ssp. *austriacus* (Palla) Aschers. et
   Graebn. 19
 hudsonianus (Michx.) Kryl. 20
 japonicus Fern. 21
 Maximowiczii C.B. Clarke 21
 palustris L. 23
 pauciflorus Lightf. 23
 quinqueflorus Hartm. 23
 radicans Schkuhr 20
 sylvaticus L. 20
 uniglumis Link 24
Smilacina Desf. 183, 199
 bifolia Desf. 200
 trifolia Desf. 199

Tofieldia Huds. 182, 183
 borealis Wahlb. 184
 coccinea Rich. 183, **185**
 nutans Willd. 185
 palustris Huds. 183
 pusilla (Michx.) Pers. 183, **184**
Trichophorum Pers. 2, 18
 alpinum (L.) Pers. 19, 20
 austriacum Palla 19
 bracteatum (Bigel.) V. Krecz. 19
 caespitosum (L.) Hartm. 19
  ssp. *caespitosum* 19
  ssp. *germanicum* (Palla) Hegi 19
 germanicum Palla 19

Uncinia microglochin Spreng. 61

Veratrum L. 181, 187
 album L. 189, 190
  floribus viridibus 190
  ssp. *Lobelianum* 190
  ssp. *oxysepalum* (Turcz.) Hult. 189
  var. *Lobelianum* 190
  var. *viridis* Rgl. 189
 dolichopetalum Loes. 187, 189
 Lobelianum Bernh. 189, 190, **191**
  var. *Mišae* Sir. 189, 190
  asiaticum Loes. 188
 Mišae (Sir.) Loes. 189, 190
 Mischae 190
 oxysepalum Turcz. 189, **190**

Zygadenus Rich. 182, 186
 sibiricus (L.) A. Gray 186

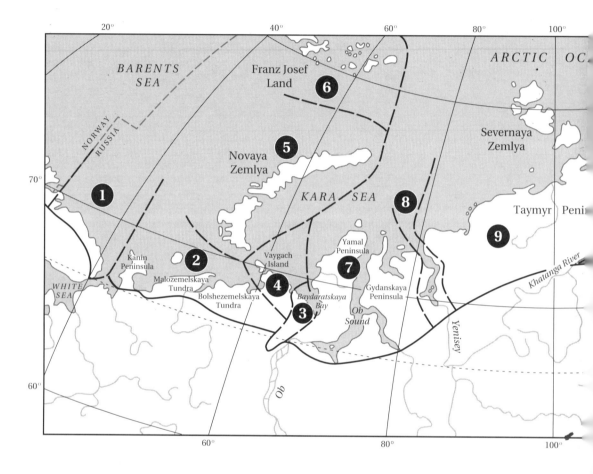

Map of geographic districts of vascular plants in the Russian Arctic